DIFFUSIVE SAMPLING

AN ALTERNATIVE APPROACH TO WORKPLACE AIR MONITORING

Diffusive Sampling

An Alternative Approach to Workplace Air Monitoring

The Proceedings of an International Symposium held in
Luxembourg, 22-26 September 1986, organised jointly by
The Commission of The European Communities (CEC, Luxembourg)

and the

United Kingdom Health and Safety Executive (HSE, London)

in cooperation with the

United Kingdom Royal Society of Chemistry

and the

World Health Organisation

Edited by

A Berlin
Health and Safety Directorate
Commission of the European Communities
Luxembourg

R.H. Brown
Occupational Medicine and Hygiene Laboratories
Health and Safety Executive
London, England

K.J. Saunders
BP Research Centre
Sunbury-on-Thames
Middlesex, England

ROYAL
SOCIETY OF
CHEMISTRY

Publication No. 10555 EN of the Commission of the European
Communities, Directorate-General Telecommunications,
Information Industries and Innovation, Luxembourg

British Library Cataloguing in Publication Data

Diffusive sampling: an alternative approach to workplace air monitoring:
   the proceedings of an international symposium held in Luxembourg,
   22–26 September 1986, organised jointly by the Commission of the
   European Communities and the United Kingdom Health and Safety
   Executive in cooperation with the United Kingdom Royal Society of
   Chemistry and the World Health Organisation.
   1. Air–Pollution–Measurement   2. Industrial hygiene   3. Air sampling
   apparatus
   I. Berlin, A.   II. Brown, R.H.   III. Saunders, K.J.   IV. Royal Society of
   Chemistry
   628.5'3   TD890

ISBN 0-85186-343-4

Published by the Royal Society of Chemistry
Burlington House, London, W1V 0BN

Printed in Great Britain by
Whitstable Litho Ltd., Whitstable, Kent

Contents

E D I T O R I A L   N O T E

The papers in these proceedings have been the subject of minor editing in order to achieve consistency. The discussion sessions have been edited substantially to eliminate undue repetition; every attempt has been made to adhere to the original meaning of the contributors. Whenever possible, English spelling and words have been used. The summary report gives an overview of the contributed papers and discussion, and contains the conclusions and recommendations for action agreed by the participants.

The editors would like to acknowlege the support given by the session chairmen and rapporteurs, as well as the organizing committee for the scientific planning of this Symposium. Thanks are also due to L. Angeletti (CEC), W. Baer (CEC), H.F. Koelbl (Morsi) and T. Jones (CEC) for their technical assistance.

ORGANIZING COMMITTEE:

A. Berlin (CEC), R.H. Brown (UK), M. Frangiadakis (Greece), J.P. Guenier (F), B. Herve-Bazin (F), K. Leichnitz (FRG), B. Miller (UK), L. Pozzoli (Italy). M. Sarivalassis (Greece), K.J. Saunders (UK), B. Striefler (FRG), J.F. Van der Wal (NL)

CHAIRMEN:

A. Berlin (CEC), J.V. Crable (USA), B. Herve-Bazin (F), R. Lidgett (B), K.J. Saunders (UK), J.F. Van der Wal (NL).

RAPPORTEURS:

R.H. Brown (UK), K. Leichnitz (FRG), B. Miller (UK), K.J. Saunders (UK), B. Striefler (FRG).

INTRODUCTORY REMARKS

A.E. BENNETT (CEC), in welcoming the participants, referred to the Communities' 1978 and 1984 Action Programmes on Safety and Health at Work. Recent developments in these Programmes had been the adoption of individual Directives, for example on lead and asbestos, and the elaboration of a proposal on a European list of Air Limit Values (for approximately 100 substances), both of which laid particular emphasis on personal monitoring.

One of the main conclusions of the Seminar on Assessment of Toxic Agents at the Workplace (Luxembourg, 1980; organized jointly by the CEC and the United States regulatory authorities) had been that personal exposure measurement was needed alongside biological monitoring for the assessment of individual exposure. In particular, there was a perceived need for continuous personal monitoring devices that would not interfere with the worker's job.

The potential of diffusive sampling in this context was stressed and the hope expressed that the present Symposium would generate advice on the future of such devices that could be transmitted to the Regulatory Authorities.

J.G. FIRTH (UK),   On behalf of the UK Health and Safety Executive (HSE), one of the sponsors and organizers of the meeting added his welcome to the participants. The HSE has also identified the potential of diffusive sampling for personal monitoring and had been actively working in the field since the late 1970's. In this work, HSE had collaborated closely with Industry and this collaboration had led to the formation of a joint working party ("Working Group 5") and the publication of diffusive methods for workplace air monitoring, which were believed to be the first to be recommended by a government body.

The potential advantages of diffusive sampling are to be considered in terms of convenience, versatility and cost-effectiveness.

The convenience of diffusive sampling was obvious. Versatility meant that samplers could be used in otherwise inaccessible environments, for example operating theatres. Cost-effectiveness meant the ability to undertake large surveys, particularly when samplers could be analysed automatically, in the context of setting exposure limits or promoting the self-regulation of industry.

D.C.M. SQUIRRELL, on behalf of the UK Royal Society of Chemistry, and expressed the particular support of the Analytical Division of that Society in the promotion of a proper evaluation of new and innovative ideas, including diffusive sampling.

DIFFUSIVE SAMPLING - A REVIEW OF THEORETICAL ASPECTS AND STATE-OF-THE-ART

Gerald Moore

SKC Ltd.
Stone Lane Industrial Estate
Wimborne, Dorset
England

ABSTRACT

The understanding of the theoretical aspects of diffusive samplers has progressed considerably in the years since 1973 when Palmes first related the application of Fick's Law. Mathematical treatments are now available which provide greater understanding of the mechanisms affecting uptake rate, sorbent effects, the need for backup systems and capacity limitations. A review of these principal theoretical aspects is provided and illustrated by examples of particular devices. The design of diffusive samplers to meet particular conditions is also considered.

A CHRONOLOGY OF DEVELOPMENTS

Although diffusive samplers using simple colorimetric principles have been available since the 1930's and have been in use since the early 1950's, it was not until the early 1970's that the first mathematical treatment was published (Palmes, 1973) which attempted to codify the factors controlling uptake rate in the application of Fick's laws of diffusion.

This early work led to the design of a sampler for NO_2 which is generally recognized to be the forerunner of most of the devices available today.

A major step forward occurred with the publication of a paper describing the development of the "Gas Badge" diffusive sampler (Tompkins and Goldsmith, 1977) which recognized the part played by external effects such as wind velocity, temperature and pressure.
Fairly rapid commercial development followed during the 1970's with the appearance of mercury vapor and organic vapor monitors from the 3M Company and an organic vapor

1

monitor from the Dupont Company. A more logical extension of the original Palmes work was the emergence of the tube type sampler in the UK sponsored by an industry group and eventually commercialized by the Perkin Elmer Company (Brown, Charlton and Saunders, 1981). During this period of development many other types of devices were developed, usually being variations on either the flat badge or tube concept, but most did not find significant commercial use during this time.

Two events of significance occurred in 1979; firstly the publication of a deceptively simple technical paper (Palmes et al., 1979) dealing with the analogies between diffusion resistance and electrical resistance and demonstrating that they could be handled in a similar fashion. The second event was the appearance of diffusive samplers having a means of backup or validation, thereby addressing the problem of sample losses or variations in sampling rate over the sampling period.

During the 1980's there has been a substantial increase in the use of both sorbent based and colorimetric samplers and a very wide variety of such devices is now available from numerous manufacturers both in the USA and Europe. A very comprehensive review of the field was presented in the American Industrial Hygiene Association Journal in 1982 (Rose and Perkins, 1982).

In 1983 there was the nearly simultaneous publication of validation protocols by NIOSH in the USA and HSE in the UK. These represent a serious attempt to provide a set of standards and test methods which can be used to compare all types of diffusive samplers against common criteria.

The period from 1983 to the present day has been particularly marked by the publication of some key papers dealing with the mathematics of various aspects of diffusive samplers. Of particular note are the mathematical treatments of the influence of sorbents on the uptake rate by two groups in the USA (Underhill, 1984; Posner and Moore, 1985); the study of transient sampling conditions (Hearl and

Manning, 1980; Bartley et al., 1983); and numerous studies of comparisons between diffusive samplers and the charcoal tube sampling method in a variety of technical journals.

An extremely intriguing recent publication (Palmes et al., July 1986) presents a further new explanation and mechanism for diffusional sampler operation. It is clear that such work is ongoing.

A REVIEW OF SOME THEORETICAL ASPECTS

The pioneering work of Dr. Palmes and his group at the New York University resulted in a definition of the basic parameters which can be used to calculate the uptake rate of most diffusive samplers from a knowledge of the geometry and the diffusivity of the gas to be sampled.

$$J = DA/L(C_o - C_e) \qquad (1)$$

and

$$Q = DA/L(C_o - C_e)T \qquad (2)$$

where

J = diffusive flux (g/sec)
D = coefficient of diffusion (cm^2/sec)
A = cross sectional area of diffusion path (cm^2)
L = length of diffusion path (cm)
C_o = external concentration being sampled (g/cm^3)
C_e = concentration at the interface of the sorbent (g/cm^3)
Q = mass uptake (g)
T = time of sampling (sec)

It is apparent from an inspection of these equations that the expression DA/L has units of cm^3/sec and therefore represents what can be considered as a "sampling rate" of the sampler when comparing to a pumped sampling system. This simple use of the sampling rate concept has proven of considerable value to users of the devices and is most often expressed in units of ml/min. Knowing the geometry of the sampler (which will be fixed for any given

type) permits the calculation of sampling rate provided the
diffusion coefficient D is known. A number of manufacturers
have published tables of sampling rates calculated in this
fashion most of whom have used the same published references
as the source for a diffusion coefficients (Lugg, 1968).
This comparison study suggested that the calculation of
diffusion coefficients proposed by Wilke and Lee in 1955 was
quite accurate in the majority of cases. More recently
this method has been computerized in order to provide a more
rapid solution to the relatively complicated mathematics
involved (GMD Systems, Inc., 1982).

Practical experience has shown that the calculation of
sampling rate by these methods can be accurate to around +/-
5% of measured values but that the use of sampling rate
measured from a dynamic gas test will always be preferable.
In particular the errors introduced by sorbent effects and
the reduction in sampling rate over time produced by these
effects can make calculated results of little value. In
addition no single method of calculating diffusion
coefficient will give accurate results over the wide variety
of compounds which can be sampled.

It is useful to explore the extremes of the original
Palmes formula in order to better understand the practical
limitations of diffusive sampler design in the real world.
Inspection of equation (2) or the simplified expression for
sampling rate (DA/L) would suggest that one can increase
uptake rate to any desired extent by simply reducing L or
increasing A. It is readily apparent that increasing A (the
surface area of the diffusion path) has some practical
limitations if the sampler is not to become excessively
large. A sampler will hardly qualify for the definition
"personal sampler" if its size becomes that of the average
dinner plate, although such a device may well be feasible
for area monitoring purposes. However, the effects of
reducing L are not so obvious and are often misunderstood.

In Palmes' original work, he constructed a number of
different experimental samplers having various lengths and

areas and performed measurements of sampling rate using both
sulfur dioxide and water vapor. His objective was to
demonstrate the correspondence of measured sampling rate
with that predicted by calculation from his formula. He
noted at the time that certain of the examples appeared to
deviate from the ideal and in particular that many of the
experimental samplers showed a shortfall in sampling
compared to the calculated rate. Largely as a result of
this data, he generally recommended the use of L/A ratios of
10 or greater when designing practical samplers. However,
he did not attempt at that time to explain the obseved
shortfalls.

    Figures 1 and 2 show his 1973 data plotted in two
different ways. It will be seen from Figure 1 that there is
little apparent trend in the results except perhaps a
general tendency for high A/L ratios to produce somewhat of
a shortfall. However, the spread of values at any given A/L
point makes it difficult to ascertain any real trend.

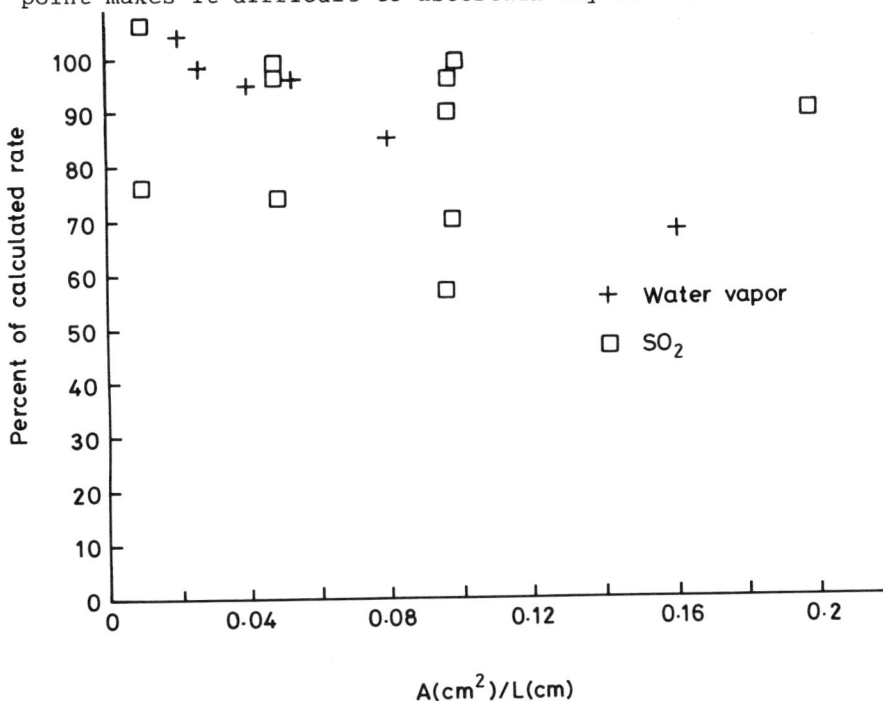

FIGURE 1   Sampling Rate vs A/L

Figure 2 shows the same data points but plotted as a
function of the diffusion path length of the various
samplers and it is immediately apparent that there is a much
more marked trend to the data. It will be noticed that the
samplers tend to approach 100% of the calculated sampling
rate when the length of the path is around 2 cm or greater.
At lengths below 2 cm the sampling rate departs from the
ideal and this deviation becomes extremely marked for
lengths less than 0.5 cm.

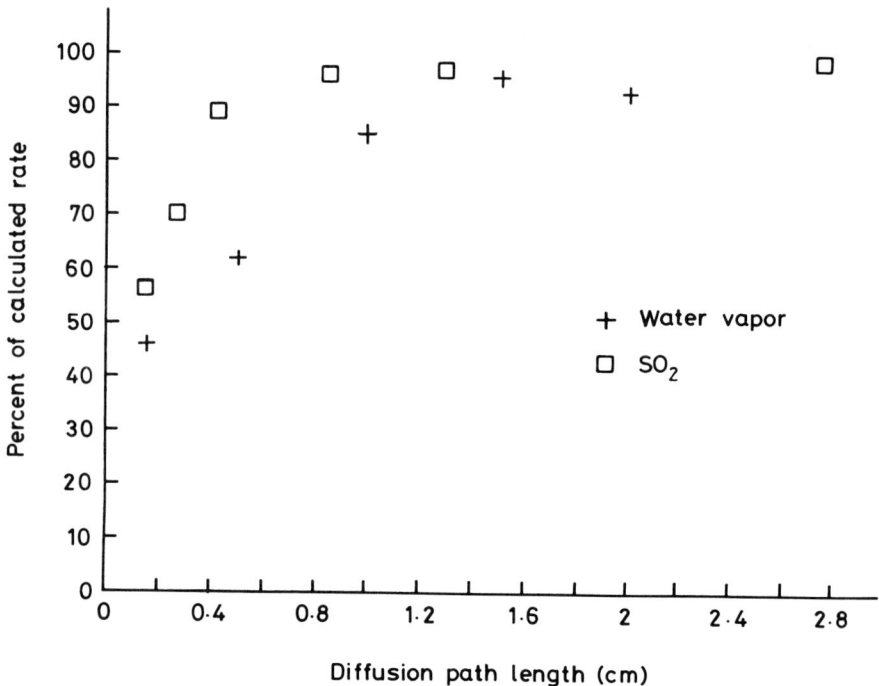

FIGURE 2   Sampling Rate vs Length

It follows, therefore, that reducing L alone will not
produce the expected increase in sampling rate. The reason
for this is quite clear if one calculates the maximum
feasible sampling rates from mass transfer considerations.

Tompkins and Goldsmith showed that a simplified expression could be used for this purpose, namely

mass uptake = kAC_o

where   k = convective mass transfer coefficient.

This coefficient can be calculated from published equations and effectively places a limit on the maximum sampling rate which can be achieved for a given face area of sampler at a given air velocity. If this treatment is applied to the original Palmes results, it immediately demonstrates that those samplers showing a shortfall are, with few exceptions, those where the sampling rate calculated from A/L considerations exceeds that permitted by kA considerations. In other words the environment at large cannot supply contaminant at the rate required by the sampler thus causing the shortfall. Another recent study also addresses this point (Persoff and Hodgson, 1985).

It should be noted that whichever method of calculation one uses, the surface area A comes into the equation thus maintaining proportionality between sampling rate and face area which can be used to advantage in the design of diffusive samplers.

Figure 3 shows some theoretical examples of samplers designed according to the various methods of calculation and the effect that will be produced in the presence of varying wind velocity. The "ideal" case shown would be independent of wind velocity down to the point where the line crosses the curve for mass transfer-based calculation. At velocities below this level, mass transfer control will take over and produce rapidly diminishing sampling rates. The third line shows test results taken from an actual device

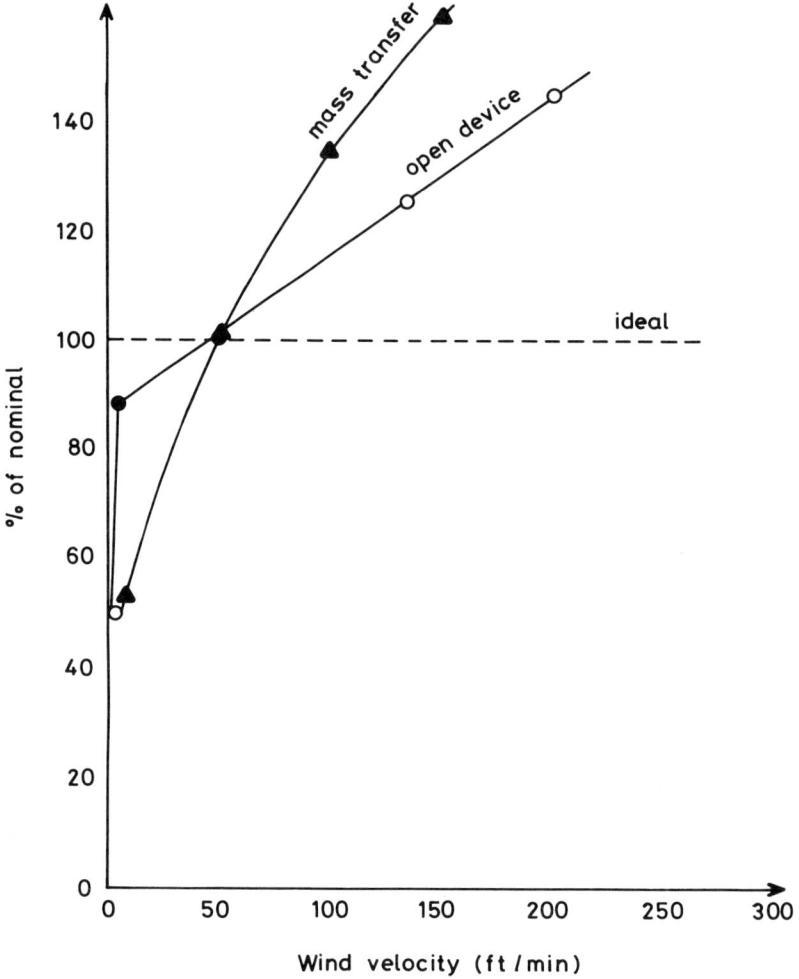

FIGURE 3   Various Samplers - Effects of Wind Velocity

where the only control on sampling rate was mass transfer, i.e. an open device.

As was stated earlier, the influence of sorbent parameters on the sampling rate can be profound. These effects will generally only be of concern for diffusive samplers relying on physical adsorption as the means of trapping. Typical systems relying on chemisorption or other reactive trapping means will not demonstrate sorbent related effects provided they are used within the overall capacity of the chemicals employed. A combination of effects takes place within the tube-type sampler whether it be of the colorimetric chemical or sorbent type since the length of the sampling path will change over the entire sampling period. This results in a more complex relationship between sampling rate and time of sampling which has been studied particularly in the case of the colorimetric diffusion tube system (McKee and McConnaughey, 1976; Gonzales and Sefton, 1985).

In the special case of sorbent-based badges, the use of validation or "backup" systems can help considerably in overcoming some of these effects. Backup systems are generally of two types, the first being a second layer of sorbent situated behind the first and separated from it by a further diffusion path. The second uses a parallel sampling technique where use is made of two similar samplers operating in parallel but at markedly different sampling rates. In this case, sorbent errors which become more apparent over time will influence the higher sampling rate system first, and inspection of the two results will quickly indicate deviations from the calculated ratio. Figure 4 shows diagramatically the effects of operating a backup system in this way.

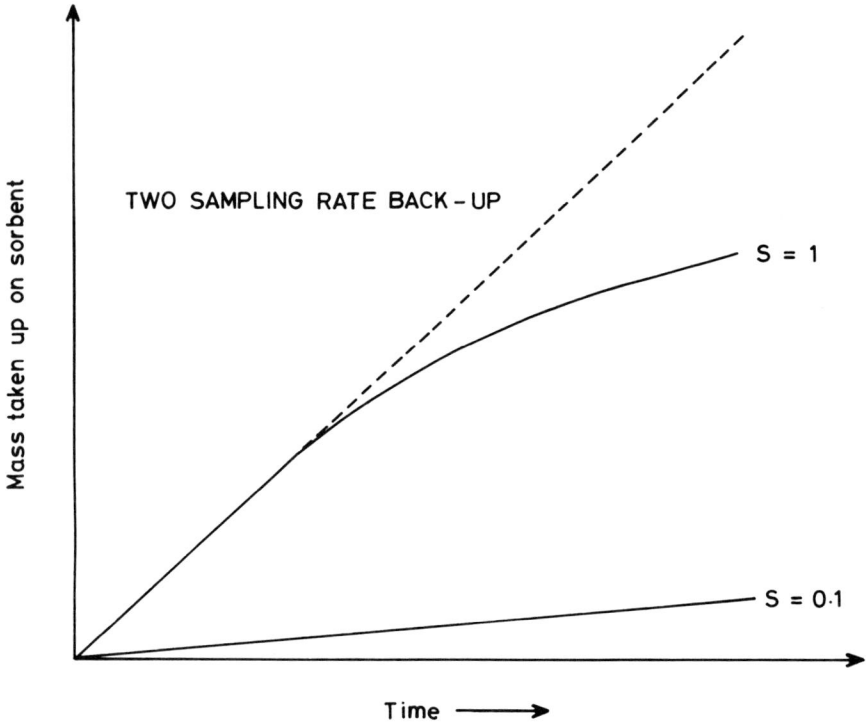

FIGURE 4   Two Sampling Rate Backup

THE ELECTRICAL ANALOGY

It has previously been shown that there is a close
correspondence between the diffusion resistance in a
diffusion path and the electrical resistance in an
electrical circuit. Diffusion resistances can be placed in
series or in parallel and the mathematical treatment of
these conditions is identical to the electrical case. The
author has carried this analogy one step further to a more
complete electrical analogy of a sorbent-based diffusion
sampler having an equivalent electrical circuit (Figure 5).

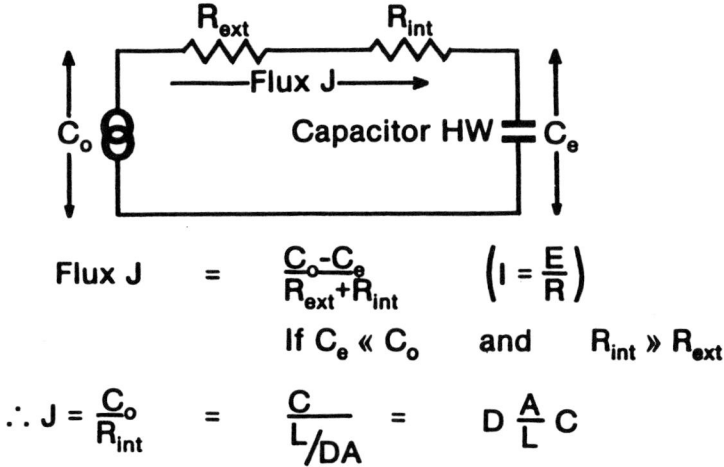

$$\text{Flux J} = \frac{C_o - C_e}{R_{ext} + R_{int}} \qquad \left(I = \frac{E}{R}\right)$$

$$\text{If } C_e \ll C_o \quad \text{and} \quad R_{int} \gg R_{ext}$$

$$\therefore J = \frac{C_o}{R_{int}} = \frac{C}{L/DA} = D\frac{A}{L}C$$

FIGURE 5   The Sampler Equivalent Electrical Circuit

The supply of contaminant from the outside environment is represented by an idealized voltage generator having the output potential C_o. Current (diffusive flux) flows from this source into the sorbent within the sampler which is represented in the electrical analogy by a capacitor having the value HW where H = specific retention volume (litres/gram) and W = weight of sorbent (grams). This current or flux flows through two resistances in series designated as R_{ext} and R_{int} representing the external mass transfer resistance and the internal diffusion resistance, respectively. The vapor concentration at the interface between the air and the sorbent (C_e) can be thought of as equivalent to the voltage appearing across the capacitor as it charges. Thus the mass taken up on the sorbent can be represented by the charge taken up on the capacitor and it is clear that the voltage (concentration C_e) will rise as the total charge (mass) rises. The equation for the derivation of the flux J clearly shows the close parallels with Ohm's Law for the electrical case. This analogy is not perfect but provides a simple framework of understanding of the interplay of some of the factors in a sorbent-based system.

Keeping in mind some of the earlier discussions presented in this paper, it might seem that the design and use of diffusive samplers is restricted by many of the basic parameters of the diffusion system. However, the ingenuity of the designers has proven equal to the task and many ingenious designs have come about as a result. Perhaps the most logical development from the original single tube approach is that followed by the Draeger Company where the sampler consists essentially of a double-ended tube thereby permitting an increase in the sampling rate or indeed an adjustable sampling rate. Higher sampling rates yet were achieved by the Dupont Company in the design of their organic vapor monitor which employs multiple Palmes tubes arranged in parallel and sampling onto a common sorbent strip. The influence of sampling face area on sampling rate has been exploited by the SKC Company in their multiple sampling rate systems.

In the field of colorimetric devices there are again multiple families available from many manufacturers, some following the flat badge type approach and others concentrating on colorimetric diffusion tubes.

The use of liquid sorbents has become quite popular in the sampling of inorganic vapors and the combination of a sorption badge with thermal desorption provides a particularly convenient analytical tool exploiting both the simplicity of a passive sampler and the ease of analysis using thermal desorption.

It is the belief of the author that European and international agreements concerning test and validation protocols for these devices will surely lead to the acceptance of diffusive samplers in standard methods published for regulatory purposes. This in turn will lead to much more widespread acceptance and use of diffusive samplers with consequent benefits to the whole industrial community in the area of safety and health.

REFERENCES

Bartley, D.L. et al. 1983. Diffusive monitoring of fluctuating concentration. Am. Ind. Hyg. Assoc. J., 44, 241-247.

Brown, R.H., Charlton, J. and Saunders, K.L. 1981. The development of an improved diffusion sampler. Am. Ind. Hyg. Assoc. J., 42, 856-869.

DIFCAL 4 - A computerized method for calculation of diffusion coefficients. 1982. GMD Systems, Inc. Hendersonville, PA   USA.

Gonzales, L.A. and Sefton, M.V., 1985. Laboratory evaluation of stain-length passive dosimeters for monitoring of vinyl chloride and ethylene oxide. Am. Ind. Hyg. Assoc. J., 46, 591-598.

Hearl, F.J. and Manning, M.P. 1980. Transient response of diffusion dosimeters. Am. Ind. Hyg. Assoc. J., 41, 778-783.

Lugg, G.A., 1968. Diffusion coefficients of some organic and other vapors in air. Anal. Chem., 40, 1072-1077.

McKee, E.S. and McConnaughey, P.W. 1986. Laboratory validation of a passive length-of-stain dosimeter for hydrogen sulfide. Am. Ind. Hyg. Assoc. J., 47, 475-481.

Palmes, E.D. and Gunnison, A.F. 1973. Personal monitoring device for gaseous contaminants. Am. Ind. Hyg., 32, 78-81.

Palmes, E.D. and Lindenboom, R.H. 1979. Ohm's Law, Fick's law and diffusion samplers for gases. Anal. Chem. 51, 2400-2401.

Palmes, E.D. et al. 1986. A simple mathematical model for diffusive sampler operation. Am. Ind. Hyg. Assoc. J.,47, 418-420.

Persoff, P. and Hodgson, A.T. 1985. Correction for external mass-transfer resistance in diffusive sampling. Am. Ind. Hyg. Assoc. J., 46, 648-652.

Posner, J. and Moore, G. 1985. A thermodynamic treatment of passive monitors. Am. Ind. Hyg. Assoc. J., 46, 277-295.

Rose, V. and Perkins, J.L. 1982. Passive dosimetry - state of the art. Am. Ind. Hyg. Assoc. J., 43, 605-621.

Tompkins, F.C. and Goldsmith, R.L. 1977. A new personal dosimeter for the monitoring of industrial pollutants. Am. Ind. Hyg. Assoc. J. 38, 371-377.

Underhill, J.W. 1984. Efficiency of passive sampling by adsorbents. Am. Ind. Hyg. Assoc. J., 45, 306-310.

VALIDATION OF PASSIVE DOSIMETRY THROUGH BIOLOGICAL
MONITORING,
AND ITS APPLICATION IN SOLVENT WORKPLACES

Masayuki IKEDA, Akio KOIZUMI and Miyuki KASHARA,
Department of Environmental Health,
Tohuko University School
of Medicine, Sendai 980, Japan

ABSTRACT

Under experimental conditions in servo-mechanized exposure
chambers, carbon cloth (CC) TF 1500 in a badge-type holder
adsorbed toluene, isopropyl alcohol, and acetone in a manner
linear to exposure duration (up to 8 hrs) as well as exposure
concentration (up to 200, 800 and 400 ppm, respectively).
Linear adsorption was also held when CC was exposed to
n-hexane, ethyl acetate and toluene either in combination
(ratio, approx. 1:4:1 in ppm) or separately, and the amount
of each solvent adsorbed after combined exposure was identical with that after single solvent exposure. When the effect
of humidity on the solvent adsorption was examined, it was
found that the same amounts of the three solvents were
adsorbed on the CC at 93% relative humidity (RH) as at 40%
RH, while decrease in the adsorbed amound at 93% RH was
observed with n-hexane (down to 20% of that at 40% RH; less
so with other two solvents) in 5 of 7 various CC preparations
tested. When packed in aluminum foil right after the exposure and kept at room temperature, loss was significant
($p < 0.05$) with n-hexane in 9 days and toluene in 14 days, but
no significant ($p > 0.05$) loss was detected with toluene for
over 6 months or with n-hexane for 2 weeks when kept at 4°C,
indicating that CC is applicable to surveys in factories
distant from an analytical laboratory. In field surveys, CC
TF 1500 in a holder was equipped on a lapel of each worker in
various solvent workplaces for a shift of the day (approx.
for 8 hrs) to measure the time-weighted average exposure
concentration. The urinary metabolite concentration was
determined in the sample collected from the worker at the end
of the shift. The comparison of the two exposure parameters
disclosed that a linear relation holds between the pairs of
solvent concentration in breath zone air (with a maximum
concentration measured in parentheses) and metabolite level
in urine (separated by a slanting mark when multiple metabolites were examined) in most cases; namely benzene vs.
phenol, toluene vs. hippuric acid/o-cresol, styrene
vs.mandelic acid/phenylglyoxylic acid, and methyl ethyl
ketone vs. unmetabolized methyl ethyl ketone. When
tetrachloroethylene-exposed workers were examined, however,
the level of urinary metabolites (i.e. total
trichloro-compounds) was linearly related to the exposure
intensity only up to approx. 100 ppm and no longer proportional at higher concentrations, suggesting the saturation of
metabolic capacity in humans. Application of CC to measure

methyl alcohol (MeOH), one of the most popular solvents, was unsuccessful. Experimental exposure of CC to MeOH up to 350 ppm revealed that the amount of MeOH recovered from CC was not proportional to exposure duration and leveling off was evident at 6 hrs. When CC in a holder was impregnated with MeOH by vapor exposure and then kept in fresh air, the decay in the MeOH amount occured with a half-time of approx. 1 hr to suggest the spontaneous desorption.

INTRODUCTION

In this presentation, experiences in our laboratory on passive dosimetry with carbon cloth (CC) is reviewed on performance of CC after experimental exposure to organic solvents 1,2/, and validation of CC through biological monitoring of exposure by means of analyses of urine from workers in various solvent workplaces 3-8/. The difficulties in application of CC to measure methyl alcohol 9/ will also be discussed.

MATERIALS AND METHODS

CARBON CLOTH

For routine study, carbon cloth K-filter TF 1500 (Toyobo Co., Osaka, Japan; to be called CC TF 1500) was employed. Other carbon cloth preparations (CC) were also examined when specified. The carbon cloth (5 cm x 3 cm) was housed in a badge-type holder for passive dosimetry.

EXPERIMENTAL CC EXPOSURE

A servomechanized exposure system with 4 chambers built in parallel was employed for experimental exposure of CC. A FID-GC was used for automatic monitoring of vapor concentration, and the determination in each chamber was repeated in every 25 min. The performance of the system was such that, in practice, the mean observed concentration was essentially the same with the ordered concentration and the coefficient of variation was less than 5% in most cases and less 10% at maximum 10,11/.

WORKERS EXAMINED AND COLLECTION OF URINE

The factory surveys were conducted either on Thursday or on Friday when the metabolite levels are expected to reach a maximum 12/. The workers participated in the studies were mostly males, and occupationally exposed to either benzene 8/, toluene 5,7/, styrene 3/, methyl ethyl ketone 4/, or tetrachloroethylene 6/. Each worker was equipped with a passive dosimeter on the lapel for an entire shift of work (usually from 08:00 to 17:00) for the determination of time weighted average of the exposure concentration. The worker was asked to pass urine at around 13:00 to 14:00 and then urine samples were collected at 15:00 - 17:00 12/.

ANALYSES FOR SOLVENT IN CC AND SOLVENT METABOLITES IN URINE

CC piece was extracted with either 10 ml carbon disulfide (containing 240 ml tert.-butylbenzene/ml as an internal standard (IS)) in general or 10 ml water (containing up to 2 ul isopropyl alcohol/ml as IS) for alcohols and some ketones, and 1-5 ul of the extract was injected into a Hitachi FID-GC (Model 163). The column used was a 3mm x 2 m in size and packed with either 10% FFAP on Chromosorb WAW-DMCS (80-100 mesh) for the former 1/ or 25% PEG-100 on Celite 545 (60-80 mesh for the latter9/, respectively. The oven was kept at 100 °C or 50°C for the former and the latter, respectively, while the injection port was heated at 120°C in both cases 1,9/. The methods of analysis of urine for various solvent metabolites were as listed in Table 1. In some cases, the urinary metabolite concentrations were corrected for a specific gravity of urine of 1.016 13/ as measured by means of refractometry, or creatinine concentration 14/ as determined by a micromodification 15/.

Table 1   Relationship between exposure concentrations as monitored by passive dosimetry and metabolite levels in urine among solvent workers

Solvent[a]	Metabolite[b] (unit)	Method[c]	Relation[d]	Regression line[e]	Cited from
Benzene (200 ppm)	Phenol (mg/g creatinine)	GC	Linear	Men(108): $Y=4.10X+16$ (0.86); Women(175): $Y=4.50X+15$ (0.89)	Reference 8
Toluene (130 ppm)	HA (mg/g creatinine)	HPLC	Linear	Men(74): $Y=15.8X+361$ (0.60); Women(56): $Y=17.4X+262$ (0.79)	Reference 5
	o-Cresol (ug/g creatinine)	GC	Linear	Men(74): $Y=5.1X+344$ (0.40); Women(56): $Y=9.3X+168$ (0.61)	
Styrene (200 ppm)	MA (mg/g creatinine)	HPLC	Linear	Men(118): $Y=15.4X+55$ (0.86)	Reference 3
	PhGA (mg/g creatinine)	HPLC	Linear	Men(118): $Y=4.0X+38$ (0.82)	
MEK (100 ppm)	MEK (ug/ml, uncorrected)	GC	Linear	Men(62): $Y=26.3X+53$ (0.86)	Reference 4
TETRA (600 ppm)	TTC (ug/ml, corrected for a specific gravity of 1.016)	SP	Curvilinear; linear only up to 100 ppm and leveling-off thereafter. Men (36) and women (25) combined.		References 6 and 20

NOTES TO TABLE 1

a/ Figures in parentheses are the maximum concentrations studied.
   MEK: Methyl ethyl ketone.   TETRA: Tetrachloroethylene.

b/ HA: Hippuric acid.   PhGA: Phenylglyoxylic acid.
   MA: Mandelic acid.   MEK: Methyl ethyl ketone, unmetabolized
   TTC: Total trichloro-compounds

c/ GC: Gaschromatography
   HPLC: High performance liquid chromatography.
   SP: Spectrophotometry.

d/ Relationship between exposure concentrations of the solvent as monitored by passive dosimetry for the shift of work and levels of the metabolite in urine collected at near the end of the shift.

e/ Sex of workers examined (number of the workers): The equation of the regression line in which X is the time-weighted average solvent concentration (ppm) and Y is metabolite concentration followed by the correlation coefficient in parentheses.

STATISTICAL ANALYSIS

Student's t-test, either paired or unpaired depending on the cases, was employed.

RESULTS AND DISCUSSION

PERFORMANCE OF CC UNDER EXPERIMENTAL CONDITIONS

When CC TF 1500 and its sister preparation CC TF 1600 were exposed to various solvent vapors, it was found that both CC TF 1500 and CC TF 1600 adsorbed toluene, isopropyl alcohol, and acetone in a manner linear to the exposure duration (tested up to 200 ppm, 800 ppm and 400 ppm, respectively)1/. These CC preparations were exposed for 4 hrs to n-hexane, ethyl acetate and toluene either in combination (ratio; approximately 1:4:1 in ppm) or separately up to the levels 1.5 times then the occupational exposure limits (OEL) of 100 ppm, 400 ppm and 100 ppm, respectively 16/ or 4.5 times OEL under combined exposure when additiveness was considered 1/. The combination of the three solvents was selected as typical of the solvent mixture in industrial adhesives 17/. The adsorbed amount was linear to concentrations and the amount of each solvent adsorbed after combined exposure was identical with that after individual solvent exposure as far as vapor concentrations were the same 1/. In contrast, leveling-off was evident with the third CC preparation when it was exposed to the vapor mixture at 1.5 times OEL and above (additiveness considered) 1/.

When exposed to n-hexane/ethyl acetate/toluene vapor mixture under very humid conditions (i.e., 93% relative humidity

(RH)), 5 CC preparations out of 7 tested failed to adsorb n-hexane quantitatively. Typical results are summarized in Table 2.

Table 2   *Disturbed adsorption of one CC preparation due to high humidity*

Solvent	Relative humidity (%)			
	40	63	86	93
n-Hexane	100	90	45	21
Ethyl acetate	100	107	98	102
Toluene	100	118	119	107

The values in the table are

$$\frac{\text{the observed amount}}{\text{the expected amount}} \quad X \ 100 \ (\%)$$

where the expected amount was the amount detected at 40% relative humidity.

Although the effect of humidity was not remarkable at 63% RH or less, adsorption of n-hexane was reduced to a half at 86% RH or less than a quarter at 93% RH, while the efects were nil with ethyl acetate and toluene. CC TF 16 00 could adsorb essentially the same amount of n-hexane at 95% RH as at 40% RH 1/.

The recovery of the solvent after the storage of the exposed CC was examined under four conditions; CC TF 1500 exposed to n-hexane and toluene were wrapped in either aluminum foil or polyethylene sheet and stored either at 4 C or at room temperature. Storage at 4 C resulted in no loss in n-hexane nor toluene in both package of aluminum foil and polyethylene sheet for 2 weeks 2/. No significant reduction was observed in adsorbed amount of toluene for 6 months when stored at 4 C in aluminum foil (unpublished data). When kept at room temperature, howeverr loss became significant (p<0.05) on Day 9 of storage with n-hexane and on Day 14 with toluene both packed in aluminum. The loss was slightly but insignificant-ly (p>0.05) more when packed in polyethylene than in aluminum 2/. Thus, CC TF 1500 was considered to be applicable to field studies in factories distant from an analytical laboratory, in case the CC is packed in aluminum foil immedi-ately after exposure and kept refrigerated as soon as possible till analysis.

VALIDATION OF CC THROUGH BIOLOGICAL MONITORING OF EXPOSURE BY MEANS OF ANALYSES OF URINE FROM WORKERS IN VARIOUS SOLVENT WORKPLACES

Field surveys 3-8/ were conducted in shoe-making (using benzene in glue), printing (toluene, methyl ethyl ketone), fabric degreasing/cleaning (tetrachloroethylene), FRP boat

production (styrene) and other (toluene or methyl ethyl
ketone in glue) workshops, in which CC TF 1500 in a holder
was equipped on a lapel of each worker for a shift of the day
(approximately 8 hrs) to measure the time-weighted average
exposure concentration. The metabolite concentration was
determined in the urine sample collected from the worker at
near the end of the shift. The comparison of the analyses
results disclosed (Table 1) that a linear relation holds
between the solvent concentration in breathing zone air and
the corresponding metabolite level in urine (separated by a
slanting mark when multiple metabolites were examined) in
most cases studied; namely, benzene versus phenol, toluene
versus hippuric acid/o-creasol, styrene versus mandelic
acid/phenylglyoxylic acid, and methyl ethyl ketone versus
unmetabolized methyl ethyl ketone 3-5, 7,8/. The correlation
coefficients were generally large and statistically significant (p<0.01 to 0.05) in all of the cases with these four
kinds of solvents, indicating that passive dosimetry employed
is a reliable method to monitor solvent exposure when variation in physical load 18/ is essentially neglibible. Two
typical cases, one with toluene and the other with styrene,
are depicted in Fig. 1 and 2, respectively. In both cases,
the 95% confidence range was narrow enough to allow biological monitoring at least on group basis.

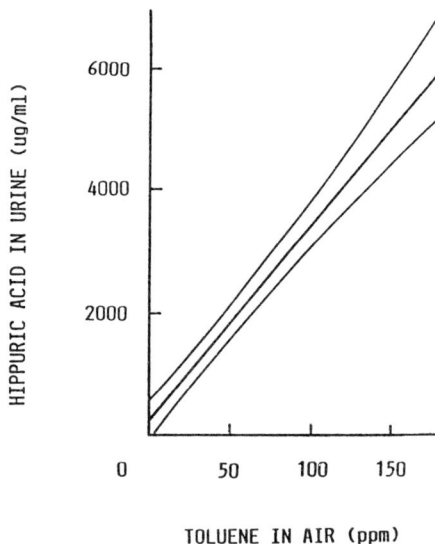

TOLUENE IN AIR (ppm)

Fig. 1 The relation between toluene concentration in breathing zone and hippuric acid level among male workers.
Toluene-workers, 74 in total, were examined. The calculated
regression line and the 95% confidence ranges of the regression line are shown. Hippuric acid levels in urine were
uncorrected (Cited from Reference 5).

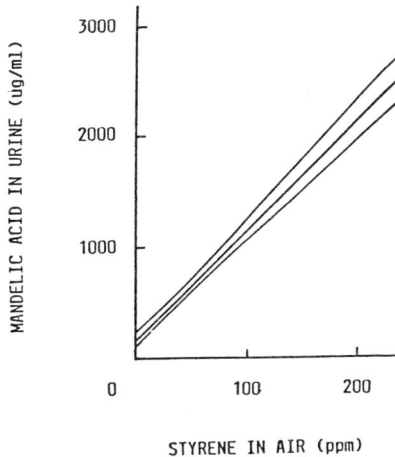

STYRENE IN AIR (ppm)

Fig. 2 The relation between styrene concentration in breathing zone and mandelic acid level among male workers.
Styrene-workers, 118 in total, were examined. The calculated
regression line and the 95% confidence ranges of the regression line are shown. Mandelic acid levels in urine were
uncorrected (Cited from Reference 3)

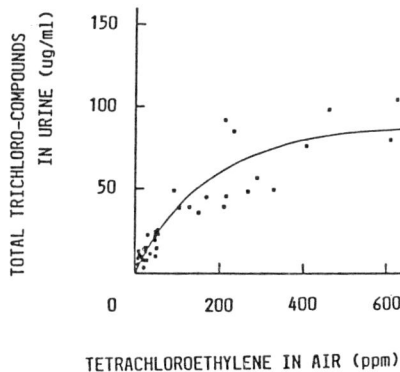

TETRACHLOROETHYLENE IN AIR (ppm)

Fig. 3 The relation between tetrachloroethylene concentration in breathing zone and level of total-compounds among
male and female workers. Tetrachloroethylene-workers, 36 men
and 25 women or 61 in total, were examined. The relation was
linear only up to approx. 100 ppm. leveling-off was evident
at higher tetrachloroethylene concentrations (Cited from
Reference 6)

When tetrachloroethylene-exposed workers were examined, however, the level of urinary metabolites (i.e., total trichloro-compounds) was linearly related to the exposure intensity only up to approximately 100 ppm and no longer proportional at higher concentrations (Fig. 3)6/. Such findings not only suggest that the capacity of humans to metabolize tetrachloroethylene is rather limited and readily saturated after relatively low exposures 6/ but indicate that passive dosimetry can offer reliable data of exposure even when biological monitoring by means of urinalysis for metabolites is no longer effective.

DIFFICULTIES IN APPLICATION OF CC FOR MEASUREMENT OF METHYL ALCOHOL

Recent survey on solvent constituents in various commercial solvent preparations for industrial use disclosed that methyl alcohol is one of the most popular solvents and present in thinners, paints, inks, degreasers and even adhesives 16,19/. The trials to apply CC TF 1500 to measure methyl alcohol is, however, so far unsuccessful; experimental exposure of the CC to methyl alcohol at up to 350 ppm revealed that the amount of methyl alcohol recovered from the CC was not proportional to the exposure duration and leveling-off was evident as early as 2 hrs after initiation of the exposure, although proportionality up to 500 ppm was observed when the exposure duration was fixed, e.g., at 4 hrs 9/.

CC TF 1500 in a holder was exposed to the vapors of several water-miscible solvents including alcohols at about 1000 ppm for 30 min and then kept in fresh air to examine possible spontaneous desorption. It was found that the amount recovered from the CC became less as a function of time after termination of the solvent vapor exposure, and that the half-time of the decay was as short as less than 1 hr for methyl alcohol and about 14 hrs for acetone. Much longer half-time was observed with other solvents tested (Table 3; unpublished data). Such findings indicate that methyl alcohol and, to a lesser extent, acetone are selectively problematic in passive dosimetry with CC TF 1500 9/.

Table 3   *The velocity of spontaneous desorption of some water-miscible solvents from CC TF 1500*

Solvent	Half-time (hrs)
Methyl alcohol	0.95
Ethyl alcohol	20.1
Isopropyl alcohol	68.4
Acetone	8.8
Methyl ethyle ketone	>100
Ethyl acetate	28.4

REFERENCES

1. Hirayama T and Ikeda M (1979) Applicability of activated carbon felt to the dosimetry of solvent vapor mixture. Am Ind Hyg Ass J 40:1091-1096

2. Ikeda M, Kumai M and Aksoy M (1984) Application of carbon felt dosimetry in field studies distant from analytical laboratory. Ind Health 22:53-58

3. Ikeda M., Koizumi A, Miyasaka M and Watanabe T (1982) Styrene exposure and biologic monitoring in FRP boat production plants. Int Arch Occup Environ Health 49:325-339

4. Miyasaka M, Kumai M, Koizumi A, Watanabe T, Kurasako K, Sato K and Ikeda M (1982) Biological monitoring of occupational exposure to methyl ethyl ketone by means of urinalysis for methyl ethyl ketone itself. Int Arch Occup Environ Health 50:131-137

5. Hasegawa K, Shiojima S, Koizumi A and Ikeda M. (1983) Hippuric acid and o-cresol in the urine of workers exposed to toluene. Int Arch Occup Environ Health 52:197-208

6. Ohtsuki T, Sato K, Koizumi A, Kumai M and Ikeda M (1983) Limited capacity of humans to metabolize tetrachloroethylene. Int Arch Occup Environ Health 51: 381-390

7. Ikeda M, Watanabe T, Kashahara m, Kamiyama S, Suzuki H, Tsunoda H and Nakaya S (1985) Organic solvent exposure in small scale industries in north-east Japan. Ind Health 23:181-189

8. Inoue o, Seiji K, Kashara M, Nakatsuka H, Watanabe T, Yin S-G, Li G-L, Jin C, Cai S-X, Wang X-Z and Ikeda M (1986) Quantitative relation of urinary phenol levels to breathzone benzene concentrationa; a factory survey. Br J Ind Med, in press

9. Koizumi A and Ikeda M (1982) Dynamics of methanol adsorption on carbon felt with special referrence to interaction with co-existing toluene. Ind Health 20:259-269

10. Koizumi A and Ikeda M (1981) A servomechanism for vapor concentration control in experimental exposure chambers. Am Ind Hyg Ass J 42:417-425

11. Kumai M, Koizumi A, Morita K and Ikeda M (1984) An exposure system for organic solvent vapor. Bull Environ Contam Toxicol 32:200-204

12. Ikeda M and Hara I (1980) Evaluation of the exposure to organic solvents by means of urinalysis for metabolites. Jpn J Ind Med 22:3-17 (In Japanese)

13. Rainsford SG and Lloyd Davies TA (1965) Urinary excretion of phenol by men exposed to vapour of benzene. A screening test. Br J Ind Med 22:21-26

14.   Jackson S (1966) Creatinine in urine as an index of urinary excretion rate. Health Phys 12: 843-850

15.   Ikeda M and Ohtsuji H (1969) Hippuric acid, phenol and trichloroacetic acid levels in the urine of Japanese subjects with no known exposure to organic solvents. Br J Ind Med 26:162-164

16.   Japan Association of Industrial Health (1978) Recommen- ded occupational exposure limits.   Jpn J Ind Health 20:290-301 (in Japanese)

17.   Kumai M, Koizumi A, Saito K, Sakurai H, Inoue T, Takeuchi Y, Hara I, Ogata M, Matsushita T and Ikeda M (1983) A nationwide survey on organic solvent components in various solvent products: Part II. Heterogeneous products such as paints, inks and adhesives. Ind Health 21:185-197

18.   Astrand I, Ehrner-Samuel H, Kilbom A and Ovrum P (1972) Toluene exposure: I Concentration in alveolar air and blood at rest and during exercise. Work Environ Health 9:119-130

19.   Inoue T, Takeuchi Y, Hisanaga N, Ono Y, Iwata M, Ogata M, Saito K, Sakurai H, Hara I, Matsushita T and Ikeda M (1983) A nationwide survey on organic solvent components in various solvent products: Part I. Homogeoeous products such as thinners, degreasers and reagents.   Ind Health 21: 175-183

20.   Tanaka S and Ikeda M (1968) A method for determination of trichloroethanol and trichloroacetic acid in urine. Br J Ind Med 25:214-219

DISCUSSION - SESSION I

BLOME (FRG)

What is the training required for someone who carries out diffusive sampling in industry?

MOORE (UK)

Most of the training comes before a diffusive sampler is used; in making sure that the device you specify is the correct one. There is very little on-the-job training required once the sampler has been selected. It's not like pumped sampling, where one has to know how to set flow rates and make sure that the batteries are charged. the simplest of instructions: how to open the device, how to close it at the end of the day, how to record the sample data are all that are required. The only practical factors that could cause problems would be the wearing and the positioning of the device which have been shown to be quite critical. There are a number of field trials that show the differences in results of samplers placed on one lapel compared to the other lapel. Workers who are left-handed or right-handed approached their work from a different direction. So this would bias the result. In summary, training for the proper selection of the samplers is essential, some on-the-job training for the handling of the samplers is necessary and some recommendations to the wearer are desirable.

BERTONI (Italy)

From my experience the positioning of a sampler is a problem which is common to any kind of sampler whether it's diffusive or not. What is important is that the manufacturers should provide instructions with any limitations of use of the sampler.

AUFFARTH (FRG)

Referring particularly to the geometry of a diffusive sampler, do you think that the specific characteristics, amount of adsorbent, cross-section, and path-length, should be standardized so that the user can be confident that the apparatus will work?

MOORE (UK)

If one takes the simple case that I was talking about where the geometry of the sampler controls the uptake rate, then obviously it's necessary to control the length and the area in manufacture. The next point would be the sorbent itself. First, one has to choose the correct sorbent and it would be nice to see some independent recommendations of optimum sorbents for particular applications. This is perhaps a job for government laboratories of Universities as manufacturers' resources are limited. Second, the amount (mass) of sorbent should also be standardized, as it is clear that a certain minimum of sorbent is necessary to achieve a good result.

On the question of accuracy, some (pumped) colormetric devices are only capable of achieving plus or minus 25% with a 95% confidence level at the TLV. They are not particularly accurate but they are certainly usable. Most industrial hygienists feel that a device with an overall accuracy of plus or minus 10% would be very acceptable and are prepared to work with devices of plus or minus 20% or greater. My own experience is that plus or minus 20% would be very easy to achieve, but that plus or minus 10% requires a lot more care, plus or minus 5% is not achievable and not necessary.

COKER (UK)

Have you looked for a correlation with exhaled breath and if so, was the difference between the ethnic groups and the sexes due to the differential between excretion and by the exhalation route or by the conversion to metabolites?

IKEDA (Japan)

We have never examined exhaled breath. It was easier in our field conditions to examine urine rather than breath.

SCHALLER (FRG)

In order to achieve significant correlations between ambient monitoring and biological monitoring, skin adsorption mustn't play a significant role. Are you sure that skin adsorption was minimal in this case?

IKEDA (Japan)

Skin penetration under usual occupational exposure conditions does not exceed 10%, so I do not think that it plays a big role in metabolitic excretion.

BERLIN (EEC)

In your correlations between biological monitoring and diffusive sampling, the bands looked surprisingly narrow. Normally, in such a correlation, the dispersion is considerable. Is your good correlation a consequence of your having taken a very large number of samples?

IKEDA (Japan)

The number of examinees was betwwen 60 and 170 and naturally the larger the number, the narrower the distribution; but please do not confuse the 95% confidence limit for the group with that for the individuals. The line shown is for the group.

SAMINI (USA)

Did you ever look at the effect of air velocity on exposure measurements?

IKEDA (Japan)

No, we have never measured air velocity in occupational settings. but in most situations, the air velocity would be expected to be high because of general or local exhaust ventilation.

MOORE (UK)

Would you like to express an opinion as to whether the diffusive monitor would have been better or worse than an active pump method?

IKEDA (Japan)

I am in favor of passive dosimeters. They are light and small and very easy for workers to use. I think that they are as valid as active sampling.

TRIEBICK (FRG)

I think that your results are very important, and our work in FRG largely confirms what you have found. Do you think that in the future it will be possible to carry out monitoring at the workplace of, for example, solvents, either through diffusive sampling or through biological monitoring, or would you recommend using both methods simultaneously in order to achieve a more reliable result?

IKEDA (Japan)

At present, we use both methods, in order to establish which is the best. In some cases, for example tetrachloroethylene, where man has limited capacity for its metabolism, passive dosimetry is definitely better, especially for measuring high concentrations. But in general, air monitoring does not take account of individual physiological variability.

MEYER (NL)

In the triangle, active monitoring, diffusive monitoring and biological monitoring, the first two measure exposure and the last dose. The two measurements are complementary and not alternatives.

DIFFUSIVE SAMPLING: AN OVERVIEW

D.C.M. Squirrell

9 Graysfield, Welwyn Garden City,
Hertfordshire, AL7 4BL, England

ABSTRACT

Some criteria set in the 1970s for the ideal personal
monitor are outlined. An overview is presented of the
diffusive samplers and monitors produced since that time.
Such monitors rely on diffusion, permeation or a
combination of both principles. Reagent type monitors with
direct or indirect read-out are reviewed together with
diffusive samplers operated with solvent or thermal methods
of desorption of the sorbed analyte. An assessment is
given of the advances made.

INTRODUCTION

In the 1970s with the increased awareness of the need
for personal monitoring and the realisation that the
requirement to monitor might become mandatory at a fairly
high frequency in several industries, thought was given to
the criteria which would be met by the ideal system of
sampling and analysis.

In my experience, the criteria set by different groups
within society overlapped in many respects. For example,
in the pursuit of safety they all required that the
monitoring system be reliable in operation and give results
of sufficient accuracy and precision to facilitate valid
conclusions and prompt correct actions. There were however
differences in emphasis, influenced by the nature of the
activity and the availability of other monitoring systems
on the site.

The manufacturer of a toxic volatile material would
often require continuous monitoring and, in addition,
systems for detecting leaks and for the monitoring of

clearance of major spillages. Management required time
weighted average figures to satisfy personal, medical and
legislative requirements. The operators themselves
expressed the need for an immediate warning of "short term"
exposure above a safe limit as well as an integrated dose
measurement over a period of time. The primary user of the
toxic product stated similar needs but placed less emphasis
on continuous area monitoring. The down stream user had
well defined areas of plant over which contamination might
occur and only particular operations required personal
monitoring. Outside the industrial boundaries
contamination was likely to be infrequent and at extremely
low levels and long term accumulations were the main
interest. Facilities had to be available however for
routine and emergency monitoring of the environment.

Within the above categories, different groups of
people suggested different factors or objectives which
could influence the technical specification of a personal
monitor suitable for use in large numbers, should that
become necessary. Clearly everyone insisted on good
performance. Personnel who had to wear the monitors
required them to be small, unobtrusive and to cause no
inconvenience during working operations. The analyst, of
course, required simplicity, reproducibility of
performance, ease of calibration and as far as possible
uniformity of proceedure with options for automation for
the handling of large numbers of samples. The occupatioal
hygienists, medical officers and management expressed the
wish for a monitoring system not subject to abuse by
intentional or accidental exposure of the sampler to unreal
high levels of contaminant not actually encountered by the
operator. In this respect, a monitor to be worn near to
the nose and mouth but which would be covered by any
breathing apparatus being worn was clearly desirable so
that only true exposure was registered. In addition, they
preferred to have a monitoring system capable of measuring
several contaminants simultaneously for survey purposes.

There were, of course, other aspects which had to be considered by management. Capital outlay and the cost of administering an extensive monitoring programme could be significant and simplicity and uniformity are conducive to reducing these costs. Personnel and Union acceptance and cooperation were considered essential as was the recognition by medical and regulatory authorities that the results obtained were meaningful.

In summary therefore our ideal monitor would fit the following general specification:-

Reliable, precise and well validated.
Operable under all normal environmental and working conditions.
Very small — can be covered by breathing apparatus.
Simple to use.
Fast response to high levels with direct read-out.
Instant warning of high levels.
Integrated collection/assessment of levels over minimum of 8 hour shift with direct read-out.
Not easily subject to abuse.
Where direct read-out not possible or more than one analyte present, then a simple and uniform system of analysis.
Automated analysis possible.

The intention of this paper is to explore how far the above criteria have been met by the diffusive monitoring systems so far developed and to overview their fields of application. The monitors will be examined in two groups — those based on a chemical reagent collection/analysis principle and those based on a sorption collection basis with subsequent desorption for the analysis stage. Both systems can work with either a diffusion geometry, a permeation principle or a combination of both, (Saunders 1981). Many ingenious and different designs of both tube and badge configurations have been described and most are available commercially but for the sake of brevity only

simplified outline diagrams are presented in figures 1-7 of
this paper. Full details will no doubt be provided in
other contributions to this Symposium.

SAMPLER EVALUATION
    In the 1970s several factors were shown to have a
possible effect on the efficiency of diffusive samplers and
it is now commonplace for the following to be investigated
as routine:-

Pre-use
    Storage of unused sampler - shelf life.

Sorption/reaction stage
    Sorption efficiency and range.
    Deviation from predictions from Fick's law.
    Temperature.
    Pressure.
    Humidity.
    Air velocity/face velocity.
    Saturation level of sorbent/reactant.
    Response time.
    Effect of short exposures at high concentrations.
    Evidence of back diffusion/permeation during
exposure.
    Reproducibility of chemistry.

Storage before desorption
    Evidence of loss or decomposition of sorbed
pollutant.

Desorption
    The reproducibility and efficiency of the desorption
    process, whether this be by a solvent or
    thermal method.
    Direct reuse of sampler after thermal desorption.

Other factors which received attention were calibration
methods, the varibility of calibration with range and the
provision of back-up when the primary sorbent became
saturated.   In the discussion which follows the above
points are termed "standard variables" although clearly not
all are investigated by every researcher.

REAGENT TYPE MONITORS : DIRECT READ-OUT
     Direct-reading reagent type monitors, shown
diagramatically in figures 1 & 2, rely on the pollutant
reacting chemically with the reagent system dispersed on a
suitable support to produce a colour change, the length or
tint of which is proportional to the pollutant
concentration.

Direct Reading Reagent-type Monitors

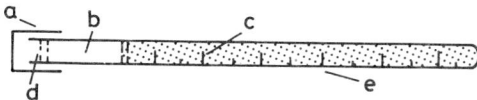

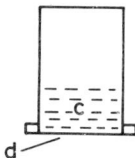

Figure 1   Tube   Figure 2   Badge

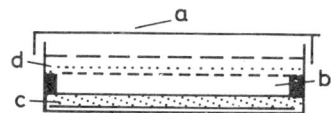

stain length measured   stain tint compared
(a) removable cap   (b) diffusion layer or gap
(c) supported reagent   (d) membrane (optional)
(e) calibrated scale

Indirect Reading Reagent-type Monitors
Figure 3   liquid reagent   Figure 4   Badge

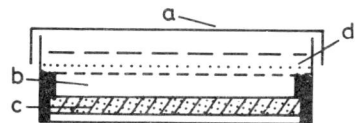

(a) removable cap   (b) diffusion layer or gap
(c) supported reagent/solution   (d) permeation membrane
     (optional in Figure 4)

Lead chloride on acid washed diatomaceous earth was used (Sefton et al, 1982) to determine H_2S, and iodine pentoxide in sulphuric acid as the reagent for benzene. These authors investigated the necessity for a silicone membrane in their tube monitors and discussed the relatively short stain length obtained for benzene in the range up to 120 ppm-hrs. Tube monitors manufactured by MSA (McConnaughey et al. 1985) for NH_3, CO, CO_2, H_2S, SO_2 and NO_2 relied on diffusion only and no permeation membrane was used. The reacting chemical was supported on a paper strip or inert gel held in a plastic tube. The effects of some standard variables were investigated and interferences quantified. Pannwitz (1984) described in trade literature the Dräger direct-reading diffusion tubes for NH_3, HCl, H_2S, NO_2, SO_2, CO, CO_2, and HCN. These devices again relied on diffusion only and the supported chemicals were contained in glass tubes opened for exposure by the breaking method used in the well known detector tubes. The effects of several variables were documented and interference details given. A similar tube for CO is supplied by Gastec. Dräger diffusion tubes operating on the reagent principle are also available for organic vapours, including acetone, acetic acid, butadiene, ethyl acetate, and olefines. All these tubes operate over a wide humidity range but require a small correction for pressure. With the exception of acetone, a correction is also recommended for use at temperatures other than 20°C.

Moore et al. (1980) developed a badge monitor for GMD Systems Inc. for the determination of phosgene. One exposed area of reagent coated paper gave an approximate value for the dose received and hence an immediate warning. As the sampler was affected by HCl, a second area was protected by a filter barrier unit and could be used as a measure of total dose at the end of the shift.

Two methods of colour comparison were available; a pocket sized comparator containing two colour wheels permitting rapid on-site measurement and a Dosimeter Reader designed to give more accurate measurement of the colours by reflectance. The system has been well evaluated against the standard variables.

A badge-type monitor was developed by the UK Health and Safety Executive (HSE. 1985) which incorporated both direct and indirect read-out facilities for the determination of mercury vapour. The diffusive sampler consisted of a Porton Down type badge (Bailey et al, 1977) containing a paper impregnated with cuprous iodide to collect the Hg vapour in the concentration range $0.0025-0.10mg/m^3$. for an exposure time of 8 hours and pro rata for 4 hours etc. For on-site read-out, the pink orange colour produced was compared with synthetic colour standards on paper and for more accurate laboratory evaluation the Hg was desorbed with acid and determined by cold vapour atomic absorption spectrometry at 253.7nm.

REAGENT TYPE: INDIRECT READ-OUT

Reagent type monitors with indirect read-out rely on the instrumental measurement of the weight changes, colour changes or changes in conductivity (Merino, 1979), due to the chemical reaction/adsorption of the diffusing pollutant with an appropriate reagent. As early as 1973, Palmes and Gunnison (1973) described diffusion devices for the determination of SO_2 and water vapour and in the same year, Reiszner and West (1973), a permeation monitor also for SO_2

The chemistry here was the West-Gaeke procedure and the absorption solution was supported by a silicone membrane at the bottom of a 41mm glass tube, (fig 3.). Bell, Reiszner and West (1975) continued with this principle in a monitor for CO using silver p-sulfamoylbenzoic acid solution as reagent. The absorbance due to the yellow silver sol produced with the CO was measured at 380nm. Good response to standard variables, temperature, humidity etc. were reported.

As well as the REAL Minimonitors for organic pollutants, mentioned elsewhere in this paper, the same company markets a series of Biobadges for Cl_2, HCN, SO_2, and NH_3. These systems use chemical collection media and rely on chemical analysis after the exposure period, (fig.4.).

The use of an indirect read-out system in monitors in a badge configuration was reported by Kring et al.(1981) in describing the Pro-tek range for NH_3, SO_2, and NO_2. The pollutants diffused through a multicavity diffusion layer to be absorbed in the selected solution prior to colorimetric reaction with specific reagents contained in plastic blisters. After colour development the absorbance was measured directly in the read-out instrument. The effects of some standard variables were studied and a temperature correction recommended for NH_3 and NO_2.

A Pro-Tek badge is also available for formaldehyde. Here the absorption solution is sodium bisulphite and the analysis is by the standard chromotropic acid-sulphuric acid method. Response time is short and storage stability good. No temperature corrections are required between 15 and $40°C$. Ethanol and higher alcohols cause a reduction in colour and phenol also interferes.

Chemical reagents impregnated onto appropriate substrates provided the collecting elements for the range of Walden GAS-BADGES (Tomkins and Goldsmith 1977). After exposure, these elements were removed from the badges (cf fig.4) for chemical analysis. The authors showed that the mass of analyte collected was virtually independent of temperature and pressure variation and that the air velocities in use were, for most practical purposes, sufficient to ensure satisfactory results with short response times. Dosimeters are available for SO_2, NO_2, and H_2S.

Purnell et al.(1981) compared the Porton Down badge sampler with a Health and Safety Executive tube system using carbon cloth or reagent impregnated paper sorbent elements for the collection of Hg, H_2S and several organic vapours containing characteristic elements which could be

determined by X-ray spectrometry. They concluded that the
XRS finish was good and that the use of a reagent
impregnated paper for inorganic gases would enable on-site
analysis based on colour change to be made and the results
confirmed later by the more precise XRS method. The
badge-type sampler proved the more efficient for organic
vapours. Olin (Cohen, 1981) also produced and evaluated a
diffusive sampling dosimeter for mercury.

SORPTION TYPE MONITORS

Descriptions of diffusive samplers based on the
reversible sorption principle and in both badge and tube
configurations (figs. 5 & 6) have been numerous.
Desorption of the pollutant has been mainly by solvent
extraction but the use of thermal desorption techniques is
gaining favour mainly because of the ease of automation of
this stage and direct coupling with gas chromatography,
particularly with samplers in the tube format.

Solvent Desorption

The 3Ms Organic Vapour Monitor (Merino, 1979) also
operates on a diffusion principle with sorption onto a
charcoal wafer and desorption carried out within the
monitor itself by the addition of carbon disulphide prior
to gas chromatographic analysis. In a more sophisticated
version of this badge, a second layer of sorbent is
included, separated from the first by a small spacer as in
fig. 7. This portion can be analysed separately thus
coping with any over saturation of the primary layer.

The Pro-Tek G-AA diffusion type badges for organic
vapours utilise a multicavity diffusion layer and a 300mg.
activated charcoal sorption element in a strip
configuration. Lautenberger et al.(1980), described how
sampling times could be up to 16 hours, after which the
strip was removed from the case for solvent desorption and
gas chromatographic analysis. Sampling rates and the
effects of the standard variables were evaluated including

air face velocity, range, humidity, sorption temperature,
storage stability and desorption efficiencies under
different conditions. A badge was also available with
back-up facilities. The badge testing protocol and
evaluation results including satisfactory comparisons with
conventional pump and tube methods have been described by
Kring et al.(1982)

Sorption Samplers

Figure 5   Tube configuration

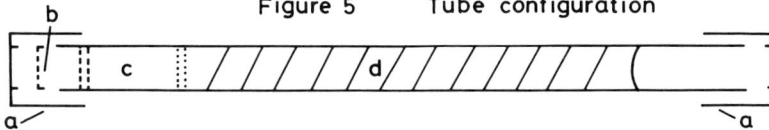

(a) removable cap   (b) membrane (optional)
(c) diffusion layer or gap (d) sorbent

Figure 6   Badge   **Figure 7   Badge with back-up**

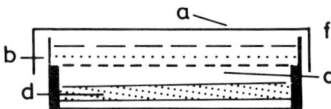

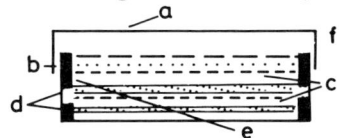

(a) removable cap   (b) membrane (optional)
(c) diffusion layer or gap   (d) sorbent
(e) spacer   (f) wind shield

Badges for organic vapours are also available in a
similar configuration from MSA as VaporGard organic vapour
dosimeters. These badges include the usual windshield,
diffusion gap and primary sorption strip separated from a
back-up strip by a special separator. The badges are not
recommended for high molecular mass or high vapour pressure
compounds.

Three sorption discs of carbon cloth were housed in a
circular badge similar to the Porton Down design, in a
system described by Pozzoli et al (   ) which also
incorporated a porous polymer diffusive layer. These
authors described their test rig used for calibration and

the determination of the effect of the standard variables. Particular attention was paid to surface area exposed and the height of the diffusion barrier. After exposure the three discs were separately solvent desorbed and analysed by gas chromatography. The design features minimised the effects of short exposure times and air turbulance and satisfactory performance was reported for a range of solvents.

A compact tube sampler having a diffusive barrier at each end and a 300mg loading of coconut shell carbon at the centre is available from Dräger as the ORSA 5. A special capping and holder arrangement is provided for the tube which is solvent desorbed after exposure, for chromatographic analysis. The standard variables have been examined and no major problems found with the large range of solvents investigated. Care must however be taken to avoid losses during transfer of the sorbent from the tube to the desorption vial. Pannwitz 1984) has published a comparison of the ORSA 5 sampler with other methods including detector tubes, pumped tubes and liquid absorption systems. For the examination of a large range of solvents he concluded that there was no essential difference between the results obtained by the active and diffusive methods of sampling.

Thermal Desorption

Although thermal desorption can be used in conjunction with samplers relying on both diffusion and permeation principles, it so happens that most applications described have related to diffusion type samplers. However, because the weaker sorbents, permitting effective thermal desorption, require greater protection from the effects of humidity, a silicone membrane is also employed so that both diffusion and permeation principles apply.

West and Reiszner (1978) described a permeation type sampler for vinyl chloride. This REAL badge used a charcoal sorbent and a silicone membrane,the permeation

characteristics of which were reasonably constant over the
temperature range 0-40'C. Changes in humidity also had
little effect and calibration studies in the UK gave
excellent results. Many of the standard variables were
examined in the Author's own laboratory using several
different designs of permeation sampler. Provided
calibration factors were established for each design and
for each batch of silicone membrane the devices proved very
reliable for vinyl chloride. The REAL badge sampler,
marketed as the Minimonitor can also be used for other
organic analytes and both solvent and thermal desorption
can be used.

The Monsanto Dosimeter, described for acrylonitrile
monomer uses the diffusion principle in conjunction with a
porous polymer sorbent and thermal desorption. In their
paper Benson and Boyce (1981) confirmed the advantages
found by other workers in using a desorption system with a
cold trap to concentrate the desorbed pollutant for a rapid
revolatilisation as a sharp band into the gas
chromatograph. The standard variables were examined and
side by side comparisons with other methods made. The
device and analytical method proved satisfactory for
working in the region of 4ppm acrylonitrile and clearly
demonstrated its potential for the determination of other
contaminants.

The advantage of a diffusion/permeation sampler which
could be thermally desorbed, preferably on an automatic
basis, and then be reused without further treatment was
first seen by Brown, Charlton and Saunders (1981). As a
first stage a tube-type sampler was developed containing a
charge of Porapak Q and fitted with a silicone permeation
membrane above a short diffusion gap. The final design
developed in conjunction with an ad-hoc committee of other
researchers is shown in fig.8 and an evaluation against the
standard variables gave very encouraging results. This
design was the forerunner of the tubes now used in the
automated thermal desorber ATD 50, marketed, after a

collaborative development programme, by Perkin Elmer. The
ad hoc committee, now Working Group 5 of UK Health and
Safety Executive, Committee on Analytical Requirements have
subsequently evaluated this sampler, amongst others. As a
result, the HSE has published diffusive sampling methods in
its MDHS series for styrene (MDHS 43), benzene (MDHS 50)
and acrylonitrile (MDHS 55).

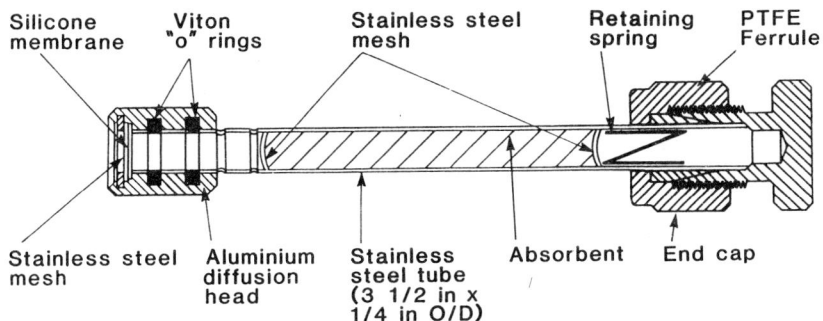

Fig. 8 Permeation/diffusion version of Perkin Elmer
tube sampler. (Saunders, 1981)

The tube configuration with thermal desorption is also
preferred by Bertoni et al.(1982) who used Pyrex glass
tubes packed with graphitized carbon black. The diffusion
barrier was made with glass wool and a short length of
Carbopak B. The weak sorption properties of graphitized
carbon black permitted high efficiency thermal desorption
and these authors described very satisfactory performance
with a range of solvents including butanol, toluene and
butylacetate, determined during field trials.

The sampler described above (Brown, Charlton and
Saunders, 1981) has been used for sampling nitrous oxide
prior to thermal displacement for infra-red absorption
measurement (Cox and Brown, 1984). Molecular sieve was
used as sorbent and as would be expected the uptake rate

for this sorbent/pollutant combination was dose dependent and a series of calibration curves was required for low concentrations. The authors have investigated the standard variables and shown that humidity had no great effect and there was no problem with back diffusion. There was a loss of 27% nitrous oxide during tube storage for two weeks indicating analysis should be carried out as soon as possible after sampling. Comparative field trials indicated that the performance obtained was similar to that from conventional pump and tube methods.

Several papers (Underhill, 1984, Bartley, 1983, Bartley et al., 1983, Koizumi and Ikeda, 1982) have given theoretical considerations to the errors which could arise due to large fluctuations in pollutant concentration, particularly over short periods of time, when weaker sorbents were being used the sorption efficiency of which may change with total up-take of pollutant. The message would seem to be, use the strongest sorbent possible, consistent with the desorption system to be employed and use a high enough capacity to suit the application in hand.

CONCLUSIONS

Some of the important concepts for the ideal personal monitor have been met by the diffusive samplers described in this overview. They are all reasonably small and cause minimal inconvenience to the wearer during normal working operations. Most however, have been designed to wear on the lapel and would require capping or removal whilst breathing apparatus was being worn. None are free from intentional abuse but it is realised that prevention of this would be difficult (expensive) to attain. The designs are, in general, easy to operate in terms of capping and uncapping or activating by means of break-seal tubes. Simple wearer identification by labelling facilities eases administration problems. Most samplers will respond to short term high exposures and integrate over a collection period of 8 hours. Only those reagent-type tube or badge

monitors giving direct read-out of accumulated dose will,
however, warn of short term high exposure and then only by
visual examination of the length or tint of a chemical
stain. Frequent observation of the monitor is thus
required and this of course is not desirable for a busy
operator. As distinct from monitors in the tube
configuration, only one badge monitor has a facility for
very rapid response to short term concentrations. An
audible warning would be preferable but it is realised this
would add considerably to cost. Most of the reagent type
monitors , because of their chemistries, can only be used
for one compound or one group of compounds. This is an
advantage in providing some specificity but a disadvantage
in not quantifying any actual interferences, although
likely interferents and their effects are usually well
documented by the manufacturers. The indirect read-out of
reagent type monitors requires a variety of read-out
systems and/or chemistries. The sorption/desorption
systems particularly for organic compounds have the
advantage that most analyses can be completed by gas
chromatography.

Diffusive samplers are very much cheaper for extensive use
than the conventional pump-tube samplers and because of
their small size find greater wearer acceptance.
Administration of an extensive monitoring programme or
survey has been made easier by manufacturers by their
clever design and in some cases by the provision of a
complete analysis evaluation service. With the sorption
systems the emphasis to date has been on solvent
desorption, probably because it was the first method used
and usually permits more than one analysis per desorbed
sample. Tube configuration monitors with automated thermal
desorption are now finding greater favour and will no doubt
continue to do so as the number of samplers in use
increases making the capital cost of automated systems
worthwhile.

ACKNOWLEDGEMENTS
     The Author acnowledges the assistance provided by Dr
R.H. Brown and Dr K.J Saunders in the preparation of this
paper.

REFERENCES
Bailey, A., Hollingdale-Smith, P,A. 1977. A personal
     diffusion sampler for evaluating time weighted
     exposure to organic gases and vapours. Ann.
     Occup. Hyg., 20, 345-356.
Bartley, D. 1983. Passive monitoring of fluctuating
     concentrations using weak sorbents. Am. Ind. Hyg.
     Assoc. J. 44(12), 879-885.
Bartley, D., Doemeny, L.J., Taylor, D.G. 1983. Diffusive
     monitoring of fluctuating concentrations. Am. Ind.
     Hyg. Assoc. J. 44(4), 241-247.
Bell, D.R. Reiszner, K.D., West, P.W. 1975. A permiation
     method for the determination of average concentrations
     of carbon monoxide in the atmosphere. Anal. Chim Acta.
     77, 245-254.
Bertoni, G., Perrino, C., Liberti, A. 1982. A graphitized
     carbon black diffusive sampler for the monitoring of
     organic vapours in the environment,. Anal. Letters
     15(A12), 1039-1050.
Brown, R.H., Cox, P.C., 1984. A personal sampling method
     for the determination of nitrous oxide exposure,
     Am. Ind. Hyg. Assoc. J. 45(5), 345-350.
Benson, G.B., Boyce, G.E., 1981. A thermally desorbable
     passive dosimeter for personal monitoring for
     acrylonitrile. Ann. Occup. Hyg. 24,55-75.
Brown, R.H., Charlton, J., Saunders, K.J., (1981), The
     development of an improved diffusive sampler. Am. Ind.
     Hyg. assoc. J. 42(12), 865-869.
Dräger, Trade literature, 81 01 291, 81 01 071, 81 01 161,
     81 01 241, 81 01 171
Gastec, Trade literature, CO.
HSE: Occ. Med. & Hyg. Lab., 1985. MDHS, Mercury vapour in
     air.
Kring, E.V., Lautenberger, W.J., Baker, W.B., Douglas,
     J.J., Hoffman, R.A., 1981. A new passive colorimetric
     air monitoring badge system for ammonia, sulphur
     dioxide and nitrogen dioxide. Am. Ind. Hyg. Assoc.
     42(5), 373-381.
Kring, E.V., Thornley, G.D., Dessenberger, C.,
     Lautenberger, W.J., Ansul, G.R., 1982. A new passive
     colorimetric air monitoring badge for sampling
     formaldehyde in air. Am. Ind. Hyg. Assoc. J. 43(10),
     786-795.
Kring, E.V., Graybill, M.W., Morello, J.A., Ansul, G.R.,
     Adkins,J.E., Lautenberger, W.J. 1982. Pro-tec organic
     vapour air monitoring badges. ASTM STP 786. 85-103.
Lautenberger, W.J., Kring, E.V., Morello, J.A. 1980. A new
     personal badge monitor for organic vapours. Am. Ind.
     Hyg. assoc.J. 41(10), 737-747.

McConnaughey, P.W., McKee, E.S., Pritts, I.M. 1985. Passive colorimetric dosimeter tubes for ammonia, carbon dioxide, carbon monoxide, hydrogen sulphide, nitrogen dioxide and sulphur dioxide. Am. Ind. Hyg. Assoc. J. 46(7), 357-362.

Merino, M. 1979. Passive Monitors. National Safety News. 56-58.

3M Trade Literature, Organic Vapour Monitors, 3500/3510.

Moore, G., Matherne, R.N., Self, C., 1980. The development of a disposable passive dosimeter for phosgene detection and experiences with field application. Am. Ind. Hyg. Assoc. Paper to conference May 1980.

Olin Corporation, Trade Literature. 1983

Palmes, E.D., Gunnison, G.D. 1973. Personal monitoring device for gaseous contaminants. Am. Ind Hyg. Assoc. J. 34, 78-81.

Pannwitz, K., 1984. Direct-reading diffusion tubes. Drager Review. 53, 10-14.

Pannwitz, K., 1984. Comparison of active and passive sampling devices. Drager Review. 52, 19-28.

Pannwitz, K., 1981. ORSA 5, A new sampling device for vapours of organic solvents. Drager Review, 48, 8-13.

Pozzoli, L., Cottica, D., Ghittori, S., Universita di Pavia Fondazione Clinica del Lavoro. A new passive sampling device.

Pro-tec, Trade Literature. The new colorimetric air monitoring badge system.

Purnell, C.J., West, N.G., Brown, R.H., 1981. Colorimetric and X-ray analysis of gases collected on diffusive samplers. Chem. & Ind. 5th. September.

Reiszner, K.D., West, P.W., 1973. Collection and determination of sulphur dioxide incorporating permiation and West-Gaeke procedure. Env. Sci. & Tech. 7(6), 526-532.

REAL, Trade Literature. Minimonitor.

Saunders, K.J., 1981. An alternative approach to personal monitoring. International Environment and safety, August, 54-56.

Sefton, M.V., Kostas, A.V., Lombardi, C., 1982. Stain length passive dosimeters. Am. Ind Hyg. Assoc. J. 43, 820-824.

Tompkins, F.C., Goldsmith, R.L., 1977. A new personal dosimeter for the monitoring of industrial pollutants. Am. Ind. Hyg. Assoc. J. 38(8), 371-377.

Underhill, D.W., 1984. Efficiency of passive sampling by adsorbents. Am. Ind. Hyg. Assoc. J. 45(5), 306-310.

MSA, Data sheet 08-00-35 VaporGard inorganic vapour dosimeter tubes. Data sheet 08-00-36 VaporGard organic vapour dosimeter badges.

West, P.W., Reiszner, K.D., 1978. Field tests of a permiation-type personal monitor for vinyl chloride. Am. Ind. Hyg. Assoc. J. 39(8), 645-660.

HOW GOOD ARE PASSIVE SAMPLERS?

A PRACTISING HYGIENIST'S VIEW OF FIELD TRIAL DATA

David T Coker
EXXON Company International
ESSO Research Centre
ABINGDON, OXON OX13 6AE. UK

ABSTRACT

Passive samplers have been available as a replacement for pumped samplers for around 15 years now. But they have only been partially successful in displacing pumps in spite of the very considerable amount of work that has been put into their evaluation. This paper reviews the body of evaluation data available from the user - the field Industrial Hygienist - viewpoint, re-examines the conclusions and recommendations in published papers and summarises their general advantages and disadvantages.

INTRODUCTION

Most of the evaluation studies on passive samplers have concentrated primarily on their accuracy, however there are two other important factors which are often overlooked; the integrity of the method and the lower limit of its reliable range.

The integrity of exposure measurements must be high because;

o   they must be acceptable as proof of compliance with exposure limits
o   they must be acceptable to employees as proof that their working conditions are "healthy"
o   they may indicate the need for investment in expensive control measures
o   they may be required as evidence in legal actions, possibly retrospectively
o   they may be used to assess the suitability of exposure limits

The monitoring method should work reliably at levels well below the current exposure limit because;

o   exposure levels must, to be in compliance, be well below the exposure limit
o   airborne concentrations in the workplace commonly fluctuate over many orders of magnitude above and below the mean concentration
o   statistical analysis of data requires all data values to be valid
o   exposure limits are frequently lowered, and past exposure data will be needed to assess the impact of new lower limits.

These factors must be born in mind when assessing any new method, particularly when it is to replace a proven technique. The incentive to justify changing must be high.

The currently used pumped sampler is the accepted proven technique for personal sampling and, in many cases, exposure limits have been based on exposure data obtained using this technique.

Many of the published assessments of passive samplers are from laboratory trials in exposure chambers, comparing data from sets of both pumped and passive samplers. This type of trial is a useful preliminary step in the evaluation of a new technique but it can in no way be considered a sufficient validation in itself for a personal monitoring method, because it does not reproduce the conditions and variables that occur in real use.

For instance,

- The temperature is fixed and stable
- The air velocity is fixed and stable
- The samplers are immobile and in fixed orientation
- The airborne concentrations are fixed and stable
- Humidity is fixed and usually lower than outdoors

In real situations most of these parameters can change, often widely and rapidly, and although some can be varied in a lab trial, it is almost impossible to simulate real conditions.

In laboratory trials the performance of passive samplers has usually been found to be adequate, with precision often better than pumped samplers. However, as these are not a sufficient assessment of performance in real use, this paper will only consider the data from field trials.

In practice, sampling is carried out with a single sampler per subject, and so one passive sampler would be replacing one pumped sampler. The main consideration for accuracy will thus be how well a passive sampler measurement agrees with a pumped measurement and so paired data of pumped/passive are the most appropriate to consider.

PERFORMANCE CRITERIA

A major problem with assessing a new method is that there is no criterion of acceptability. The NIOSH standard which is often applied, specifies that in lab trials the bias of the method plus twice the repeatability relative standard deviation should not be more than 25%. However, this was originally intended for the lab evaluation of substances using the charcoal tube method and was not for applying to field tests of different techniques.

When applied to paired field samples, the statistical approach in the NIOSH standard is a more understandable test for assessing the degree of difference to be expected by replacing a pumped sampler with a passive than either the paired t-test or linear regression analysis. These are both commonly used tests, but are often misapplied (17,18), do not give a direct indication of the accuracy of the method and have sometimes been observed to give conflicting indications of acceptability.

Many papers on field trials give only statistical data to support the authors conclusions. Few give the raw data to allow readers to make their own assessment.

REPEATABILITY

It is often said that the repeatability from replicate samples in a field situation is better for passive than pumped samplers.

In our own field trials (1) we found that this was true for passive samplers of the same type, but not true between different types of passive sampler. i.e.

	Repeatability (1 S.D.)
Pumped (solvent and heat desorption)	6%
Same passive	3%
Different passives	24%

A similar study (7) recently also contained data to confirm this phenemonon.

	Repeatability (1 S.D.)
Pumped (all solvent desorption)	10%
Same passive	7%
Different passives	22%

These differences, in our trials, did not appear to be due to uptake rate errors as the scatter of data was evenly distributed.

RELIABILITY

It is often said that passive samplers are more reliable than pumped, because they eliminate pump failure. However passives do appear to suffer from the occasional "wild" result. In field trials these can be seen as values which differ by more than a factor of 2. In our evaluation we found 2% of "wilds". In other trials there were up to 10% of "wilds".(2)

It should be appreciated that in use, a pump failure can be detected. A "wild" passive sampler measurement would not be detected.

PAIRED TESTS

The analysis of 25 sets of field data covering 8 substances, comprising 8 sets of our data and the rest, largely unpublished from various sources has previously been reported.(2)   The NIOSH type analysis of these sets showed;

bias averaged	20%, range	-75% to +35%
S.D's averaged	25%, range	8% to 43%
95% C.L. averaged	70%, range	33% to 115%

A recent report (5) quoted 95% C.L.'s of 87% and 201% in field trials of more than 200 pairs.

Some of these field trials were sampled at fixed locations and some were personal samples. It was noticeable that the level of agreement on personal sample pairs tended to be worse than at fixed locations. this may be due to local variations where the samplers were not close enough together. However one aspect of passive samplers which does not appear to have been investigated is the affect of the wearers movement. This could conceivably cause turbulence of the air within the sampler's diffusion layer,which must be immobile where the uptake rate is controlled by molecular diffusion

These trials were all made on outdoor plant where it has been shown that concentrations vary widely and rapidly within time periods much shorter than the time constants of passive samplers (9) . There are some studies which indicate better agreement when used in an indoor situation. (10)

INTERPRETATION OF RESULTS

Where raw data has been published the conclusions expressed can often be questionable. For example:

A field trial of 69 paired samplers (4) reported 95% confidence limits of 18% overall, whereas examination of the data (15) showed that 60% of the pairs differed by more than 18%. The NIOSH calculation gave a figure for 95% C.L. of well over 100%, although by breaking this down into concentration ranges the accuracy was found to be concentration dependant. But the accuracy was only acceptable above the exposure limit, thus severely limiting the usefulness of the method. This appeared to be the case in another recent field trial of 241 paired samplers (3) which also concluded that passive samplers were suitable because their accuracy at the exposure limit was said to be 12%. Examination of the data (16) showed less than five values were around the exposure limit and the accuracy of the bulk of the data was 75%.

Quoting accuracy at inappropriate levels appears to be not uncommon. A recent paper (6) showed a plot of pumped and passive results for 15 field test pairs and concluded that the coefficient of variation was 24%. However, the plot showed that 11 out of 15 were above the exposure limit, the values below the limit were crammed into the area around the axis of the plot and all appeared to be outside the coefficient of variation quoted.

LOWER RANGE

Most of the data available on field test pairs indicates that the agreement between pumped and passive samplers is concentration dependant, with wide discrepencies apparent at levels below 0.5 to 1 ppm, but more acceptable agreement at high concentrations. This indicates that passive samplers are most suited to higher concentrations, and less useful for substances with exposure limits of only a few ppm, or where exposures are commonly below one ppm. For instance the new UK MDHS passive method for acrylonitrile (8) has a lower limit of use of 1 ppm (the pumped method is 0.1 ppm). With the current exposure limit of 2 ppm it seems likely that this method will be of limited use.

GROUPED DATA

When exposure values are assessed in sets rather than singly the 95th percentile of the data distribution is a commonly used parameter for "compliance" with an exposure limit. Calculating this from our field data for four different types of passive sampler gave predictions of the 95th percentile which varied on average by 21% from the pumped value. A value of 20% was found in another study (5).

CONCLUSIONS

o   The overall difference in individual exposure measurements arising
    from using a passive instead of a pumped sampler is, on average 70% at
    95% confidence level for outdoor plant use.

o   This difference is probably lower for indoor workplace situations and
    is less than the repeatability of pumped samplers in laboratory trials
    under steady conditions.

o   The differences between pumped and passive samplers are concentration
    dependant and limit the reliable use of passives to levels above about
    0.5% to 1 ppm in outdoor situations over full shifts.

o   The differences between pumped and passive sampler pairs is also
    evident between different types of passive sampler, but not between
    the same types.

o   It appears probable that rapid and wide fluctuations in concentration
    have a fundamental affect on passive samplers, different designs being
    affected differently. There are some theoretical explanations for this
    phenomenon, (11,12) but lab trials have not been able to simulate
    sufficiently rapid and extensive changes to enable it to be assessed
    experimentally. (13,14).

o   Differences between pumped and passive samplers appear to be greater
    with personal sampling than at fixed locations. This may be because
    concentration is locally more variable on a person than in free space.
    Another possibility which does not seem to have been investigated is
    that the movements of the wearer may cause turbulence of the air
    within the samplers diffusion gap due to its inertia, this air must be
    immobile for a molecular diffusion controlled uptake rate.

o   When data are statistically analysed in groups, the random differences
    tend to become averaged out and the inaccuracies are less apparent.

o   Passive samplers show occasional "wild" results with errors of more
    than a factor of two. These would not be detectable in normal useage.

o   Passive samplers main functional advantage is that the use of the
    sampling pump is eliminated
    Their main functional disadvantage is that there is no direct
    indication of the sample "volume", which has to be estimated from
    laboratory measurements of uptake rate.

    It is against this advantage and disadvantage, and the conclusions
    discussed above that the decision to replace pumped sampling by
    passive must be judged.

REFERENCES

1   COKER,D.T. JONES,A.L. SIMMS,M.C. 1981. A practical and theoretical assessment of passive monitors. British Occupational Hygiene Society Annual Conference. Nottingham UK

2   COKER,D.T. 1981. Use of passive samplers in the field. Royal Soc. Chem. Conference. December 1981 London

3   WOOD,I. 1986. Measurement of organic nitrile concentrations. Ann. Occup. Hyg. 29, 399-413

4   BENSON,G.B. BOYCE,G.E. 1981. A thermally desorbable passive monitor for personal monitoring of acrylonitrile. Ann. Occup. Hyg. 24, 55-75

5   TINDLE,P. 1983. The pros and cons of diffusion sampling. Inst. Occup. Hyg. Ann. Conf. Report No.4

6   KRING,E.V. et al 1984. Laboratory validation and field verification of a new passive air monitoring badge for sampling ethylene oxide in air. Am. Ind. Hyg. Assoc. J. 45, 697-707

7   STOCKTON,S.D. et al 1985. Field evaluation of passive organic vapour samplers. Am. Ind. Hyg. Assoc.J. 46, 526-531

8   UK Health & Safety Executive. Methods for the Determination of Hazardous Substances.

9   COKER,D.T. 1984. Bodily uptake rate differences between real life fluctuating concentrations and steady state lab. conditions. 21st Int. Congress on Occup. Health. Dublin, Republic of Ireland

10 VAN DER WAL,J.F. MOERKERKEN,A. 1984. The performance of passive diffusion monitors for organic vapours for personal sampling of painters. Ann. Occup. Hyg. 28, 39-47

11 HEARL,F.J. MANNING,M.P. 1980. Transient response of diffusion dosimeters. Am. Ind. Hyg. Assoc. J. 41,778-783

12 BARTLEY,T.L. et al 1983. Diffusive monitoring in fluctuating concentrations. Am. Ind. Hyg. Assoc. J. 44, 241-247

13 EINFELD,W. 1983. Diffusional sampler performance under transient exposure conditions. Am. Ind. Hyg. Assoc. J. 44, 29-35

14 COMPTON,J.R. 1984. The effect of square wave exposure profiles upon the performance of passive organic vapour monitoring badges. Am. Ind. Hyg. Assoc. J. 45, 446-450

15 COKER,D.T. JONES,A.L. 1981. Letter to Editor. Ann. Occup. Hyg. 24, 399-402

16 COKER,D.T. JONES,A.L. 1986. Letter to Editor. Ann. Occup. Hyg. 30, 263-265

17 TUGGLE,R.M. HAWKINSON,R.W. 1981. Incorrect use of t-tests. Am. Ind. Hyg. Assoc. J. 42, 325-326

18 TILEY,P.F. 1985. The misuse of correlation coefficients. Chem. in Brit. Feb. 162-163

COMPARISON OF ACTIVE AND DIFFUSIVE SAMPLING FOR THE MEASUREMENT
OF ORGANIC SOLVENTS IN PRACTICE - CONSIDERATIONS ON THE
APPLICABILITY OF DIFFUSIVE SAMPLERS

H. Blome*), D. Wolf*)

*) Berufsgenossenschaftliches Institut für Arbeitssicherheit
   (BIA) Postfach 2043, Lindenstrasse 80, D-5205 Sankt Augustin-2
   (FRG)

Abstract

Diffusive samplers, type Monitor 3500, ORSA 5 und PRO-TEK G-AA,
were simultaneously used with active sampling instruments in
order to determine the concentrations of harmful substances at
workplaces and to compare the results. Toluene, ethyl benzene,
xylene, styrene, acetone, 2-butanone, hexone, ethyl acetate,
butyl acetate, dichloromethane, 1,1,1-trichloroethane, and
acrylonitrile occurred individually or as mixture within a great
range of concentrations. Due to extensive trials, air velocity,
substance type, kind of material mixture, air humidity, storage
stability of loaded sampling media, and analytical recovery have
to be regarded, among others, as decisive parameters which are
able to influence the results of diffusive sampling. First
recommendations regarding the use of diffusive samplers can be
summarized as follows:

1) Diffusive samplers for use in practice should be approved only
   after laboratory and field tests referred to the materials to
   be sampled. According to the present knowledge, special
   consideration is required if low-boiling halogenated
   hydrocarbons alcohols and ketones are concerned.

2) Material mixtures require an applicability check-up of the
   corresponding diffusive sampler for each individual case.

3) The manufacturers of diffusive samplers have to quote minimum
   air velocity, analytical recovery, the method to determine the
   analytical recovery, capacity, and storage stability of loaded
   sampling media.

4) The manufacturer has to guarantee the invariable quality of
   diffusive samplers by appropriate control measures.

Diffusive samplers with activated carbon, types Monitor 3500, ORSA 5 and PRO-TEK G-AA, were used for our measurements of organic vapours in the air at workplaces. Field measurements were made at fixed points and on persons. The diffusive samplers were used in parallel with recognized, adapted or tested active sampling instruments (1,2,3). In the first phase of the investigations a diffusive sampler made by each manufacturer was positioned in the immediate vicinity of the active sampling system in measurements on persons; a second diffusive sampler was positioned on the other half of the body in the respiratory area. There were therefore seven sampling systems per person. The point of this arrangement was to establish any differences in concentration between the right and left half of the body. For the stationary measurements a gas chromatograph was also used as a double check. Without anticipating the individual results, it was possible to make the following general observations:

- the results of the active sampling using fixed sampling instrument did not deviate from those of the gas chromatograph by more than 30% at any stationary measurement point. The vast majority of the active sampling measurements were within +- 15% of the average observed with the gas chromatograph system.

- the deviations from their average of the results of the two diffusive samplers made by one manufacturer per measurement point were always under 10% and in most cases under 4%. It can therefore be taken that at the workplaces we investigated there is generally speaking no difference between measurements on the right and left sides of the body.

Measurements were made in a variety of branches of industry and works. The areas covered were: the spray painting of metal components, dipping, wood varnishing, paint and varnish manufacturing, the production of corrugated tubes, the glucing of shoe soles, fabric and expanded plastic components, the manufacture of fibreglass-reinforced plastics and polyacrylonitrile.

The following substances were analysed during these
investigations: toluene, ethyl benzene, xylene, styrene,
acetaone, 2-butanone, 4-methylpentane-2-one, ehtyl acetate, butyl
acetate, dichloromethane, 1,1,1-trichloroethane and
acrylonitrile.

There were relatively high concentrations of some of these
substances in some parts of a number of the works under
investigation - sometimes in conjunction with other substances.
The data were evaluated by means of a point correlation diagram
showing the regression line and correlation coefficient. The
following graphs show the results for toluene for the three
diffusive samplers mentioned:

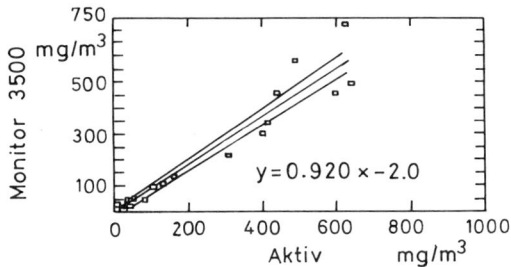

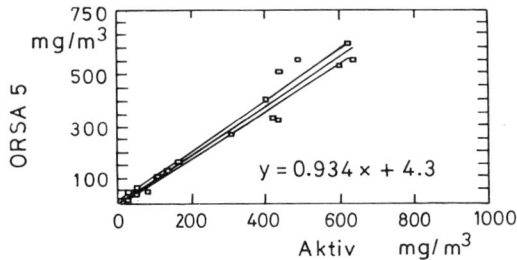

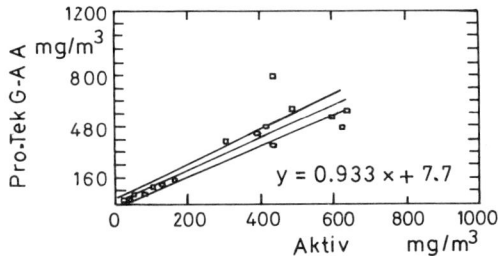

Measurement of toluene in practice

The results for the other substances are given in table form. In
addition to the substance and the diffusive sampler used, the
number of samples, the regression line and the correlation
coefficient are given.

Table 1: Results of field measurements

Substance	Diffusive sampler	Number of samples	Regression equation	Correlation coefficient
Ethyl benzene	Monitor 3500	28	y = 0,904 x + 3,8	0,94
	ORSA 5	28	y = 0,835 x + 11	0,92
	PRO-TEK G-AA	28	y = 0,902 x + 1,4	0,94
xylene	Monitor 3500	28	y = 0,821 x + 9,0	0,98
	ORSA 5	28	y = 0,842 x + 17,9	0,95
	PRO-TEK G-AA	28	y = 0,941 x + 9,4	0,95
acetone	Monitor 3500	11	y = 0,875 x + 11,2	0,85
	ORSA 5	11	y = 1,132 x + 59,8	0,85
	PRO-TEK G-AA	11	y = 1,146 x + 17,1	0,89
2-butanone	Monitor 3500	14	y = 0,637 x + 29,5	0,85
	ORSA 5	14	y = 0,785 x + 24,0	0,94
	PRO-TEK G-AA	14	y = 0,678 x + 32,0	0,85
4-methylpentane-2-one	Monitor 3500	5	y = 0,622 x + 6,9	0,71
	ORSA 5	5	y = 0,923 x + 20,7	0,83
	PRO-TEK G-AA	5	y = 1,080 x - 9,4	0,72
ethyl acetate	Monitor 3500	15	y = 0,769 x - 4,7	0,98
	ORSA 5	14	y = 0,970 x - 2,3	0,99
	PRO-TEK G-AA	15	y = 0,927 x - 15,3	0,98
butyl acetate	Monitor 3500	10	y = 0,805 x - 4,6	0,89
	ORSA 5	9	y = 0,872 x + 18,3	0,94
	PRO-TEK G-AA	8	y = 1,168 x + 9,4	0,87
acrylonitrile	Monitor 3500	6	y = 0,974 x - 0,03	0,99
	ORSA 5	6	y = 1,557 x - 0,3	0,99
	PRO-TEK G-AA	6	y = 0,612 x + 0,4	0,93
dichloromethane	Monitor 3500	14	y = 0,855 x - 7,5	0,95
	ORSA 5	14	y = 0,853 x - 7,8	0,94
	PRO-TEK G-AA	14	y = 1,299 x - 7,0	0,94

Although correlation is generally good (coefficient almost 1), the diffusive samplers do in some cases tend to produce assessments which are too high or too low, as can be seen from the regression equations. With some substances the measurement value do not correspond or relatively low correlation coefficients are observed.

Expressing measurement values exclusively in terms of regression lines and correlation coefficients means that parameters affecting the results of the diffusive or active sampling may well be overlooked.

In the case of dichloromethane for example, some of the ratios resulting from the diffusive and active sampling differed significantly in two works. In both works the dichloromethane concentrations at the workplaces in questions were 10-270 mg/m3. In the first care 1,1,1-trichloroethane in concentrations of 50-80 mg/m3 was also observed and in the second there was a continuation of methyl and ethyl acetate in a concentration of up to 480 mg/m3.

The lower conversion factors were observed when ester was present. Any escape of dichloromethane could be excluded in the active sampling system where a 1000 mg activated carbon tube was used, in both works. a possible explanation for the observations could be an escape in the diffusive samplers and/or different absorption rates for the diffusive samplers of a given manufacturer according to the actual situation.

Further programmes with selected mixtures of substances are required. It would appear that the measurement values supplied by the diffusive samplers were too low for the toluene/dichloromethane mixtures and selected halogenated hydrocarbons when the air humidity was high (4,5).

The parameter environmental air speed is also very important. At some points at a number of workplaces the environmental air was around 5 cm/s and below.

Here the absorption rates of the diffusive samplers of some manufacturers for 2-butanone were lower than at other workplaces where the environmental air speeds of 15-100cm/s measured were regarded as decisive. These field observations are supported by laboratory investigations. For air speeds in the test gas section of around 2 cm/s some lower absorption rates were observed (4).

It is possible that some organic substances, which, according to the manufacturers, can be determined by diffusive samplers are also found in particle form. The active samplers must in these cases ensure not only quantitative separation but also specific types of measurement. Diffusive samplers are not usually suitable for sampling substances in particle form.

On the basis of the results of our comparative measurements we would make the following recommendations for the use of diffusive samplers:
1) Diffusive samplers for use in practice should be approved only after appropriate laboratory and field tests. Given the present state of the art, special care should be taken if low-boiling halogenated hydro-carbons, alcohols and ketones and substances in particle form are likely to be present.

2) The suitability of the diffusive sampler must be checked in each individual case where mixtures of substances are present.

3) The manufacturers of diffusive samplers must quote minimum air speed, analytical recovery, the method used to determine the analytical recovery, capacity and storage stability of loaded sampling media. It is particularly important to determine and check analytical recovery in the laboratory. The effect of environmental air speed should be investigated in the field.

4) The manufacturer must guarantee maintenance of the quality of the diffusive sampler by suitable control methods.

Bibliography:

(1) Henschler, D. (1985). Analytische Methoden zur Prüfung gesundheitsschädlicher Arbeitsstoffe, Band 1-3, Senatskommission zur Prüfung gesundheitsschädlicher Arbeitsstoffe der Deutschen Forschungsgemeinschaft, Verlag Chemie, Weinheim

(2) Von den Berufsgenossenschaften anerkannte Analysenverfahren zur Feststellung der konzentrationen krebserzeugender Arbeitsstoffe in der Luft in Arbeitsbereichen, ZH 1/120, ausgabe Dez. 1983, Carl Heymanns Verlag KG, Köln

(3) Blome, H. et al. (1985). Messen gesundheitsgefährlicher Stoffe in der Luft am Arbeitsplatz. BIA-Handbuch. Ergänzbare Sammlung der sicherheitstechnischen Informations- und Arbeitsblätter für die betriebliche Praxis. Herausgeber: Berufsgenossenschaftliches Institut für Arbeitssicherheit - BIA- des Hauptverbandes der gewerblichen Berufsgenossenschaften. Erich Schmidt Verlag GmbH, Bielefeld

(4) Blome, H. und Hennig, M. (1985, 1986). Leistungsdaten ausgewählter Passivsammler. Staub-Reinhalt.Luft 45 (1985), teil 1, 505-508; Teil 2, 541-546; Staub-Reinhalt.Luft 46 (1986), Teil 3, 6-10

(5) Gregory, E.D. and Elia, V.J. (1983). Sample retentivity properties of passive organic vapor samplers and charcoal tubes under various conditions of sample loading, relative humidity, zero exposure level periods and a competitive solvent. Amer.Ind.Hyg.Ass.J.44,88-96

DISCUSSION - SESSION II

MOORE (UK)

Mr. COKER commented on the amount of what he called "wild data" in field trials of varying passive samplers and pumped systems and I agree that such data is very common in those situations. But I'm not so sure that I can agree with the conclusion that this is some inherent deficiency of the passive sampler as compared to the pump sampler. Certainly most field trials show the effect and there are far too many data points for us to dismiss them as out-liars or wild. I think that as they are real data points - perhaps Mr. COKER could speculate as to why they have come about.

COKER (UK)

As to speculating as to why they come about, I don't know. I think that they probably are wild data points. The only thing that you have to decide is whether they are going to effect the conclusions which you come to on your sampling.

MOORE (UK)

I shall think the results are real rather than wild. Wild is, after all, the comparison between the two. I don't believe the wild points are the result of error mechanisms as such, but simply different readings taken from a changing environment.

COKER (UK)

The trials we did were actually on fixed locations. The monitors were put into real workplace situations and consequently I can't see any reason other than error why a wild result should have occurred.

VAN DEN HOED (The Netherlands)

We have conducted some field tests on the 3M organic vapour badge, in which we duplicated the side-by-side comparisons. We also observed so-called "wild data", but you could see where this wild data had occurred because you would get three similar data points and one different. In all cases, the outlying data was pumped.

BERTONI (Italy)

My experience of comparing active and passive samplers in the field has led to excellent results. I think that the tests which Mr. COKER illustrated are too different to lead to any conclusions. When you compare very diverse methods. It's quite usual to find differences of 50 to 100%, particularly in the field tests and especially if you are comparing passive monitors which are manufactured very differently. Regarding cost, Mr. COKER pointed out that the personnel costs are higher that the material costs. This may be true although a pump, even a rather mediocre one, costs about one thousand dollars. But staff costs also exist when you are using active samplers. The pumps have to be calibrated and maintained.

COKER (UK)

I take your point about the differences between the two types of passive samplers. We get a very good agreement between pump methods in the field. However, the passive sampler ones which were all side-by-side, didn't all agree with each other. So which one do you take as the right result? On other issue, we pay about five hundred dollars for a pump and we have pumps now which we've had for about twelve years. So the actual capital cost is almost nothing. The cost is in employing me and the technicians who analyse the samples.

SAUNDERS (UK)

How many surveys are your results based on?

COKER (UK)

The ones that I quoted were twenty-five survey. This is including eight of ours and some for some other people. A lot of these came from the original work which was done by Working Group 5. But some are quite recent ones in the literature.

HARPER (UK)

There has been some work done recently in the United States which has shown that there is quite a high degree of variability in NIOSH standard charcoal tube methods and I wonder just what is the validity of comparing two methods, both in which may be highly variable.

COKER (UK)

There is a lot of data around on how good an agreement you can get with pump methods, and I think that you will find the level of agreement you get depends on how much care you take when sampling. I would think that would apply to passive samplers as well as active samplers. It all gets back to the situation you find yourself in. We don't see a lot of benefit in changing our samplers, but other people who are doing a lot of sampling will come to a different conclusion and I think that each view point is valid. BERTONI (Italy)

I wanted to make another point which is very important with regard to the difference between active and passive samplers. The passive sampler makes a selection between what exists in the vapour phase and what exists in the condensed phase. The diffusion coefficients are at least one order of magnitude different between aerosol and molecule. Active samplers can not make this distinction.

CHALVIDAN (France)

In my experience with ethylene oxide, we did not see any difference in results for different types of samplers but we did see major differences in the behaviour of individuals. The active samplers were not very well accepted. The pumps were heavy and we noted in many cases what one might call problems of sabotage because the people didn't like these samplers. This type of drawback is practically non-existent for passive samplers which are far easier to wear and which people can forget about.

POZZOLI (Italy)

We looked at a factory producing shoes where there was a lot of solvent used. We considered the respiratory area of the worker. We built a kind of a wall which had five passive samplers: two at head level, two at the respiration level, and one on the forehead of the worker. The sampler data seemed different but in fact it was consistent with patterns of ventilation, direction of air-flows, the expiration speed and localised breathing. For example, for the worker gluing the shoe and then putting it on the conveyor belt to her right, we saw that the samplers on the right gave higher values than the samplers placed on her left.

LEICHNITZ (FRG)

There has been recently a publication in the American Industrial Hygiene Association Journal presenting results found by TNO in the Netherlands regarding the testing of an ISO draft method for chlorinated hydrocarbons and these results have been obtained exclusively by active sampling. The highly qualified analysts involved in this testing programme knew about the importance of this test. Even in this situation 10% of the results were "wild", the precision being much higher than 25%.

FIRTH (UK)

We have been talking about diffusive monitors as though they were a homogeneous group of devices. Of course some diffusive monitors will give better results than others. Also, I would like to comment from the point of view of a large user of pumps. I think within HSE we've got well over a half a million pounds of investment in pumps and they are costing us about twenty to forty thousand pounds a year to maintain. So this is one of the reasons why we are attracted to diffusive samplers, because the cost savings for a large user are considerable.

COCHEO (Italy)

Passive samplers have all the advantages of active samplers and some additional ones. They are very easy to handle, they can do a very large amount of monitoring and can often be analysed in a centralized laboratory. But it is very important to have protocols on the sampling and analysis which would be valid for everyone. The analysis protocol ought to indicate maximum storage time of the sampler prior to use. It should give precise information on how it should be stored. It should give information on how to carry out a blank check before doing the analysis. The purpose of this Conference I think is to provide the Commission with information on how to draft recommendations not just for sampling but for analysis.

BROWN (UK)

Both the NIOSH and the HSE protocols do include clauses which suggest that those undertaking the evaluations should look at storage before and after use with samplers of this type. With respect to contamination, part of any diffusive sampling method should contain a clause with normal analytical practice to examine "blank" samples and this should avoid the use of contaminated samplers. Regarding the compression of points at the origin on correlation plots of diffusive against pump systems certainly the HSE protocol recommends that you shouldn't do this. You should do a log or some other suitable transformation to spread the points out equally. One last point - I think there is a specific problem with acrylonitrite which forms the basis of some the Mr. COKER's correlation plots. For this compound, particularly with thermal desorp-tion methods, most of the sorbents that are commonly used have acrylonitrite as part of their composition monomers. There is thus a tendency for a larger than usual background level of that particular analyte.

MOORE (UK)

I think that it is unreasonable to expect the manufacturers to write an encyclopedia for you on exactly how to do every possible sampling task. It's simply not reasonable. We try very hard as manufacturers to have good quality control and to prevent such things as the contamination of the blanks, but from time to time it will happen. It's a normal precaution in this type of measurement to run blanks. I welcome the advent of standardized testing protocols so that we can say to you that this particular device will meet the requirements of this particular organization. It has been tested against a set of standard tests to prove the point.

WOOLFENDEN (UK)

We manufacture a tube-type sampler which is packed by the individual user with whichever sorbent they require for the job. We haven't any control over what they put in the tubes or what sort of sampling they do with them. But it may be a good thing for us and for all manufacturers to emphasise more the precautions that should be taken.

BROADWAY (UK)

If you use thermal desorption you can always check the sampler before you use it to make sur it's clean.

MILLER (UK)

We have done some laboratory and field trials on carbon tetrachloride. We used pumped charcoal tubes. 3M organic vapour monitors and Perkin-Elmer tubes with and without membranes. We followed the HSE Protocol. MDHS 27. We found no effect of face velocity on the Perkin-Elmer tubes in the range 0 to 156 ft/min. In field trials, in comparison of paired samplers and replicates of six. The agreement between all four samplers was excellent. We did experience a problem in one particular plant, however, where there was a lot of airborne dust containing occluded carbon tetracloride. In this case, Perkin-Elmer samplers without membranes occasionally gave high results due to the ingress of dust, but this was easily avoided by exposing the samplers face downwards.

THERMALLY DESORBABLE PASSIVE SAMPLERS: Use of Graphitized carbon black in working sites evaluation.

G. Bertoni, R. Fratarcangeli, A. Liberti and M. Rotatori

Istituto sull'Inquinamento Atmosferico del C.N.R.
Area della Ricerca di Roma
Via Salaria Km 29,300 - C.P. 10
00016 Monterotondo Stazione (Roma) ITALY

ABSTRACT

The use of tubular passive samplers packed with graphitized carbon black for the evaluation of the organic vapours concentration in various sites is described.

Investigations carried out under different conditions and with different composition of the polluted atmosphere, indicate that, this device, allow a direct determination of volatile compounds when thermally desorbed.

Results obtained with this procedure have been compared with those carried out with other conventional devices.

INTRODUCTION

Recently passive samplers consisting of a glass tube filled with a light adsorbent (Carbotrap - Supelco Inc. Bellefonte, Pa. U.S.A.) have been carried out. They are suitable for the TLV-TWA determination of organics in working site (Bertoni et al., 1982).

Performances and parameters affecting sampling capacity and rate of such passive samplers have been previously studied (Bertoni et al., 1982, 1985). It has been shown that when passive samplers are exposed to a varying polluted atmosphere (like a worker assigned to various jobs during the working day) retro–diffusion phenomena and changing in sampling rate may occur. Such a phenomena are minimized when a suitable adsorption material quantity is filled into a "pencyl type" device which also offers the advantage that its sampling rate is independent of the air velocity.

In this work we have compared results obtained using tubular passive samplers and other classic devices. For this purpose we investigated different working sites using passive samplers filled with graphitized carbon black, active samplers packed with active charcoal and other instruments dedicated to the continous monitoring of given pollutants working in parallel. So that in various investigations we measured the personal exposure to halogenated anaesthetics of people working in operating theatres, the exposure of peltry artisans to volatile solvents and the exposure to the styrene vapours of some boat yard operators making glass fiber reinforced plastic ships.

EXPERIMENTAL

In the above-mentioned working sites we employed:

a) Tubular passive samplers, 15 cm x 4 mm i.d. glass tubes with 1 cm of diffusional path length, packed with about 500 mg of Carbotrap-(Supelco Inc. Pa. U.S.A.);

b) Active charcoal passive samplers (Gasbadge from Abcor Dev. Corp. Wilmington, MA - U.S.A.);

c) Active charcoal active samplers (S.K.C. Inc. Pa. U.S.A.);

d) MIRAN 1A Infrared Spectrometer (Foxoro Anal. Division, U.S.A.);

e) Auto-injecting gaschromatograph (Perkin-Elmer - U.S.A.).

All the analysis have been carried by a DANI 3600 gaschromatographic system modified for thermal desorption.

RESULTS

Table I shows concentration levels of halo-organics measured in six operating theatres in Rome. The amounts evaluated with active and passive samplers are consistent within the instrumental error while you can sometimes see values which do not fit with those measured by means of a single beam infrared spectrometer MIRAN 1A. In this case differences can be attributed to the fact that the I.R. continous monitoring device stayed on a fixed place while the anaesthetist left his position for a short time.

TABLE I   Concentrations (p.p.m.) of halogenated anaesthetic gases measured by I.R. spectrometry and adsorption devices.

Theatre	compounds	meas.device	average conc. (p.p.m.v.)	measurement time (min)
1	Ethrane	I.R. Spectrometer	9.0	61
1	"	active sampler	3.6	300
1	"	passive sampler	3.8	300
2	"	I.R. Spectrometer	10.4	40
2	"	active sampler	10.6	300
2	"	passive sampler	10.2	300
3	"	I.R. Spectrometer	13.5	40
3	"	active sampler	10.4	280
3	"	passive sampler	11.2	280
4	"	I.R. Spectrometer	10.6	88
4	"	active sampler	6.7	310
4	"	passive sampler	7.2	310
5	Fluotane	I.R. Spectrometer	4.4	73
5	"	active sampler	4.3	250
5	"	passive sampler	4.8	250
6	"	I.R. Spectrometer	0.6	38
6	"	active sampler	0.3	180
6	"	passive sampler	0.3	180

Table II compares concentrations evaluated with different sampling devices in a peltry's laboratory. All the results are in good agreement but the thermally desorbable passive sampler shows the best sensitivity in quantization of organics at very low concentrations.

TABLE II Comparison of volatile solvents concentrations (p.p.m.v.) measured with: A) active sampler of active charcoal; B) passive sampler of active charcoal; C) passive sampler of graphitized carbon black.

compounds	A	B	C
Methyl-ethyl-cetone	0.15	–	0.10
2-2 dimethyl-butane	0.07	–	0.07
2-3 dimethyl-butane	0.45	0.45	0.42
Cyclohexane	0.35	0.33	0.35
Ethyl-acetate	0.26	0.25	0.25
Methylcyclopentane	0.14	0.12	0.17
3-methylpentane	1.30	1.20	1.20
2-methylpentane	1.40	1.50	1.50
1-2 dichloropropane	0.05	–	0.06
trichloroethylene	–	–	0.14
n-hexane	0.80	0.80	0.78
Toluene	0.05	–	0.05
Total	5.00	4.65	5.09

Table III shows measurements of personal exposure to the styrene vapours carried out in a boat yard where glass fiber reinforced plastic ships were built. In this case the results obtained with passive samplers were in agreement with discontinous determinations performed by an auto-injecting gaschromatographic system while values carried out by the MIRAN 1A I.R. spectrometric system were over-estimated because of the interference of acetone used to wash tools.

TABLE III Average concentrations of styrene in a boat yard. Personal expsoure measurements: A) by means of passive samplers; B) with an auto-inject ing gaschromatographic device; C) with an I.R. spectrometric device.

| Sample n° | styrene average (mg/m^3) | | | notes |
	A	B	C	
1	6.18			
2	13.21	15.3	33	opencast work
3	26.72			
4	27.97			
5	31.01			
6	31.32	33.2	51	indoor work
7	31.48			
8	35.62			

CONCLUSIONS

Field tests confirm reliability of passive sampling devices employed in very different environmental situations.

Thermal desorption of a passive sampler filled with light adsorbents (Carbotrap) can be considered an inexpensive and easy alternative to the other classic devices.

REFERENCES

1) Bertoni G., Perrino C. and Liberti A., 1982. A Graphitized carbon black diffusive sampler for the monitoring of organic vapours in the environment. Anal. Lett., 15(A12), 1039-1050.
2) Bertoni G., Perrino C., Fratarcangeli R. and Liberti A., 1985. Critical parameters for the adsorption of gaseous pollutants on passive samplers made of low specific area adsorbents. Anal. Lett., 18(A4), 429-438.

USE OF A PASSIVE SAMPLER FOR MONITORING VAPOURS FROM A COMPOSITE SOLVENT IN WORKPLACE AIR

R. Bertrand, P. Berthier

Commissariat a l'Energie Atomique
Institut de Protection et de Surete Nucleaire
Service d'Hygiene Industrielle

26701 PIERRELATTE CEDEX, FRANCE

Abstract

A passive sampler was used to investigate the concentrations of vapours from a composite solvent (aromatics from C7 to C12) in workplace air.

The experimental method used to determine the constant of the sampler is described and its validity is discussed.

The performance of passive samplers under various conditions was studied: air direction in relation to the surface of the dosemeter, variations of concentrations in space and time, minimum sampling duration, saturation of the dosemeter, sample retentivity.

The conclusions are highly favourable towards the use of the passive sampler studied, subject to certain restrictions.

INTRODUCTION

We decided to monitor the airborne concentrations of vapours from a solvent composed mainly of aromatic compounds from C7 to C12 by means of personal sampling.

The personnel involved did not work at fixed posts but it premises where the concentrqtions differed widely, and in a variety of postures (standing, seated, bent over, etc.)

The sampling method involving an activated-carbon tube linked to a personal portable pump proved to be impractical because of the wide variety of postures adopted by the workers. We therefore decided to use the passive sampler, the only means of obtaining a representative sample of the toxic substances inhaled by the worker without hampering him in his work.

The practical aspects of the passive sampler are known; it remained for us to demonstrate its reliability in sampling the vapours from the solvent used, which was made up of a mixture of constituents with different vapour tensions and boiling points.

We had to set up a study protocol which would enable us to achieve, in experiments, the constants of the dosimeter selected (the 3 M - 3 500) and to measure its reliability under various conditions of use.

EQUIPMENT AND METHOD

The airborne concentration of a vapour collected by the passive sampler is given by the formula:

$$C = \frac{Qd \quad A \quad 1000}{P_D \cdot t}$$

where:

C : Concentration in mg.m $^{-3}$
Qd : Mass of vapour trapped by the dosimeter
A : Dosimeter constant vis-a-vis the vapour
P_D : Vapour extraction efficiency of the dosimeter
t : Sampling time in minutes
If we know the concentration C and the efficiency PD, we can calculate the constant A.

We set up a calibration standard which enabled us to create within an enclosed space, an atmosphere with a constant concentration. We measured the concentration by means of gas chromatography, and at the same time we monitored the rise in equilibrium of the exposure chamber by means of a total hydrocarbons analyser.

In addition, in order to confirm the concentrations given directly by the chromatograph we sampled using pumped-activated-carbon tubes (100 + 50 mg) for the same sampling time as the passive samplers. This method indicates the mean concentration during the tests.

DIAGRAM OF CALIBRATION STANDARD GENERATOR

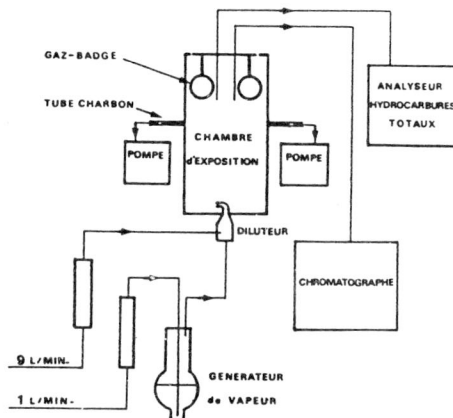

The vapour generator consists of a bottle containing the solvent to be studied. Air is fed into the bottle at a rate of 1 litre/min. The diluter mixes and homogenizes the vapour-filled air with pure air fed in at a rate of 9 litres/min.

The exposure chamber has a volume of 2 500 ml. This low volume enables the atmosphere to be brought rapidly into a state of equilibrium.

The extraction efficiency of the dosimeter, Pd is determined by depositing a known mass of solvent in the dosimeter. The solvent is left in contact with the dosimeter for a minimum of two hours and then extracted with a known volume of carbon disulphide. The extraction efficiency is given by the formula:

$$Pd = \frac{\text{quantity extracted}}{\text{quantity deposited}}$$

The value of the constant of the passive sample (under the temperature and pressure conditions obtaining in the experiment) is given by the formula:

$$A = \frac{C \cdot Pd \cdot t}{Qd \quad x \quad 1000}$$

STUDY OF SAMPLER PERFORMANCE UNDER VARIOUS CONDITIONS

We studied the response of the sampler under various sampling conditions:

- Air direction in relation to the surface of the sampler

Our calibration standard is dynamic and the airflow is unidirectional.

We placed the samplers in the exposure chamber at angles ranging from 0 to 90o, and found no difference in the results obtained.

- Variation of the concentration in space.

We placed two batches of samplers in the same atmosphere. After a given sampling time we withdrew one batch from the chamber and we kept the second batch in solvent-free air.

There was no evidence of any loss of solvent from the second batch when the results for the two batches were compared. In contrast, when the sampler is exposed to a relatively high concentration for a short period it under-records the concentration.

We conclude that the passive sampler can be used for measurements of atmospheres where there are concentration fluctuations but not for measurements at workplaces where there are sudden bursts of vapour, such as may occur, for example, when taking a liquid sample from an extraction installation. For the solvent in question we established that a minimum sampling time of three minutes was needed to ensure a correct result.

- Saturation of the passive sampler

We verified that when the sampler is exposed to a constant concentration the quantity of vapour found is a linear function of the time up to saturation point.

- Sample retentivity

A series of four samplers was exposed at the same time. Analyses were made on days D, D+8, D+64 and D+68. No significant variation in the results was found.

CONCLUSIONS

These tests enabled us to use the passive sampler, a more practical system than the activated-carbon tube linked to a pump. The results of fixed-location measurements of ambient air using carbon tubes and passive samplers corresponded closely.

In the case of the particular solvent whose vapour concentrations we had to measure in a workshop, we concluded that:

- there is no desorption of the trapped vapours on transferring from a polluted place to a vapour-free place;

- at least three minutes' exposure time is needed in order to ensure a correct response;

- the passive sampler is poor at assimilating sudden bursts of vapour;

- the saturation point of the sampler is relatively high (7500 mg/m3 for an exposure period of eight hours);

- samplers can be stored in a cold room for at least two months with no risk of desorption of the vapours.

A STUDY OF PAINTERS OCCUPATIONALLY EXPOSED TO WATER AND
SOLVENT BASED PAINTS

J. Kristensson*, H. Beving**

*Analytical Chemistry Department
University of Stockholm
S-106 91 Stockholm, Sweden
**Department of Experimental Surgery
Thoracic Clinics, Karolinska Hospital
S-104 01 Stockholm, Sweden

ABSTRACT

A study was performed in which health effects on painters mainly exposed to water based paint were compared to
painters mainly exposed to solvent based paint.
An unexposed group was used as reference. Each group consisted of 40 test persons, all men.
In the investigation occupational exposure was measured
by diffusive sampling and gas chromatographic analysis. The
ATD-system (Perkin-Elmer Ltd.) was used. Ten compounds out of
the complex mixture was chosen. These ten compounds consisted
of some typical solvents used in water respectively solvent
based paint. Qualitative and quantitative analysis was performed on these ten compounds.
The results showed that none of the two exposed groups
were purely exposed to solvents from either water or solvent
based paint. Both groups were mixed, but the "water-group"
was more exposed to acrylates than the "solvent-group".
The quantitative analysis showed that the ten compounds
chosen represented only 15 - 20 % of the total peak area.
The sum of the concentrations of the ten compounds was
usually low and the calculated hygienic effects were
approximately 0.2 but in some casese the hygienic effect was
as high as 3.
Health effects were studied by investigation of blood
cells, regeneration time of platelets, uptake rate of serotonin in platelets, solvent metabolites in blood, function of
the kidneys and effects on the nervous system.
Blood cells were analyzed with Contraves Autolyzer 801.
The amount and size distribution of the cells were studied.
Significant differences of several investigated parameters
were found between the groups. Of special interest was the
regeneration time of the platelets. A new HPLC method for the
studies of regeneration time was used (Olsson et.al., 1986).
Some of the other investigated biological parameters
also showed differences between the groups.

EXPOSURE MEASUREMENTS

Exposure measurements were performed during 6 - 8 hours periods. No short time measurements were done. The work carried out during the exposure measurements was indoor repainting in houses.

Diffusive samples were taken at two different times on each of the test persons. The ATD-50 system and a Sigma 2B gas chromatograph were used in the analysis. Parameters for the ATD-50 are given in table 1 and parameters for the gas chromatographic system are given in table 2.

Table 1.   ATD-50 parameters.

Primary adsorbent:	Tenax GC 60/80 mesh
Membrane:	No
Desorption temp.:	250°C
Desorption time:	5 min
Trap packing:	Tenax GC 60/80 mesh
Trap temp. low:	-30°C
Trap temp. high:	300°C
Heated line temp.:	150°C

Table 2.   Gas chromatographic parameters.

Column:	CP Wax 57 CB, 4 um, 25 m x 0.32 mm (Chrompack)
Carrier gas:	N_2, 1.5 ml/min (8 psi) Split 30 ml/min
Detector:	FID
Temperatures:	Injector 100°C Column. 60°C, 2 min - 4°/min - 100°C - 8°/min - 200°C, 5 min. Detector 350°C

The ten compounds used as probes were ethyl acetate, methyl ethyl ketone, methyl acrylate, methyl isobutyl ketone, n-butyl acetate, ethylbenzene, m-xylene, n-butanol, 1,3,5-trimethylbenzene and styrene. Particularly methyl acrylate was used as probe for water based paint and m-xylene for solvent based paint.

Usually the measured concentrations were well below the TLV-values.

BLOOD CELL MEASUREMENTS

A 5 ml blood sample was taken from each of the test persons. The blood samples were automatically analysed in Contraves Autolyzer 801.

Histograms showing the amount and the size distribution of platelets, red and white blood cells were obtained, figure 1.

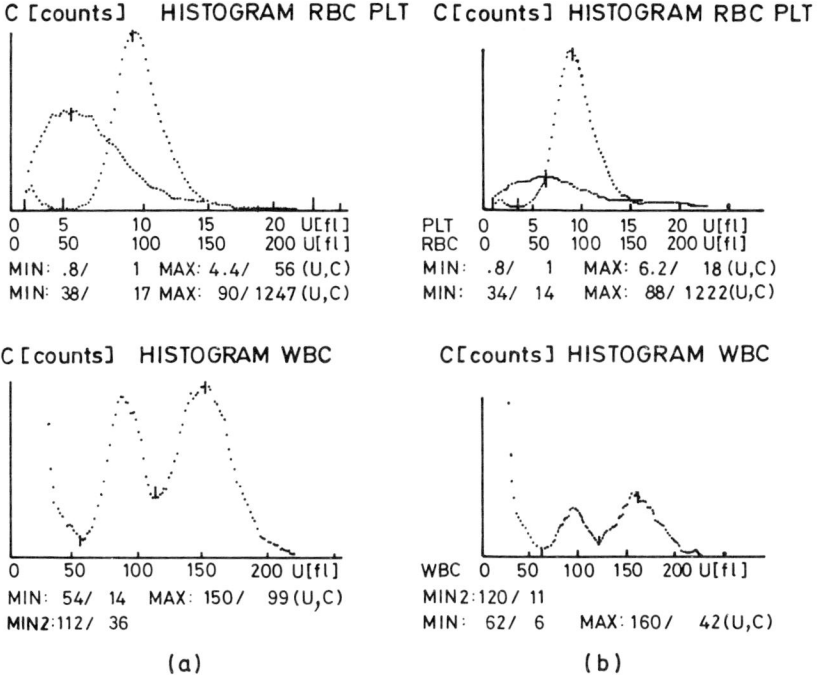

C [counts]   HISTOGRAM RBC PLT   C[counts] HISTOGRAM RBC PLT

```
0     5      10      15      20  U[fl]
0     50     100     150     200 U[fl]
MIN: .8/    1  MAX: 4.4/    56 (U,C)
MIN: 38/    17 MAX:  90/ 1247 (U,C)
```

```
PLT  0     5     10    15    20  U[fl]
RBC  0     50    100   150   200 U[fl]
MIN: .8/   1    MAX: 6.2/   18 (U,C)
MIN: 34/   14   MAX:  88/ 1222(U,C)
```

C [counts]   HISTOGRAM WBC   C[counts] HISTOGRAM WBC

```
0     50    100    150    200 U[fl]
MIN: 54/   14  MAX: 150/   99 (U,C)
MIN2:112/  36
```

```
WBC  0    50    100   150   200 U[fl]
MIN2:120/  11
MIN:  62/   6   MAX: 160/   42(U,C)
```

(a)   (b)

Figure 1.   Histograms of blood cells obtained from Conteaves Aotulyzer 801.
(a) = test subject exposed to water based paint
(b) = test subject exposed to solvent based paint.

The obtained results showed that painters exposed to solvent based paint, compared to both the painters exposed to water based paint and the reference group, had lower amount of platelets but the size of the platelets was bigger. The amount of white blood cells was lower. The uptake of serotonin was higher (Beving et.al., 1984).

The regeneration time of the platelets was significally changed (lowered) in blood from painters exposed to solvents.

A new method based on HPLC analysis of a metabolite of arachidonic acid was used (Olsson et.al., 1986). The results obtained from the regeneration times are shown in figure 2.

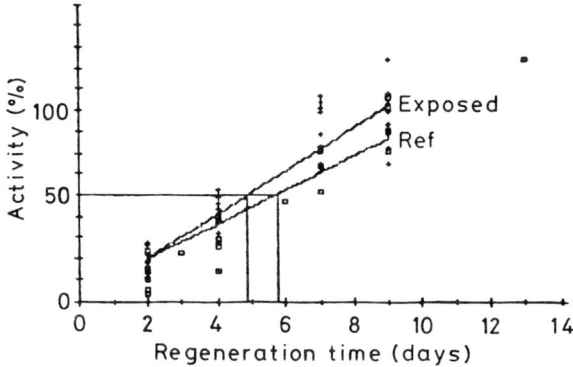

Figure 2.   Regeneration times of platelets.

Investigation of the kidneys showed that painters exposed to water based paint had more proteins in the urine and investigation of the transmitting speed in the peripheral nervs showed lower speed in painters exposed to solvent based paint.

All results are not yet fully evaluated. Correlation studies between exposure and biological effects will be carried out.

REFERENCES

Olsson, U., Beving, H., Kristensson, J.,Malmgren, R.,
    Schmidt, S., 1986.
    J. of Chrom., Biomedical Applications, in press.
Beving, H., Kristensson, J., Malmgren, R., Olsson, P.,
    Unge, G., 1984.
    Scand. J. Work Environ Health, 10, 229-234.

USE OF DIFFUSIVE SAMPLERS IN MONITORING ORGANIC VAPOR
EXPOSURES IN OIL MIST CONTAMINATED WORK ENVIRONMENT

Bjørseth, O., Børresen, E., and Malvik, B.

SINTEF, Div. Applied Chemistry,
N-7034 Trondheim-NTH, Norway

ABSTRACT

Sampling for hydrocarbon vapors in oil mist contaminated areas using diffusive samplers has been studied for three different samplers (3M, DuPont and SKC) and charcoal tubes. The main purpose has been to verify whether or not diffusive samplers give different results to those from the charcoal tubes. The study has been performed both under field conditions and in the laboratory. The variation in sampling conditions and atmospheres has not given the opportunity to treat the results with suitable statistical methods. The results obtained so far can be summarized as follows:

- Use of diffusive samplers for hydrocarbon sampling in oil mist contaminated areas gives higher concentrations than charcoal tubes.
- Storing diffusive samplers at room temperature before analyzing may increase the amount of hydrocarbons on the absorbent by a factor of ten.
- Reuseable samplers may contaminate later samples by transport of hydrocarbons from the front membrane.

INTRODUCTION

Monitoring of solvent vapor concentrations in working atmospheres by use of diffusive samplers (passive monitors) has increased in Norway during recent years. The diffusive samplers have been tested against the charcoal tube method in the laboratory and in real situations. Fjeldstad, P.E. et al. have suggested special sampling rate constants for actual organic pollutants in the workplace for three passive monitors (3M; 3500 Organic vapor monitor, DuPont; Organic vapor G-AA and SKC; Anasorb CA).

In mud processing areas offshore, the work environment can be contaminated with a complex mixture of organic vapors evaporated from hot drilling fluid, and oil mist produced by mechanical agitation (shale shakers) and high pressure cleaning equipment. The oil mist production is normally during short periods, but the concentrations in these periods can be substantial.

The practical advantage of diffusive sampling in these areas is obvious. In this work, the influence of oil mist on the analyzed samples has been investigated. Diffusive samplers (3M, DuPont, SKC) and charcoal tubes have been used simultaneously in mud processing areas during drilling operations at offshore installations and in test situations in the laboratory. Both the sampling period and the handling of the passive monitors between sampling and analysis have been studied.

METHODS

In the laboratory the tests have been carried out in an exposure chamber. Vapor and oil mist have been generated by evaporation and using a spray nozzle. The oil used was a low aromatic mineral oil with boiling point range from 240^0C to 450^0C.

In the test atmosphere two of each of the following samplers: 3M, DuPont, SKC, and charcoal tubes (SKC) were utilized. To verify a homogenous vapor distribution in the exposure chamber, toluene was used as a reference.

During field sampling on an integrated drilling and production platform, the same types of diffusive samplers and charcoal tubes were used. The samplers were placed close to the breathing zone of a person working in the drilling fluid processing area.

The samplers were analyzed according to the NIOSH manual of analytical methods. The sampling rates used in this study are taken from Fjeldstad et al. Before analyzing, the samples were treated in different ways:

1. Procedure according to manufacturer.
2. Removed absorbent from sampler and stored in freezer.
3. For SKC dual samplers, analyzing one capsule immediately after exposure while the other was capped and stored at room temperature.

For the samplers with a front membrane, this membrane was extracted and analyzed separately.

RESULTS

The precision in sampling and analysis was verified by the toluene reference which showed a standard deviation less than 10%. Using sampling rates given by the manufacturers, the diffusive samplers normally gave slightly higher concentrations than the charcoal tubes.

The same samplers analyzed with respect to oil vapor gave higher standard deviations and typically higher concentrations. In laboratory tests the three diffusive samplers indicated approximately twice the concentration of charcoal tubes. Dual SKC samplers, where one part was analyzed immediately after exposure and the other stored overnight in a freezer at -22^0 C, showed an average increase of 43% in the parts that were capped and stored. Storing SKC samplers at room temperature for 5 days gave an increase of 420%.

Field sampling using dual SKC samplers that were stored for several days at room temperature, gave values up to 10 times those of simultaneously taken charcoal tubes. Analysis of the membranes showed high oil concentrations, especially for the SKC sampler.

DISCUSSION

This study has shown that diffusive sampling of hydrocarbon vapor is more greatly influenced by oil mist contaminated air than are charcoal tubes. The reason is due to the construction of the samplers and the handling procedures of the adsorbent between sampling and analyzing. In particular, the SKC sampler which has a thick porous membrane which is not removed after sampling, seems to be sensitive to oil mist contamination. If the membrane is contaminated with oil droplets, the storage time after sampling will allow transfer of hydrocarbons to the adsorbent.

Reuseable samplers may contaminate later samples through transfer of adsorbed material from the front membrane. Therefore, disposable dosimeters seem to be preferable when sampling of vapors has to be performed in atmospheres contaminated with oil mist.

REFERENCES

Fjeldstad, P.E. et. al.: Diffusion samplers-determination of samling rates during practical use and simulated practical use, Nordic Occupational Hygien Conference 1986, Tammerfors, Finland.

DEVELOPMENT OF A PASSIVE SAMPLER FOR MONITORING AMBIENT

LEVELS OF ORGANIC VAPOURS PARTICULARLY BENZENE

C. Choo Yin, G. Layton-Matthews

National Smokeless Fuels Limited,
Central Laboratory, Cardiff,
CF4 7YH, United Kingdom.

ABSTRACT

The aim of this project, which was partly funded by ECSC, was to assess the effect of emissions on the benzene, toluene and xylene levels found in the general environment around coke works.

In order to collect the large number of samples required, a passive sampler, designed for use with the EMS single stage thermal desorber, has been developed to replace the VA tube and pump system for the static sampling of low levels of benzene, toluene and xylene in the workplace and general atmosphere. Analysis in both cases was by gas-liquid chromatography using a flame ionisation detector. Laboratory tests were carried out to compare the background contamination levels of the sampler with those obtained using VA tubes and to determine up-take rates.

Field trials, in which the passive samplers were paired with the VA tube/pump samplers have been carried out. The results showed good agreement between the two systems over the range 2-600 ppb (10^{-9}).

INTRODUCTION

For many years the coking industry has been a major producer of aromatic hydrocarbons including benzene, toluene and xylene (BTX). Because of the potential hazard to health of these materials, particularly benzene, National Smokeless Fuels Limited (NSF) and a number of other organisations within the European Coal and Steel Community (ECSC) undertook a cooperative project to measure the levels of BTX, both within the workplace and in the environs of coke works, with a view to assessing the effects of emissions from a Works on the BTX levels in the general environment. The work was partly funded by the ECSC.

Active systems, e.g. adsorption tubes packed with polymeric adsorbents (VA tubes) and low flow rate pumps, have been used to determine ambient levels of BTX. Simultaneous monitoring of a large number of sites presented the obvious problems of high cost and manpower resources. The major technical problem was the interference with the analysis caused by water collected in the VA tubes when sampling in conditions of high humidity or during periods of heavy precipitation.

To overcome these problems a passive sampler (EVM sampler), compatible with the available analytical equipment (EMS thermal desorber and gas-liquid chromatograph fitted with a flame ionisation detector) was developed.

CRITERIA

    The most important criteria were that the EVM sampler should be thermally desorbable using the existing equipment and should have similar performance to the VA tube/pump with regard to blank interference, desorption efficiencies, sensitivity and accuracy and that it should be relatively inexpensive.

DESIGN AND OPERATION

    To meet the criteria outlined above, a sampler consisting of two co-axial cylinders of different internal diameters (15.9 and 5 mm) was prepared as shown in Fig. 1, the larger diameter end being the sampling end with its large exposed surface.

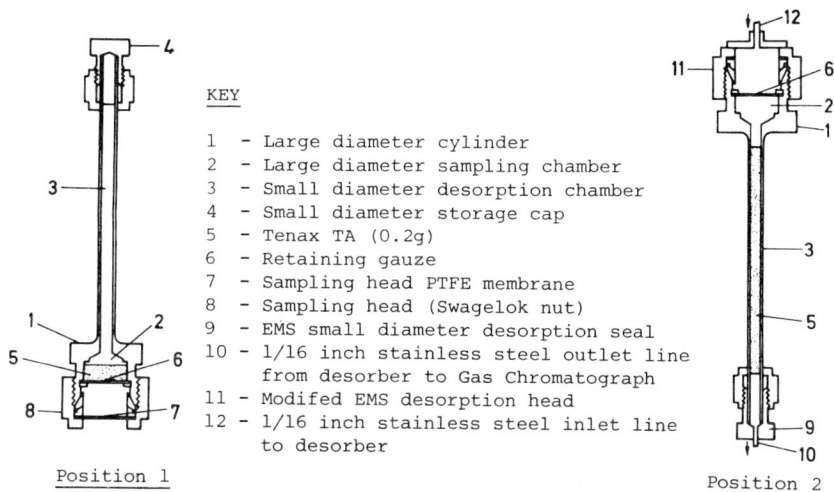

KEY

1 - Large diameter cylinder
2 - Large diameter sampling chamber
3 - Small diameter desorption chamber
4 - Small diameter storage cap
5 - Tenax TA (0.2g)
6 - Retaining gauze
7 - Sampling head PTFE membrane
8 - Sampling head (Swagelok nut)
9 - EMS small diameter desorption seal
10 - 1/16 inch stainless steel outlet line from desorber to Gas Chromatograph
11 - Modifed EMS desorption head
12 - 1/16 inch stainless steel inlet line to desorber

Position 1

Position 2

Fig 1.   NSF environmental vapour monitor (EVM)

    Sampling heads and storage caps were prepared from standard Swagelok nuts. Tenax TA (0.2 g) was used as the polymeric adsorbent.

Sampling and Desorption

    For sampling, the larger end is pointed downwards (Fig. 1, pos. 1) and vibrated for a few seconds. In that orientation a sampling head is fitted and the sampler suspended from a suitable support. At the end of the sampling period (8 hours) a storage cap is fitted before storing over charcoal as for VA tubes.

For desorption the sampler is inverted and again vibrated for a few seconds (Fig. 1,pos. 2). In that orientation the sampler is inserted into the desorber, fitted with a modified head, flushed with carrier gas and placed in the heated jaws of the desorber. The analysis was completed by a standardised glc technique, an aliquot of a standard solution being used for calibration.

LABORATORY TESTS

Blank Interference

After storage over charcoal for 72 hours the mean background levels of B, T and X found on 26 EVM samplers were 2.7, 2.3 and 1.8ng respectively, the standard deviations being 1.34, 2.63 and 3.24. Under the same conditions similar levels were obtained for the VA tubes.

Desorption Efficiencies

Similar desorption efficiencies were obtained for both the EVM samplers and VA tubes ranging from $96\pm0.6\%$ for benzene to $100\pm0.3\%$ for o-xylene.

Up-take rates (R)

Pairs of EVM samplers were exposed to a standard atmosphere for two hour periods, desorbed and the mass of each hydrocarbon was determined as above. The up-take rates were calculated as follows:-

$$R = \frac{\text{Mass in ng (m)}}{\text{Conc in ppm (C) * Time in min (t)}} \quad (1)$$

The mean up-take rates so determined were 19.7 ± 0.7, 22.7 ± 1.3 and 23.0 ± 0.9 ng ppm^{-1} min^{-1} for B, T and m-X respectively, the units ng ppm^{-1} min^{-1} being used for convenience. These values were used to calculate all subsequent measurements of BTX concentrations.

Field trials

Paired sampling, EVM versus VA tube/pump samplers, was carried out at locations within the curtilage of a coke works. The masses of B, T and X were determined as above and the concentrations obtained with EVM samplers calculated as follows:-

$$C = \frac{1000.m}{R.t} \text{ (ppb)} \tag{2}$$

The levels of B, T and X obtained with both samplers were comparable. A typical set of values for benzene is shown in Table 1.

TABLE 1   Results of paired sampling at various locations in the benzole plant : EVM Sampler versus VA Tube/pump

		Benzene Concentration (ppb)			
EVM	VA	EVM	VA	EVM	VA
4	5	112	132	3	2
24	23	62	60	59	58
11	10	15	15	122	132
11	11	3	6	3	5
46	52	92	109	71	65
33	35	243	253	565	611
583	576	139	148	135	152

Subsequent experience has shown that, with the equipment available, the lowest detectable level for each hydrocarbon was 2 ppb.

Cost

The cost of each unit was estimated to be about £30 which was less than a tenth of the cost of a suitable pump and VA tube.

CONCLUSIONS

The EVM passive sampler has met the overall criteria applied.

It is a suitable replacement for the VA tube and pump for the sampling of low levels of BTX.

The views expressed are those of the authors and not necessarily those of National Smokeless Fuels Ltd.

USE OF PASSIVE DOSIMETERS IN THE MONITORING OF

OCCUPATIONAL EXPOSURE TO SOLVENTS

E. De Rosa*, G.B. Bartolucci*, G.P. Gori*, L. Perbellini**, F. Brugnone**

*Istituto di Medicina del Lavoro, Universita' di Padova, Italy.
**Istituto di Medicina del Lavoro, Universita' di Verona, Italy.

ABSTRACT
    Environmental monitoring of three groups of subjects exposed to styrene, toluene and n-hexane respectively, was carried out using passive dosimeters. Excellent correlations were found between daily time-weighted average (TWA) exposures to solvents examined and corresponding urinary metabolite levels. The data showed that passive dosimeters turn out to be particularly advantageous in studies of contemporaneous environmental and biological monitoring, because they are less expensive and easier to use than active samplers, and allow the collection of more data.

INTRODUCTION

    A current research trend tends to associate environmental and biological monitoring, with the aim of identifying biological limit values for solvents or their metabolites which may be used to estimate the level of exposure to risk. The ACGIH (1985) has recently emphasized the usefulness of such studies and proposed biological exposure indices (BEIs) for some solvents. Clearly the exposure levels must be accurately measured (so that to calculate the daily TWA exposure for each subject) if they are to be correlated with biological indicators. In this context passive dosimeters are particularly advantageous because they are less expensive and easier to use than traditional active samplers, so that large numbers of exposed workers can be monitored at the same time.

    The aim of the present study is to measure the solvent concentrations by passive samplers, and to correlate the values thus obtained with urinary metabolite levels in exposed workers.

MATERIALS AND METHODS

TK-200 passive dosimeters (from Zambelli, Milan, Italy) placed in the breathing area of the workers were used. We monitored 13 subjects exposed to styrene, 22 to toluene and 14 to n-hexane. Fractionated samples in sequence over the whole workshift were taken from all subjects (each sampling lasting from 2 to 3 1/2 h), so as to evaluate their TWA exposure. Analyses of passive dosimeters were carried out by gas chromatography after desorption with CS_2 according to the NIOSH method.

Contemporaneously with environmental monitoring, biological monitoring was carried out. As regards styrene, a gas chromatographic method (Bartolucci et al., 1985) was used to determine levels of mandelic (MA) and phenylglyoxylic (PGA) acids in urine collected both at the end of the workshift (EW) and the next morning (NM). For toluene, a HPLC method (De Rosa et al., 1985) was used to measure hippuric acid and ortho-cresol in EW urine samples. For n-hexane, a gas chromatographic method (Perbellini et al., 1981) was used to determine the amounts of 2,5-hexanedione in EW urine samples.

RESULTS

Table 1 shows the results of environmental monitoring: the values exceeded the TLV-TWA only in one subject exposed to n-hexane. Table 2 shows the relationships between exposure to solvents and urinary metabolites.

Table 1: Daily TWA exposure in mg/m^3 in subjects examined.

Solvents	No of workers	$\overline{X}$	$\pm$	SD	Range
Styrene	13	95.7		59.0	15-210
Toluene	22	139.2		58.4	36-263
N-Hexane	14	61.9		55.5	22-219

Table 2: Results of linear regression between TWA exposure to styrene, toluene and n-hexane (x) and urinary metabolite levels (y) in exposed workers. The BEIs obtained from regression parameters are also reported.

Urinary metabolites	Cases	slope	intercept	r	BEIs
MA + PGA EW	13	5.40	-88.00	0.81	1073 mg/g creat
MA + PGA NM	13	1.86	- 3.87	0.89	396 mg/g creat
Hippuric Acid EW	22	0.008	0.24	0.89	3.24 g/g creat
Ortho-Cresol EW	22	2.48	79.08	0.70	1009 µg/g creat
2,5-Hexanedione EW	14	0.025	0.43	0.87	4.93 mg/l SW1024

CONCLUSIONS

In our study we found highly significant correlations between environmental data (such as TWAs) measured with passive dosimeters and biological indicators for styrene, toluene and n-hexane. Correlations between exposure levels and urinary metabolites turned out to agree with data in the literature; in particular, the levels of the urinary metabolites of styrene, toluene and n-hexane corresponding to the relative TLVs agree with those recently reported by various Authors or proposed by ACGIH (1985). These results confirm the utility of passive dosimeters, because they may give as good an indication of exposure as traditional active samplers, and allow the collection of more data with limited resources.

REFERENCES
ACGIH. TLVs-Threshold Limit Values and Biological Exposure Indices for 1985-86, Cincinnati, Ohio, 1985.
Bartolucci GB, De Rosa E, Gori GP, Chiesura Corona P, Perbellini L, Brugnone F. Biomonitoring of occupational exposure to low styrene levels. Ann Am Conf Ind Hyg, 12, 275-282, 1985.
De Rosa E, Brugnone F, Bartolucci GB, Perbellini L, Bellomo ML, Gori GP, Sigon M, Chiesura Corona P. The validity of urinary metabolites as indicators of low exposures to toluene. Int Arch Occup Environ Health, 56, 135-145, 1985.
Perbellini L, Brugnone F, Faggionato G. Urinary excretion of the metabolites of n-hexane and its isomers during occupational exposure. Br J Ind Med, 38, 20-26, 1981.

DIFFUSIVE SAMPLING ONTO SOLID ADSORBENTS FOR THE ANALYSIS OF BENZENE AND
1,3 BUTADIENE IN AIR BY GAS CHROMATOGRAPHY

Bernard Fields
ICI Petrochemicals and Plastics Division
PO Box 90
Wilton
England

ABSTRACT

Diffusive sampling methods are used at ICI Petrochemicals and Plastics
Division for the atmospheric sampling of benzene and 1,3 butadiene prior to
analysis by gas chromatography. Perkin Elmer type sampling tubes are used,
worn in the breathing zone of workers, in order to provide time weighted
average personal exposures, usually of 8 hours duration. An extensive
laboratory and field comparison of pumped versus diffusive sampling methods
for benzene was undertaken before these methods were introduced. The
decision to move to diffusive sampling where large numbers of measurements
are required (about 200/yr for each) was based on the accuracy and
reliability of the diffusive sampling technique perceived from these trials
as well as on the obvious practical advantages.

Less than 0.1 ppm (8 hr TWA) of benzene and butadiene can be determined and
above 0.1 ppm the precision of analysis based on repeatability of uptake in
the laboratory and on field comparisons of pumped versus diffusive samplers
is better than 10% for benzene and about 15% for 1,3 butadiene (1).

INTRODUCTION

This paper describes a programme of work carried out to assess the suitability of Perkin Elmer type diffusive tubes for the analysis of organic
vapours in workplace air. The samplers were evaluated for benzene with a
programme of work which satisfies the principles of the Health and Safety
Executive (H&SE) Protocol for Assessing the Performance of Diffusive
Samplers (1) although much of the data were obtained before the protocol was
published. In addition two extensive field comparisons of pumped versus
diffusive samplers were undertaken. The first, on an aromatics production
plant involved a 6 week trial during which 186 employees were monitored. 60
of these measurements were made using both diffusive and pumped samples.
This covered 12 occupational groups. The second survey assessed exposures
on an Olefins complex, an open-air structured plant. The survey which took
place covered all of the normal operations of the plant and included a unit
where 80% benzene streams were present. A significant number of employees
were chosen at random from each occupational group such that with 90% confidence at least one individual from the higher 10% exposure would be contained in the sample (2).

The development and evaluation of a diffusive sampling method for 1,3 butadiene took place after satisfactory results were obtained in the benzene trial. The laboratory validation of the method was carried out in pursuance of the H&SE protocol requirements and a less detailed field study was performed since it was considered that sampler performance would be similar for most organic vapours providing a well behaved absorption and desorption process can be demonstrated with laboratory data.

METHODOLOGY

During the benzene trials a thermal desorption-cold trapping-GC system was used to determine benzene. This comprised of an Environmental Monitoring System's (EMS) thermal desorber heated to 180°C, a 12 cm cold trap packed with 5% OV101 on 60-80 Chromosorb W capable of being cooled to -78°C with Drikold (CO_2) or directly heated. Separation and quantitation were achieved with a Pye 104 GC and an integrator. The system was subsequently replaced with Perkin Elmer ATD50 system and a GC which has now provided routine analyses for a number of years.

The dimensions of the Perkin Elmer type tubes used are approximately 6 mm OD, 5 mm ID and 90 mm long and these are worn on the workers' lapel. Preparation of the tubes is as follows. Analysis is performed as soon as is practicable after sampling.

	Benzene	1,3 Butadiene
Silicone membrane	Yes	Yes
Air gap	14 mm	14 mm
Packing	Poropak Q (60-80)	13X molecular sieve
Uptake rate 8 hr	1.33 ng/ppm/min	1.32 ng/ppm/min
Analytical column	TRIS	Picric acid/Carbopak

The pumped methods for comparisons used the same tubes with an Accuhaler model 808 pump attached operating at 2 ml/min. All of the butadiene work was performed on the ATD50 system.

RESULTS

Benzene

Results of the initial field survey on the aromatics plant comparing the results of pumped versus diffusive sampling are given in Figure 1. Of the 60 paired analyses three pairs could not be used because of pump failure and spoilt analyses and three pairs gave discrepancies of more than a factor of 10. These three were rejected as outliers. Two of the high results were

obtained with diffusive samplers and one with a pumped tube. The remaining
54 pairs when subjected to a test for bias using the Students'
t-distribution showed no bias at the 95% significance level. If only the 18
results below 0.1 ppm are considered then the passive sampler gives higher
results than the pumped tube.

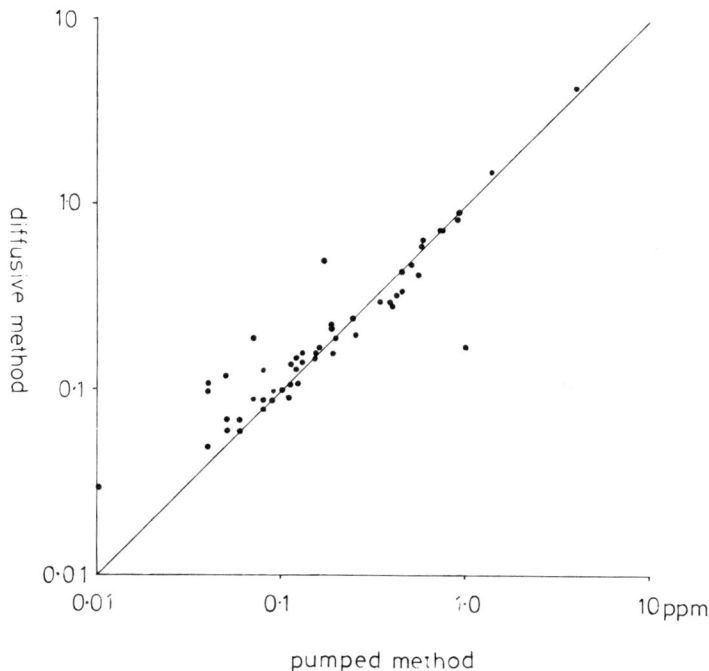

Figure 1   Log–Log plot of concentration given by diffusive sampler versus
concentration given by pumped sampler for paired analyses of
benzene.   The line drawn is the theoretical relationship (y–x)

In the second survey one pair of the 44 pairs of results were rejected
because of a discrepancy greater than a factor of 10 in the results.   The
remaining results subjected to the t-distribution test showed a significant
difference between the results at 95% significance level with the pumped
sampler giving higher results.   This bias however was small compared with
recommended exposure limits.   Again using results below 0.1 ppm a positive
bias in favour of the diffusive sampler was found.

Laboratory uptake rate determinations were made by analysing diffusive
tubes exposed to known concentrations of benzene with a view to providing a

means of calibration of the method, an indication of repeatability of the
sampling process under fixed conditions and variations in uptake rate under
the different conditions of concentration, humidity and sampling time which
may be experienced on the plant. The work included 65 determinations of
8 hr uptake rates over the concentration range 0.09–15.3 ppm and a
coefficient of variation of 6.5% was found under these conditions.

1,3 Butadiene

Laboratory determinations of diffusive uptake rates were made by
exposing tubes to 1,3 butadiene concentrations between 1 ppm and 50 ppm
under conditions of different humidity and for periods of exposure varying
from 30 mins to 8 hrs in accordance with the H&SE protocol already referred
to. The uptake rate for 1,3 butadiene is independent of sampling time and
humidity providing a silicone membrane is used. The repeatability of
determination of uptake rate is poorer than for benzene with a coefficient
of variation of 15%. The sensitivity of the method is the same as for
benzene with detection limits of about 0.01 ppm and reasonable precision of
analysis is obtained above 0.1 ppm.

DISCUSSION

The results presented demonstrate that a Perkin Elmer type sampling
tube operated diffusively provides a means of sampling which is sufficiently
sensitive to analyse at 0.1 ppm or less with a precision which is similar to
that obtained when pumped samplers are used. As a result of the ease of
portability, the reliability and lower cost of the devices they are now used
routinely in air analysis at this location.

Much of the method development and validation work on both of these
methods has been done alongside and with the cooperation of other members of
Working Group 5 of the H&SE Committee for Analytical Requirements. The
results presented here form part of and are consistent with other data
produced by Working Group 5 members in support of method validation. The
method for benzene is now published in the Methods for the Determination of
Hazardous Substances Series as MDHS50.

REFERENCES

Health and Safety Executive. 1983. Protocol for assessing the performance of
    a diffusive sampler (MDHS27).
Leidel, N. A., Busch, K. A. and Lynch, J. R. An occupational exposure
    sampling strategy manual. NlOSH Report No. 77-173.

THE USE OF DIFFUSIVE SAMPLERS TO MONITOR

OCCUPATIONAL EXPOSURE TO WASTE ANAESTHETIC GASES

W M Gray*, J O'Sullivan**, H B Houldsworth**, N Musgrave**

*West of Scotland Health Boards
Department of Clinical Physics and Bio-Engineering
11 West Graham Street
Glasgow G4 9LF, U.K.
**Barnsley Health Authority
Barnsley District General Hospital
Barnsley S75 2EP, U.K.

ABSTRACT

We have developed methods for monitoring the exposure of health service staff to waste anaesthetic gases, using a commercially available diffusive sampler. This paper describes the adsorbents used, conditioning procedures, desorption and analysis methods and the uptake characteristics of the samplers. We have also assessed the performance of the samplers in the laboratory and during routine use, paying particular attention to the repeatability of measurements, storage characteristics and ease of use. We conclude that the samplers can provide satisfactory performance, but only if special care is taken during conditioning, storage and transport.

INTRODUCTION

In view of the growing evidence that occupational exposure to waste anaesthetic gases represents a health hazard (Gray, 1985), there is a need for a simple and reliable method of monitoring exposure to these gases. Following preliminary studies by Houldsworth et al. (1982) and Cox and Brown (1984), we have developed methods to enable a commercially available diffusive sampler to be used for anaesthetic gases and have assessed its suitability for routine monitoring.

DETAILS OF SAMPLER

The sampler, manufactured by Perkin-Elmer, is a tube of length 3.5 inches and o.d. 1/4 inch (Brown et al., 1981). It is supplied with two push-fit storage caps, each of which is sealed with a single O-ring, and one of which is replaced by a special open-ended cap during sampling. The tubes for nitrous oxide (N_2O) adsorption were packed with molecular sieve 5A and conditioned overnight at 225°C in nitrogen before each use. The tubes for the adsorption of volatile anaesthetic agents (halothane, enflurane, isoflurane) were packed with Tenax TA, previously conditioned by heating in a stream of helium using a temperature programme of 7°C min^{-1} from room temperature to 250°C and maintaining this temperature for 4 hours.

MATERIALS AND METHODS

Analysis conditions
    A Perkin-Elmer ATD 50 thermal desorption unit and a Perkin-Elmer Sigma 3B gas chromatograph were used. The N_2O tubes were desorbed at 250°C for 3 min onto a cold trap packed with Carbosieve S at -30°C which was fired to 300°C. Chromatography was performed with helium (40 ml min^{-1}) as carrier gas on a 2 m long, 4 mm i.d. glass column packed with 80-100 mesh Porapak QS. The oven was run at 40°C for 4 min (to separate N_2, CO_2 and N_2O) and then at 100°C for 10 min (to clear H_2O) and a hot wire detector at 100°C was used. The desorption and analysis conditions for the volatile agents differed in the following respects: the primary desorption temperature was 225°C and the cold trap was packed with 60-80 mesh Chromosorb 102 and was fired to 225°C; the chromatographic column was packed with 60-80 mesh Chromosorb 102, the helium flow was 30 ml min^{-1}, the oven temperature was a constant 150°C and a flame ionisation detector at 250°C was used.

Test atmospheres
    These were obtained by flow dilution of N_2O or volatile agent in air (Gray and Burnside, 1984). N_2O and air were obtained from medical grade cylinders, and their flows measured with rotameters. The volatile agents were dispensed from a 25 µl syringe driven by a syringe pump.

Uptake rates of samplers
    16 tubes were exposed to test atmospheres of 290, 565 and 930 ppm N_2O and batches of 4 tubes removed and analysed at 1, 2, 4 and 8 hours. For the volatile agents, preliminary studies

showed that the uptake rate did not depend on exposure time. The uptake rates for each agent were obtained by exposing batches of 6 to 10 tubes to atmospheres of 0.5, 2 and 5 ppm for 4 hours and then analysing them.

Assessment of samplers

(1) For N_2O, the sampling repeatability was assessed by exposing batches of tubes to a variety of test atmospheres and also by monitoring the exposures of operating theatre staff with duplicate samplers. In each case, a pooled estimate of the coefficient of variation (CV) was obtained. For the volatile agents, a pooled CV was obtained from the results of the measurements of uptake rate.

(2) The storage performance of the N_2O samplers was investigated by exposing 18 tubes to 470 ppm N_2O for 1 hour and analysing batches of 6 tubes at 0, 14 and 28 days. Between exposure and analysis, the tubes were stored in a desiccator containing 1/8 inch pellets of molecular sieve 5A.

(3) The sealing efficiency of the storage caps was assessed by exposing batches of N_2O samplers with storage caps on each end to a test atmosphere of 930 ppm N_2O for 1 hour. This was done using caps fitted with new O-rings and also using caps with O-rings subjected to accelerated wearing by removing and replacing the caps on a tube a number of times.

(4) The performance of the tubes use has been assessed during the collection and processing of more than 5000 samples from locations throughout the U.K.

RESULTS

Uptake rates of samplers

For N_2O, the uptake rate was independent of concentration but decreased with increasing exposure time, as follows:
Uptake rate/(ng ppm^{-1} min^{-1}) = 1.44 (time/h)$^{-0.16}$
For the volatile agents, the values for the uptake rate/(ng ppm^{-1} min^{-1}) were: halothane: 2.59; enflurane: 2.29; isoflurane : 2.20.

Assessment of samplers

(1) The CVs for laboratory and theatre exposures of the N_2O samplers were 4.6% (78 d.f.) and 10.9% (20 d.f.) respectively. The CV for laboratory exposures of the volatile agent samplers was 5.5% (70 d.f.).

(2) The mean amounts of N_2O recovered from the tubes at 14 and 28 days were 93.1% and 90.0% respectively of the 0 day value. The 28 day value was signicantly less than the 0 day value at the 0.05 level (one-sided Wilcoxon-Mann-Whitney test).

(3) 10 tubes fitted with storage caps with new O-rings showed a mean N_2O uptake of 1.4% (range 0 to 2.9%) of the uptake expected for tubes fitted with sampling caps. For 10 tubes fitted with caps subjected to 50 cycles of accelerated wearing, the mean uptake was 6.8% (range 2.5% to 13.4%) of the sampling uptake.

(4) No major problems have been experienced during routine use of the samplers.

CONCLUSIONS

(1) The performance of the samplers is adequate for routine monitoring of occupational exposure to waste anaesthetic gases.

(2) Careful conditioning is required.

(3) The design of the storage caps is poor, leading to the possibility of leakage onto the tubes. Avoidance of contamination requires care during storage and transport.

REFERENCES

Brown, R.H., Charlton, J. and Saunders, K.J. 1981. The development of an improved diffusive sampler. Am. Ind. Hyg. Assoc. J., 42, 865-869.

Cox, P.C. and Brown, R.H. 1984. A personal sampling method for the determination of nitrous oxide exposure. Am. Ind. Hyg. Assoc. J., 45, 345-350.

Gray, W.M. 1985. Scavenging equipment. Br. J. Anaesth., 57, 685-695.

Gray, W.M. and Burnside, G.W. 1984. Calibration atmosphere generator for operating theatre pollution studies. Br. J. Anaesth., 56, 543-550.

Houldsworth, H.B., O'Sullivan, J. and Musgrave, N. 1982. Passive monitors for the determination of personal nitrous oxide exposure levels. Anaesthesia, 37, 467-468.

PRACTICAL ASPECTS OF THE DIFFUSIVE SAMPLING OF BUTADIENE

J.W. HAMLIN[1] and K.J. SAUNDERS[2]

[1]Occupational Hygiene Advisor, BP Chemicals Limited, Belgrave House,
76 Buckingham Palace Road, London SW1W 0SU

[2]Senior Chemist, Analytical Environmental Group, BP Research Centre,
Chertsey Road, Sunbury-on-Thames, Middlesex TW16 7LN

INTRODUCTION

Recent toxicological work on the effects of 1,3-butadiene on rodents has focussed attention on the need for a method of measuring exposure to low concentrations.

A pumped sample/carbon adsorption method which had been adequate for determining compliance when the TLV was 1000 ppm v/v, was not found to be completely satisfactory for measuring exposures of less than 20 ppm v/v. A diffusive sampler method based upon the use of molecular sieve was developed through cooperation between the UK Health and Safety Executive (HSE) and British Industry. This paper describes the practical work required before the method can be used in field applications and will concentrate on seven aspects, viz

    i   Scope of Method
   ii   Calibration
  iii   Conditioning and Capping
   iv   Use of Permeation Membrane
    v   Validation
   vi   Employee Consultation
  vii   Field Experience

SCOPE OF METHOD

1,3-butadiene is monitored using a membrane capped diffusive monitor tube, Figure 1, containing 200 mg of 13X molecular sieve (60-80 mesh) as sorptive substrate, over the monitoring period (usually 4 to 8 hours). At the end of the exposure period, the membrane containing diffusive head is removed and replaced with a blank end-cap. For analysis, the tube is positioned in the thermal desorber and the sorbed material is thermally displaced in a stream of nitrogen into a gas chromatographic analyser. The components are separated by the use of a suitable gas chromatographic column and detected using a flame ionisation detector.

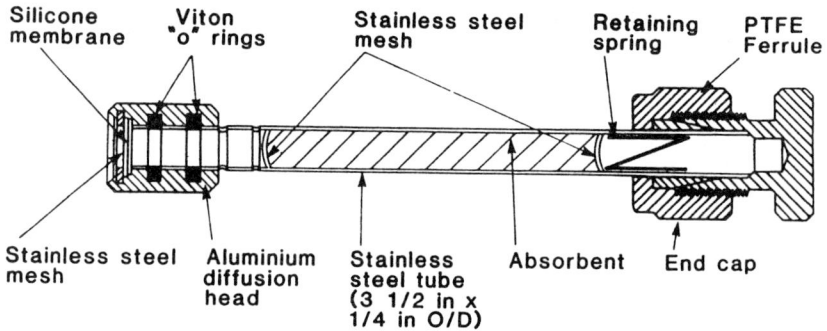

FIGURE 1.   Perkin Elmer Diffusive Monitor

The method is suitable for the time weighted average (TWA) measurement of atmospheres over a period of 4 to 8 hours at a concentration level of >0.1 ppm v/v.   The upper level has not been ascertained but it is above the TLV of 10 ppm v/v.   The method has been validated for moist atmospheres up to 95 per cent humidity.

CALIBRATION

1.   Diffusive Monitors

Prior to use the diffusive monitor constant for 1,3-butadiene must be determined as per the Diffusive Sampling Evaluation Protocol (Reference 1) by exposing the device to a known concentration (ppm) of the analyte for a fixed time period (min) and determining the mass (ng) of analyte "pick up" by thermal desorption and gas chromatography.   The experiment is repeated for various time periods and concentrations and the average diffusive monitor constant in ng/ppm/min determined.

2.   Detector System

The GC detection system also has to be calibrated.   To achieve this, the emission from a gravimetrically calibrated permeation tube of 1,3-butadiene is swept out of the thermostatted environment in a flow of nitrogen gas into a conditioned sorption trap, Figure 2 (Reference 2).

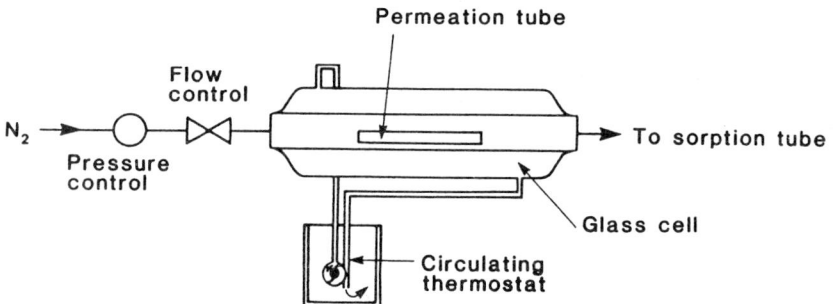

FIGURE 2.   Permeation Cell

The sorbed material is thermally displaced and measured. The procedure is repeated for various sampling periods and the area response per ng is plotted against the weight of the sample (ng). From this graph, which should closely approximate to a horizontal line, the average response for 1,3-butadiene in area units/ng is calculated.

CONDITIONING AND CAPPING

As described above, the monitoring tube is packed with 200 mg of 13X molecular sieve. This molecular sieve must be preconditioned by heating at 350°C for sixteen hours in a flow of dry nitrogen to ensure that contaminants, particularly atmospheric moisture, have been removed. The tubes are cooled back to ambient temperature in the flow of nitrogen and immediately capped with Swagelok end-caps fitted with PTFE ferrules (Figure 1).

Immediately before use the packed monitoring tube must be reconditioned by subjecting it to a normal desorption cycle of the particular thermal desorber being used. Our experience has been with the Perkin Elmer ATD-50 and the EMS thermal desorbers.

USE OF PERMEATION MEMBRANE

The silicone membrane in the diffusive sampler head is used to minimise ingress of particulates and water vapour. This is particularly important when using molecular sieve as a sorbent, because of its high affinity for water with consequent variation in uptake. More information is contained in Reference 3.

VALIDATION

Laboratory trials cannot fully simulate conditions which may be realised on a factory site. A field validation exercise was therefore undertaken, in which diffusive tubes were exposed in parallel with a conventional sampling system (ie pumped sorption tubes). On-site, the paired samplers were exposed as near to each other as practicable in the worker's breathing zone, ie on the lapel.

Results obtained in such a field trial are plotted against each other in Figure 3.

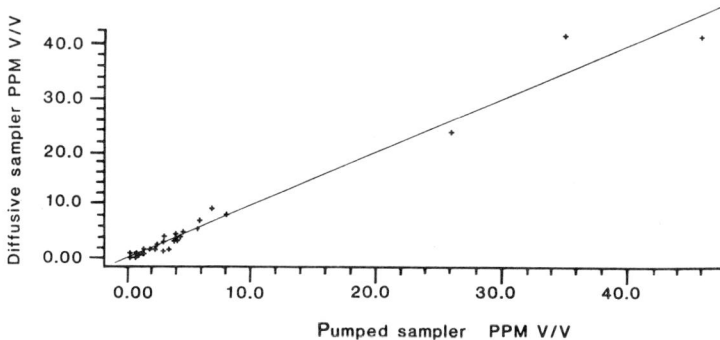

FIGURE 3. Plot of Concentration from Diffusive Sampler
Against Concentration from Pumped Sampler

The slope of the line of best fit is 0.9904 and intercept on the "X" axis is -0.2248. At the 95% confidence level the coefficient of determination is 0.9876 and the interval for the true slope is 0.9295 < slope <1.0513. This interval includes unity; it therefore indicates there is no relative systematic error in the measurements and that the agreement is good.

Butadiene is handled at a number of BP Chemicals sites and it is important that results produced at these sites are consistent and accurate. To ensure this "round-robin" comparison exercises are regularly undertaken to check both calibration and analysis. Results from a typical exercise are shown in Table 1, where tubes dosed at a Central Laboratory were analysed at two sites.

DOSED	2 µg	4 µg	13 µg
SITE 1	2.3, 2.3 2.5	5.4, 5.2 5.0, 4.8	14.6, 14.6 14.8, 16.0
SITE 2	2.0, 2.1 2.0	4.0, 3.9 3.9, 4.2	14.5, 13.6 13.8, 13.4

TABLE 1 - ROUND-ROBIN RESULTS

EMPLOYEE CONSULTATION

All BP sites have Health and Safety Committees with representation of all levels of employees. These Committees are involved with every aspect of health and safety and are kept informed of development in toxicology, analysis, etc. Particularly welcomed by these Committees has been the development of diffusive samplers. The samplers have wearer acceptability because they obviate the necessity for wearing pumps which can be cumbersome, particularly in confined spaces. The management has welcomed them on economic and practical grounds.

FIELD EXPERIENCE

The validated method described above now forms the basis for monitoring exposure to 1,3-butadiene at BP Chemicals sites involved in the production and polymerisation of butadiene. It has been used successfully for eighteen months and the results obtained have played an important part in the tripartite setting in the UK of a control limit for 1,3-butadiene. The method will shortly be published in the UK Health and Safety Executive's "Methods for the Determination of Hazardous Substance" series together with a complete back-up data report.

REFERENCES

1. "A Diffusive Sampler Evaluation Protocol" by R H Brown, R P Harvey, C J Purnell and K J Saunders, <u>Am Ind Hyg Assoc J</u> 45 (2): 67–75 (1985).

2. "Techniques for Atmospheric Analysis" by K J Saunders in "Analytical Techniques in Environmental Chemistry 2" by J Albaiges, published by Pergamon Press, Oxford and New York, 1982.

3. "The Development of An Improved Diffusive Sampler" by R H Brown, J Charlton and K J Saunders, <u>Am Ind Hyg Assoc J</u>, 42 (12): 865–869 (1981).

THE USE OF DIFFUSIVE SAMPLING TO MONITOR EXPOSURE OF PERSONNEL SERVICING OFFICE MACHINERY ON CUSTOMERS' PREMISES

David F. Houghton, D.E.W. Clarke, M. Hemming, D.J. Keech

Rank Xerox Limited
Welwyn Garden City
Hertfordshire
England

ABSTRACT

The exposure of Rank Xerox Service Representatives to Propan-2-ol from cleaning materials was studied over a five week period in London. The results were critically compared with those that have been found for simulated service calls carried out under controlled conditions. It was found that the simulated activities were representative of the exposures being experienced in the field. It is concluded that concentrations of Propan-2-ol created by Service Representatives in the course of their work, are well within occupational exposure limits. The monitoring was carried out by diffusive sampling followed by thermal desorption, and the advantages of the method for this type of activity are noted.

INTRODUCTION

It is the policy of Rank Xerox Ltd that any exposures of its employees or customers be reduced to the lowest levels commensurate with current technology. To accomplish this, regular monitoring of personnel and work places is carried out.

Prior to the introduction of any new machine product into the marketplace, Service personnel are monitored whilst carrying out simulated service activities on it in controlled chambers.

The data gathered from each of these exercises refers to a single machine type whereas, in practice, a Service Representative works on a selection of machines in the course of a day. Also the conditions in a customers' offices may vary, between themselves and from the controlled conditions of the test chambers. Thus there becomes a need to confirm that the laboratory data being produced is representative of exposures being experienced in the field.

In keeping with the policy of minimising exposures, materials are selected with lowest possible hazard ratings consistent with achieving the desired technical objectives Cleaning materials issued for field use are now formulated using Propan-2-ol as the major organic constituent. A second objective, therefore, is to ensure that exposures to Propan-2-ol do not exceed statutory occupational hygiene limits.

This paper focusses on investigations into Propan-2-ol exposures.

Testing in the controlled chambers was carried out using a portable infra-red analyser and also pumped charcoal tubes. When the decision was taken to take the experiment onto the customers' premises it was recognised that these two forms of monitoring were not acceptable for the new situation and passive diffusive sampling was adopted.

METHOD

The method of diffusive sampling onto a porous polymer followed by thermal desorption for analysis is used quite extensively for other applications within the department and is an ideal choice for this activity because of its simplicity and convenience.

Perkin Elmer diffusive sampling tubes packed with Chromasorb 102 or Tenax GC are attached to clothing near to the breathing zone. Any Propan-2-ol present in the air diffuses onto the polymer and is adsorbed by it. Analysis is performed by a Perkin Elmer ATD-50 automatic thermal desorption system interfaced with a Pye Unicam capillary gas chromatograph.

Personnel being monitored are given specific instructions on how to use the tubes together with a form to record essential data about the monitoring. Such data is required to perform the analysis and calculation.

As a preliminary study, fifty tubes were used to monitor ten French Service Representatives over one week. One tube per day was worn and opened for exposure only during service calls. Using the experience gained from this initial study, a much larger field trial was set up, monitoring personnel in the London area. Certain changes were introduced for the London trial as follows:

Separate tubes were used to monitor each individual service call. 165 service calls were monitored over a five week period covering sixty Service Representatives. Each service call was monitored by two sampling tubes, one packed with Chromasorb 102, the other with Tenax GC.

RESULTS

The results from the French field trial were quoted as 8 hour time weighted averages because of the sampling procedure used. All but one of the forty five tubes that were returned from France used, recorded concentrations of 0-8 parts per million (ppm). The one outstanding tube recorded a concentration of 48 ppm. It should be noted that twelve tubes monitored days where no solvents were used.

The results from the London field trial are more significant as far as a comparison with the monitoring of the simulated activities is concerned. They are all averaged over the particular sampling time for each tube, which was also the length of the service call monitored.

Table 1 gives a break down of the total number of results from the London field trial into parts per million area bands:

TABLE 1 Total results.

ppm	0-5	5-10	10-15	15-20	20-50	>50
frequency	219	30	20	9	16	12
% of total results	71.6	9.8	6.5	3.0	5.2	3.9

Table 2 shows the results for the twelve tubes that recorded concentrations of greater than 50 ppm, including their corresponding 8 hour time weighted averages.

TABLE 2 12 highest concentrations.

Result (ppm)	Sampling time (mins)	8 hour T.W.A. (ppm)
84	40	7
81	56	10
265	70	39
321	70	47
677	30	42
132	75	21
253	75	39
67	10	1
225	50	23
87	50	9
79	35	6
58	35	4

DISCUSSION

Since diffusive sampling was adopted for these activities, monitoring of simulated service calls within the controlled chambers has consistently given results of 0-20 ppm. The data obtained from the field trial shows 90% of results were within this 0-20 ppm range.

Of the 12 tubes (4% of the results) that gave concentrations of >50 ppm, one did indicate that the short term exposure limit may have been exceeded. However the second tube monitoring that service call gave, by comparison, an extremely low result (17 ppm). This suggests that the result was anomolous.

In all but this one result the occupational exposure limits would not have been exceeded even if the levels of Propan-2-ol had persisted for eight hours or the Service Representatives had made many calls, each time attaining a similarly high Propan-2-ol concentration.

It is unlikely that Service Representatives experience exposure to Propan-2-ol on a high percentage of the calls they make, as out of 600 tubes made available to the 60 Representatives involved over five weeks, only 306 were returned used. This indicates that the use of cleaning materials is necessary in less than half of all service calls.

It is concluded that the experiments with simulated service calls are highly representative of actual calls made in the field and that Propan-2-ol concentrations are very similar. These concentrations are well within the H.S.E. occupational exposure limits of 400 ppm (8 hour T.W.A.) and 500 ppm (short term exposure).

The diffusive sampling technique using personal monitors was very succesful. The monitors are sufficiently small and unobtrusive that they can be used in the field without inhibiting the service activities in any way, unlike techniques such as pumped charcoal tubes which require workers to wear restrictive webbing, clips and pumps. The measurements by diffusivesampling, because of the simplicity of the tubes, are made in a situation as close to reality as possible at all times with all movements. Techniques that require probes do not have this advantage.

The analyses are automatic and rapid, so that a large number of tubes can be analysed. Therefore, sufficient numbers of samples can be taken to provide statistically significant results for exposures in the field. The use of thermal desorption in the sample analysis obviates the problem of sample preparation which is time consuming, reduces accuracy due to handling steps and background solvents, and in this case would have required the use of chemicals more hazardous than the substance actually being monitored.

Overall, here is a versatile method that gives a greater freedom to take studies out of the academic field of the laboratory and firmly into workplace environments.

VENTILATION MEASUREMENTS BY DIFFUSIVE
SAMPLING TECHNIQUE

J. Kristensson*, A. Emmer*, P. Levin**

*Institute of Analytical Chemistry
University of Stockholm
S-106 91 Stockholm, Sweden

**The Royal Institute of Technology
EHUB
S-100 44 Stockholm, Sweden

ABSTRACT

The amount of pollutants in indoor air depends both
on the amount of pollutants that is emitted from the different sources and the amount that is evacuated by the ventilation system.
The pollutants can either be evacuated by ventilation
hoods at the sources or diluted by the general ventilation
system.
In occupational hygiene measurements personal monitoring
of pollutants is obvious. If one could measure ventilation in
a simple way this would be as obvious as personal monitoring
in occupatinal hygiene measurements.
The method described is a method for ventilation measurements based on the use of a tracer compound that is measured
by diffusive sampling.
The method was first described by Dietz 1982. This method has
been modified to suit the ATD-50 system.
A similar system for ATD-50 based on pumped sampling has been
described by Littler 1983 and Littler 1985.
Ventilation measurements with the tracer compound technique is usually short time measurements e.g. the ventilation
is measured during a few hours.
Ventilation is very much dependent on the outdoor climate.
Therefore measurements should be performed during longer
periods of time. With the described method measurements is
made from one week up to a month.
The ventilation can change in different places within a
room. With the described method one can also make "personal
ventilation" measurements.

DISTRIBUTION OF TRACER COMPOUND

The tracer compound was released at a constant rate by the use of a permeation tube made from teflon.
Different perfluoro compounds was used as tracers,

PMCH= perfluoro-methyl-cyclohexane

PDCH= perfluoro-dimethyl-cyclohexane

PDCB= perfluoro-dimethyl-cyclobutane

So far most experiments has been done with PMCH. ·

By changing the length and the thickness of the teflon tubing the release rate for PMCH was varied between 15 to 300 ng/min.
The permeation tubes have so far not been thermostated when used in field measurements.
The volume of the room was calculated and the ventilation rate was estimated. One or several permeation tubes with known release rates were used to achieve a concentration of PMCH of approximatly 0.1 ppb.

The permeation tube was placed in the room several days before the measurement was performed, to achieve equilibrium. The permeation tube can be permanently placed in the room because it may last for more than a year.

SAMPLING

Diffusive sampling was performed with Perkin-Elmer ATD-50 sampling tubes. The adsorbent used was Porapak QS 60/80 mesh. The adsorbent was heat conditioned at 200°C before sampling.
Sampling was continued for at least 5000 minutes (approximately 3.5 days).

So far only diffusive caps without membranes have been used. It is also necessary to use better end caps to seal the ATD tubes before and after the measurements, for instance Swagelok end caps with teflon ferrules.

ANALYSIS

The analysis was performed with the ATD-50 system and a Sigma 2B gas chromatograph. The ATD parameters are given in table 1 and the gas chromatographic parameters are given in table 2.

Table 1.   ATD-50 parameters.

Compound:	Perfluoromethylcyclohexane (PMCH)
Primary adsorbent:	Porapak QS 60/80 mesh
Membrane:	No
Desorption temp.:	150°C
Desorption time:	3 min
Trap packing:	Porapak QS 60/80 mesh
Trap temp. low:	-30°C
Trap temp. high:	200°C
Heated line temp.:	150°C

Table 2.   Gas chromatographic parameters.

Compound:	Perfluoromethylcyclohexane (PMCH)
Column:	PLOT Al_2O_3/KCl 50m x 0.32mm (Chrompack).
Carrier gas:	N_2, 1.4 ml/min (10 psi) Split 15 ml/min
Detector:	ECD, Make-Up gas 60 ml/min.
Temperatures:	Injector 150°C Column. 140°C, 3 min-5°/min-200°min, 5 min. Detector 350°C

CALIBRATION

A standard atmosphere was generated by a thermostated permeation tube and known gas flows. The permeation tube has a known rate of the release of PMCH.

From the standard atmosphere, uptake rates was calculated. Calibration of the gas chromatograph was done both by use of the standard atmosphere and pumped sampling of known volumes and by injection of diluted PMCH in pentane onto the adsorbent followed by analysis in the ATD-system.

RESULTS AND DISCUSSION

So far only laboratory tests will be reported. Only a few field measurements has been done and the results from these measurements are not yet fully evaluated.

Concentrations down to 0.01 ppb was analysed with the system. The linearity of the system has so far been tested from 0.05 to 15 ng injected PMCH.

Determination of the uptake rates showed that the uptake rate was dependent of both the time and the concentration. We have also observed that the uptake rate varies with the total volume of the room. This can be a concentration dependence because it was very difficult to generate the same concentration in a volume of 2 liters as in 40000 liters. Uptake rates was also determined by diffusive and pumped sampling in parallell.

If the concentration is > 0.1 ppb and the diffusive sampling time is > 5000 min the uptake rate can be considered as constant.

PMCH (perfluoromethylcyclohexane):	3.8 ng/ppm/min
PDCB (perfluorodimethylcyclobutane):	1.8 ng/ppm/min

REFERENCES

Dietz, R., 1982. "Brookhaven air infiltration measurements system. Manual for field depolyment".
Report BNL 31544, Dep. of Energy and Environment, Brookhaven National Laboratory, Feb. 1982.
Dietz, R., Core, E., 1982. "Air infiltration measurements in a home using a convenient perfluorocarbon tracer technique".
Report BNL 30797 R, Dep. of Energy and Environment, Brookhaven National Laboratory, May 1982.
Littler, J., Prior, J., 1983. "Development of a multitracer technique for observing air movment in buildings.
Reprt RIB/83/913/1, Research in Buildings Group, Polytechnic of Central London, 1983.
Littler, J., Prior, J., Martin, C., 1985. "Automation, extension and use of the PCL multi-tracer gas technic for measuring interzonal air flows in buildings".
Report RIB/1985/718, Research in Buildings Group, Polytechnic of Central London, 1985.

DETERMINATION OF NITROUS FUMES IN A SHIP REPAIR YARD BY MEANS OF DIRECTLY READABLE DIFFUSED COLLECTORS AND INDICATING TUBES

Herbert Muller-Wilderink,
Blohm & Voss AG, Hamburg

NITROUS FUMES

Nitrous fumes are mixture of NO, NO_2, N_2O_4, and the unstable N_2O_3. These fumes have a pungent and acrid odour which, in long concentrations, is very much like that of ozone and can hardly be distinguished from it. If stronger concentrations occur, then their colour may become conspicuous which, dependent upon the temperature, is brownish-yellowish to reddish. (Ref. 1)

Nitrous fumes are generated at temperatures of at least $1000o$ C (Ref. 2) and develop, for instance, when welding or flame-cutting metallic parts; they are somewhat heavier than air and thus they accumulate in floor areas. The investigations dealt with in this paper have mainly been carried out in connection with the flame-cutting of ships' constructional parts, the straightening of thin-gauge steel plates, and the welding of aluminized constructions components.

In shipbuilding, working in confined spaces is a necessity. However, such method of working may lead to increased concentrations of nitrous fumes if there is a lack of adequate ventilation. Confined spaces, in the meaning of an intensified endangerment by industrial fumes, are such spaces which have limited volumes of air and only a small natural change of air. (Ref. 3)

In order to be able to reliably determine the proportion of nitrous fumes, measuring techniques are applied that yield quite accurate results of measurement in relatively short a time and with little expenditure; short-time measurements by using test tubes and long-time measurements by using diffusion tubes or long-time tubes.

These devices allow to quickly ascertain any exceeding of MAK values (threshold values of concentrations of toxic substances at working places), and the requisite safety measures can be taken without endangering the workman concerned.

When treating shipbuilding parts with the acetylene-and-oxygen flame (acetylene generates a flame that is extremely hot) as is done for the purpose of straightening steel plates and flame-cutting structural components, then this process causes the concentration of nitrous fumes, if ventilation is insufficient, to rise and go up to a level that is injurious to health. (MAK value of nitrogen dioxide is 9 mg/m3) (Ref. 4)

METHODS OF MEASURING

For measuring the concentration of fumes in the breathing air of the shipyard workmen, the test tubes mentioned hereunder have been used:

-DRAGER test tubes "nitrous fumes 2/a" (CH 29401) (Ref. 5), measuring range: 10 strokes 2.0 to 50 ppm of nitrous fumes 5 strokes 5.0 to 100 ppm of nitrous fumes, change of colour towards deep blue-grey; and

-DRAGER test tubes "nitrous fumes" 0,5/a" (CH 29401) (Ref. 5), measuring range: 5 strokes 0.5 to 10 ppm of nitrous fumes, change of colour towards blue-grey.

Cross sensitivity: On the test tubes mentioned above, ozone shows a reaction similar to that of NO_2, but the indicating layer changes its colour to become light grey.

In the order to determine the proportion of nitrogen dioxide, for some of the measurements there have been used the following test tubes:

-DRAGER test tubes "nitrogen dioxide 2/c" (6719101) (Ref. 5), measuring range: 10 strokes 2.0 to 50 ppm of nitrogen dioxide 5 strokes 5.0 to 100 ppm of nitrogen dioxide change of colour towards dark grey; and

-DRAGER test tubes "nitrogen dioxide 0.5/c (CH 29401) (Ref. 5), measuring range: 5 strokes 0.5 to 10 ppm of nitrogen dioxide 2 strokes 5.0 to 25 ppm of nitrogen dioxide change of colour towards blue-grey.

Cross sensitivity of the nitrogen dioxide test-tubes: Nitrogen monoxide(NO)is not indicated. Ozone reacts in a manner similar to that of NO_2; however, the indicating layer changes its colour to become light grey.

Other measurements aimed at ascertaining the conditions of fume concentration over a longer period of time used the DRAGER test tubes "nitrogen dioxide 10/a-D" (8101111) (Ref. 5), measuring range: 10 to 200 (ppm x hours) 8 hours' time of employment = 1.3 to 25 ppm NO2 change of colour towards yellow-orange.

Cross sensitivity: Chlorine and ozone are also indicated (at approximately half the indicator sensitivity).

For the purpose of determining more precisely the cross sensitivity of ozone the DRAGER test tube "Ozone o.05/b" (6733181) (Ref.5) was used, measuring range: 10 strokes 0.05 to 0.7 ppm of ozone 5 strokes 0.01 to 1.4 ppm of ozone, change of colour towards white.

RESULTS

1. Welders' working places: Two measurements in the region

of breathing.   Characteristics: WIG welding   on   aluminium
superstructures using argon as shielding gas.

DRAGER test tubes with hand pump		Strokes/time	Measured	value
Nitrous fumes	0.5/a	5	0.5	ppm
Nitrous fumes	0.5/a	5	3.0	ppm
Nitrogen dioxide	2.0/c	10	1.0	ppm
Nitrogen dioxide	0.5/c	5	0.6	ppm
Ozone	0.05/b	10	0.7	ppm

Diffusion tube :
Nitrogen dioxide 10/a-D 2 x six hours each no indication

2. Plasma flame-cutting of 4 mm thick aluminium
   Measurement in the region of breathing.

DRAGER test tubes with hand pump		Strokes/time	Measured	value
Nitrous fumes	0.5/a	2	5.0	ppm
Nitrous fumes	0.5/a	5	over 10.0	ppm
Nitrous fumes	2.0/a	5	10.0	ppm
Nitrogen dioxide	0.5/c	5	1.5	ppm

Diffusion tube :
Nitrogen dioxide   10/a-D   5 hours   no indication

3. Flame-cutting work on a forebody in the area of the bottom, inside
   and outside.
   In the interior, small space sizes (abt. 8 m3) ;
   natural ventilation.

DRAGER test tubes with hand pump		Strokes/time	Measured	value
Nitrous fumes	0.5/a	2	10.0	ppm
Nitrous fumes	2.0/a	10	5.0	ppm
Nitrogen dioxide	0.5/c	5	0.75	ppm
Ozone	0.05/b	10	no indication	

Diffusion tube :
Nitrogen dioxide   10/a-D   2 hours   no indication
                                       70 minutes   no indication

Same work on the next day

DRAGER test tubes with hand pump		Strokes/time	Measured	value
Nitrous fumes	0.5/a	2	6.0	ppm
Nitrous fumes	2.0/a	5	5.0	ppm
Nitrous fumes	2.0/a	5	40.0	ppm

(abt.0.5m above t.flame)

Diffusion tube :
Nitrogen dioxide   10/a-D   5 hours   no indication
                                       5 hours   no indication

The above series of measuring results represents a selection of recordings from different series of measurements and has merely the nature of an example.

ASSESSMENT OF RESULTS

Experiments performed in the test laboratory of Messrs. DRAGER (Ref.6) have proved that in the case of low concentrations of nitrous fumes, i.e. in the range of parts per million, the half-life period (the time consumed by the transformation of nitrous fumes into nitrogen dioxide) will be extended to several hours.

This is one reason for the values of nitrogen dioxide as measured being low compared with the values of the nitrous fumes. (Cf. Ref.7) In most of all incidences there will be no hazard of a build-up of an atmosphere containing both nitrous fumes and nitrogen dioxide which would be of a lasting nature, that is to say, for more than one hour. This is precluded by the working place being provided with good natural or artificial ventilation. The circumstance that such atmospheric contamination of lasting nature does not occur explains why no readings on the diffusion tubes "nitrogen dioxide 10/a-D" are shown.

RESUME

At the working places of welders as well as at those of flame-cutters, the concentration of nitrous fumes can be determined by means of DRAGER short-time indicator tubes. As a guideline, reference to the threshold value of nitrogen dioxide, which according to the TLV List/USE 1986, shows 25 ppm nitrogen dioxide as a time-weighted average value. As regards the determination of harmful substances detrimental to the health of the persons involved there can be employed, in the case of the pollutant "nitrogen dioxide", the directly readable DRAGER diffusion tube "nitrogen dioxide 10 a-D". This device records concentration of nitrogen dioxide of two parts per million/length of working shift. The determination of nitrogen dioxide should be judged most cautiously as far as any conclusions are concerned with respect to the measured values exceeding or falling below the MAK values.

1) In the case of the measured values of nitrous fumes being high and those of nitrogen dioxide being low, it would undoubtedly be wrong to claim that the exposure to health imperilment is insignificant if the measured value of nitrogen dioxide falls below the relevant MAK value. (It may be referred in this context to the threshold value of nitrogen dioxide stated above.)

2) To what extent there will take place in the lungs of those having inhaled nitrous fumes a conversion from nitrous fumes into nitrogen dioxide, that is a medical problem and as such not subject of this paper.

But what can be said with certainty is that precautionary measures suited to protect workmen are recommendable in any case where the measured values of nitrous fumes exceed 25 parts per million.

BIBLIOGRAPHIC REFERENCES:

Ref 1:   Moeschlin:
            Therapie der Vergiftungen, 5. Auflage, Seite 143-144
            (Therapy of Intoxications, 5th edition, pages 143-144)
Ref 2:   Press   und   Wagner:
            Entstehen und Eigenschaften von gesundheitsschadlic
            Gasen und Stauben
            (Generation and Properties of Fumes and Dusts Injuriou
            to Health)
Ref 3:   ZH1/78, 1985
            Arbeitsgemeinschaften der Eisen-und Metall Berufs-
            genossenschaften, Seite 8 (Pamphlet ZH1/78, 1985;
            published by the Joint Committees of the Industrial
            Accident Insurance Institutions of the Iron and
            Metal Trades; page 8)
Ref 4:   Technische Regeln fuer gefaehrliche Arbeitsstoffe, 19
            Stickstoffdioxid (Code of Practice for Dangerous
            Working Materials, 1985 Nitrogen Dioxide)
Ref 5:   Draegerwerk AG Luebeck:
            Pruefroehrchen-Taschenbuch, 6. Ausgabe (Mai 1985)
            (Test Tubes Manual, 6th edition, May 1985)
Ref 6:   Mitteilung zum Drager-Gasspruehgeraet,
            22. Folge, Marz 1959;
            Nitrose-Pruefroehrchen 100/6 (Information about the
            DRAEGER Gas Spraying Device, 22nd series, March 195
            Nitrous Fume Test Tube 100/6)
Ref 7:   A. Zober
            Arbeitsmedizinische Untersuchungen zur inhalativen
            Belastung von Lichtbogen-Schmelzschweissern.
            Bau-Forschungsbericht Nr. 317 (Industrial-Medical
            Investigations on the Inhalative Stressing of
            Electric Arc Fusion Welders.
            Construction Research Report Nr. 317

ACTIVE AND PASSIVE AIR MONITORING IN COMPARISON TO
BIOLOGICAL MONITORING IN TETRACHLOROETHYLENE AND
XYLENE EXPOSED WORKERS*

K.H. Schaller, G. Triebig
Institute of Occupational- and Social Medicine
of University Erlangen-Nürnberg
(Director: Prof.Dr.med. H. Valentin)
FRG-8520 Erlangen, Schillerstraße 25/29

ABSTRACT
    Two field studies were performed to assess the relation-
ships between ambient (AM) and biological monitoring (BM) for
35 tetrachloroethylene and 28 xylene exposed workers. For personal AM passive sampling with two commercially available
systems as well as active sampling with charcoal tubes were
carried out. For BM tetrachloroethylene and xylene in blood,
tetrachloroacetic acid (TCA) and methylhippuric acid (MHA) in
urine in postshift samples were estimated.
    Significant coefficients of correlation were calculated
between active and passive AM. The AM data after passive sampling revealed a good agreement. Regarding the results of BM
significant correlations were found between AM and the solvent
and/or metabolite concentrations in blood and urine.
    On the basis of our study it could be concluded that
passive sampling is a practicable and reliable tool for the
monitoring of tetrachloroethylene and xylene concentrations in
air at the workplace.

INTRODUCTION
    An important criteria for the assessment of occupational
exposure is the measurement of toxic substances in air at the
workplace - so called ambient air monitoring (AM) - and/or in
biological materials - so called biological monitoring (BM).
The external exposure can be assessed by personal and/or stationary air sampling. In this context passive samplers have
become increasing importance.
    In the past years, several studies have been published
showing good agreements between passive and active sampling
(Gray and Thompson, 1984, Blome and Hennig, 1985/1986).
Another way to examine the validity of passive samplers is the
comparison of AM results with the individual data of BM. But

* With grants of Bundesanstalt für Arbeitsschutz, Dortmund
(Contract No. V2-539.835 (4))

this procedure is only relevant for those substances with significant relationships between external and internal exposure
and skin absorption is not relevant. Regarding this aspects,
the aim of our study was to investigate external/internal exposure relationships between AM data after passive sampling and
those of BM for the industrial important solvents tetrachloroethylene and xylene.

GROUPS AND METHODS

Thirtyfive persons occupationally exposed to tetrachloroethylene in dry-cleaning shops as well as 28 men exposed to
xylene in a lamp production unit were studied. Before the shift
a urine specimen was collected and the persons were supplied
with two brands of passive samplers (Monitor 3500, 3M Co.
Neuss, FRG and ORSA 5, Drägerwerk Lübeck, FRG) and with pumps
(Compur Electronics, München, FRG) and charcoal tubes (ACC).
Sampling of a postshift urine and blood specimens were performed. The air sampling was usually carried out over one shift
(6-8 hours).

The passive samplers were analyzed by gas chromatography
after elution with carbon disulfide according to the recommendations of the producers. The specific coefficients of desorption were established in our laboratories (Bomhard et al.,1983).
The solvents in blood and/or their specific metabolites in
urine trichloroacetic acid (TCA-U) and methylhippuric acid
(MHA-U) were determined by GC and HPLC (Schaller et al., 1983,
Angerer und Schaller, 1985).

RESULTS AND DISCUSSION

For the relationships between the results of the tetrachloroethylene determinations in air with two different passive
samplers in comparison to the active sampling with activated
charcoal tubes significant coefficients of correlations could
be evaluated (0.98-0.99). There is no intercept and the slope
of the regression line is around 1.
In table 1 the results of AM and BM are presented. Tetrachloroethylene concentrations in air (Tetra-A) ranged always below
the current MAK-value (1985) of 50 ppm.

Table 1: Results of AM and BM for tetrachloroethylene (Tetra)
exposed workers (N = 35)

Parameter		Range	Median
Tetra in air	ppm	0.1 - 31	7.5 (3500)
		0.1 - 28	6.5 (ORSA 5)
Tetra in blood	µg/l	1.0 - 134	30
TCA in urine	mg/l	0.2 - 12.7	2.7
- mg/g creatinine		0.2 - 10.1	2.4

The correlation analyses for the relationships between AM and
BM showed the following results:

Correlation	y' = a x + b	r
Tetra-A (3M) (ppm) - Tetra-B (µg/dl)	y' = 3.1 x + 6.6	0.89
Tetra-A (ORSA) (ppm)- Tetra-B (µg/dl)	y' = 3.6 x + 7.6	0.91
Tetra-A (ACC) (ppm) - Tetra-B (µg/dl)	y' = 3.0 x + 10.6	0.86
Tetra-A (3M) (ppm) - TCA-U (mg/g Cr.)	y' = 0.15 x + 1.5	0.50
Tetra-A (ORSA) (ppm)- TCA-U (mg/g Cr.)	y' = 0.17 x + 1.6	0.48
Tetra-A (ACC) (ppm) - TCA-U (mg/g Cr.)	y' = 0.15 x + 1.6	0.49

The tetrachloroethylene concentrations in air correlate significantly with the solvent levels in blood. In contrast, the
metabolite TCA in urine did not show such a good relationship.
This fact is caused by the low metabolic transformation rate
of tetrachloroethylene to TCA, which amounts about only 1 %.
Therefore tetrachloroethylene in blood is recommended for BM
of exposed workers. The parameter TCA in urine can be regarded only as an indicator for a past exposure to relatively
high tetrachloroethylene levels.
For xylene also a significant coefficient of correlation
(r = 0.97) was calculated for the comparison of the two brands
of passive monitors and the active sampling. The values found
with the monitor 3500 were lower (~15 %) than the data deter-
minated with the monitor ORSA.
    In table 2 the results of AM as well as BM as medians and
ranges are given. The current MAK-value of 100 ppm is not exceeded. For BM the levels of xylene in blood and of the main
metabolite MHA in urine were determined. The following signi-

ficant correlations between the parameters of AM and BM could be evaluated:

Correlation	$y' = ax + b$	r
Xylene-A (3M) (ppm) - Xylene-B (µg/dl)	$y' = 1.98\ x + 11.8$	0.57
Xylene-A (ORSA) (ppm) - Xylene-B (µg/dl)	$y' = 2.25\ x + 11.4$	0.58
Xylene-A (3M) (ppm) - MHA-U (mg/g Cr.)	$y' = 0.02\ x + 0.2$	0.70
Xylene-A (ORSA) (ppm) - MHA-U (mg/g Cr.)	$y' = 0.02\ x + 0.1$	0.75

Table 2: Results of AM and BM for xylene-exposed workers

Parameter	N	Range	Median
Xylene-A - Monitor 3500 (ppm)	28	2 - 60	19
- ORSA 5 (ppm)	28	1 - 53	16
Xylene-B µg/dl	28	50 - 210	30
MHA-U - g/l	28	0.1 - 3.0	0.8
- g/g creat.	28	0.1 - 1.6	0.4

In agreement with the recommendations of the German Science Foundation (DFG, 1984) our data confirm, that xylene in blood as well as methylhippuric acid in urine can be used in BM of exposed persons.

CONCLUSION

AM using passive samplers is a practicable and relatively inexpensive technique to assess the amount of longterm exposure to the organic solvents tetrachloroethylene and xylene under actual exposure conditions at the workplace.

REFERENCES

Angerer, J., Schaller, K.H. (ed.) 1985. Analyses of Hazardous Substances in Biological Materials, Vol. 1. Verlag Chemie Weinheim

Blome, H., Hennig, M. 1985/86. Teil 2/3. Staub-Reinhaltung d. Luft 45, 541-546; 46, 6-10

DFG 1984/85. Gesundheitsschädliche Arbeitsstoffe. Toxikol.-arbeitsmed. Begründung von BAT-Werten, Verlag Chemie Weinh.

Gray, C.N., Thompson, J.M. 1984. In: Harrington, J.M. (ed.) Recent Advances in Occupational Health, Churchill Livingstone, Edinburgh

Schaller, K.H., Triebig, G., Valentin, H. 1983. Arbeitsmedizin aktuell, Lieferung 13, Gustav Fischer Verlag, Stuttgart

Bomhard, A., Triebig, G., Schaller, K.H., Sulzmaier, R. 1983. Gentner-Verlag, Stuttgart, pp. 371-375

AIR MONITORING OF MERCURY-EXPOSED WORKERS WITH PASSIVE SAMPLERS IN COMPARISON TO BIOLOGICAL MONITORING

G. Triebig, K.H. Schaller

Institute of Occupational- and Social-Medicine
of University Erlangen-Nürnberg
(Director: Prof.Dr.med. H. Valentin)
FRG-8520 Erlangen, Schillerstraße 25/29

ABSTRACT

The relationships between external mercury (Hg)-exposure monitored with passive samplers (AM) and internal exposure (biological-monitoring, BM) were studied in three groups (production of thermometers (N = 30), recycling plant (N = 5) and chlorine-alkali plant (N = 20)). For personal AM we used the passive samplers 3600 and 3600 A resp. (3 M Company, FRG), which were carried about 6 to 7 hours during the workshift. Mercury in air ranged between 16 and 444 $\mu g/m^3$ (MAK-value 100 $\mu g/m^3$). Chlorine was always below 0.2 ppm (plant 3). BM comprised the determination of Hg in pre- and postshift blood-and/or urine-samples. The Hg-concentrations in blood varied between 5 and 122 $\mu g/l$ (normal below 5 $\mu g/l$), and in urine between 5 and 766 $\mu g/l$ resp. (normal below 5 $\mu g/l$). Regression analyses showed significant correlations between Hg-concentrations in air and blood (R = 0.69 and 0.82) and urine (R = 0.35, 0.72 and 0.82), resp. with better results for post-shift samples. On the basis of our study it is concluded, that AM with commercially available passive samplers is a reliable and practical but rather expensive method for monitoring of mercury vapour exposed workers.

INTRODUCTION

Air monitoring (AM) and/or biological monitoring (BM) are often used in medical health surveillance of mercury (Hg)-exposed workers (Schaller and Triebig 1984). However there is some lack of information about the correlations between AM after passive sampling and BM under field conditions. For this reason, we examined the external- and internal exposure of mercury in 55 workers at three different workplaces. Main purpose of the study was to elucidate possible correlations between Hg in air after passive sampling and Hg in blood and urine resp.

*With grants of Bundesanstalt für Arbeitsschutz, Dortmund (Contract No. V2-539.835(4))

GROUPS AND METHODS

A total number of 55 men at three different workplaces
were investigated:

  I   Production of thermometer (N = 30)
  II   Hg-recycling plant   (N = 5)
  III   Chlorine-alkali plant   (N = 20)

An exposure to metallic mercury vapour only is relevant, at
workplace III also chlorine can occur in minor concentrations.

For this reason we used two passive samplers (PS) monitor
3600 (Hg vapour) and/or monitor 3600 A (Hg vapour in presence
of chlorine gas), (3 M Co., Neuss, FRG). The PS were carried
over the whole workshift, e.g. about 6 to 7 hours. After
sampling the monitor was sealed and sent for analysis to 3 M
Co. The principle of the monitor is: mercury is absorbed on
a thin layer of pure gold depending on Hg-concentration in air.
From conductivity measurements before and after sampling the
absorbed amount of Hg is estimated indirectly.

For BM blood- and/or urine-samples were collected before
and/or after workshift. Cold vapour atomic absorption technique
with $Na(BH)_4$ was used for the measurement of total Hg content
in blood and urine (Schierling and Schaller, 1981). All methods
used were reliable and well accepted in our laboratories. The
determinations were performed under external and internal
quality control scheme.

RESULTS AND DISCUSSION

Table 1 shows the results of AM and BM for the three
groups.
In the Fed. Republic of Germany the following permissible
limit values for inorganic metallic mercury exposure are
relevant (DFG 1985):

  Hg-air   100 µg/m³ (MAK value)
  Hg-blood   50 µg/l   (BAT value)
  Hg-urine   200 µg/l   (BAT value)

Table 1: Mercury concentrations in air, blood and urine.
Given are the medians as well as the ranges, 1 = postshift
sample, 2 = preshift sample

Group	Hg in air $\mu g/m^3$	Hg in blood $\mu g/l$	Hg in urine $\mu g/l$
I Thermometer- production (N = 30)	95 (16-196)	46 (5-122)[1]	138 (5-447)[1]
II Recycling plant (N = 5)	278(173-444)	51(31-110)[1]	403(122-766)[1] 410(82-609)[2]
III Chlorine alkali plant	68(26-296) 71(47-172)	14(8-41)[1]	56 (17-224)[1] 91 (22-327)[2]

The results of AM and BM demonstrate that the Hg exposure
exceeds the permissible limit values, especially in plant II.
Furthermore there is a discrepancy between Hg-air and Hg-blood/
urine for workers in plant II and III, which is due to the
occasionally use of protective masks during workshift. There
are no significant differences in Hg-air levels after using
monitor 3600 and 3600 A resp., which is in accordance with the
low levels of chlorine at the workplaces (below 0.2 ppm) in
plant III. Mc Cammon and Woodfin (1977) found, that the 3 M
monitor for mercury is even not affected by high concentration
of chlorine. But the monitors are temperature dependent, an
increase of absolute temperature resulted in an enhancement
of the diffusion coefficient. Furthermore high levels of SO_2
and H_2S can slightly alter the results (Mc Cammon and Woodfin
1977).

In table 2 the coefficients of correlation for the relationships of Hg-air, Hg-blood and Hg-urine are given.

Table 2: Coefficients of correlation between Hg-air ($\mu g/m^3$),
Hg-blood ($\mu g/l$) and Hg-urine ($\mu g/g$ creatinine), resp. for the
three groups, 1 = postshift sample, 2 = preshift sample,
* statistical significant (p $\leq$ 0.05).

Group	Hg-air/Hg-blood	Correlated parameters Hg-air/Hg-urine	Hg-blood/Hg-urine
I Thermo- meter production (N = 30	0.69*[1]	0.80*[1]	0.68*[1]
II Recycling plant (N = 5)	0.82*[1]	-0.18[1]	-
III Chlorine alkali plant (N = 20)	0.02[1]	0.61*[1] 0.37[2]	0.19[1]

Significant correlations were found between the Hg-concentrations in air, blood and urine resp. for plant I. These results
could not be found for plant II and III. This is due to the
fact, that the workers in these plants had used protective
masks during work activities with high Hg-exposure.
On the basis of these results it is concluded, that the passive
sampling mercury-monitor allows a reliable and practicable
measurement of the mercury-vapour-concentration at the workplace over a longer period (work shift). During sampling
extreme temperatures have to be regarded. The relatively high
costs (about 200.- DM) of the monitor probably limits its
widespread use.

LITERATURE
Deutsche Forschungsgemeinschaft (Ed.) 1985. MAK-/BAT-Werte
    Liste 1985, Verlag Chemie, Weinheim.
Mc Cammon, Jr., Ch.S. and Woodfin, J.W. 1977. An evaluation of
    a passive monitor for mercury vapor. Am.Ind.Hyg.Assoc.J.,
    38, 378-386
Schaller, K.H. und Triebig, G. 1984. Personenbezogene Proben-
    nahme von Quecksilberdämpfen am Arbeitsplatz - ein Ver-
    gleich zwischen externer und interner Quecksilberexposition.
    Arbeitsmed.Sozialmed.Präventivmed. 12, 289-292
Schierling, P. und Schaller, K.H. 1981. Einfache und zuver-
    lässige Methoden zur atomabsorptionsspektrometrischen Be-
    stimmung von Quecksilber in Blut und Urin. Arbeitsmed.
    Sozialmed.Präventivmed. 3, 57-61

AN OVERVIEW OF THE EFFECTS OF TEMPERATURE, PRESSURE

HUMIDITY, STORAGE AND FACE VELOCITY

POZZOLI L. - COTTICA D.

Fondazione Clinica del Lavoro - Laboratorio di Igiene Industriale

PAVIA - ITALIA

The elements which compose a passive sampler can be classified as follows:

- outer covering
- substratum for collection
- diffusion chamber
- primary resistance

The concentration of the substance to be examined, expressed in mgs/cubic m. present in the environment and coming from outside the sampler, is calculated from the application of the following formulas derived from Fick's law:

$$\frac{Q.X.}{S.t.D.}$$

where:

Q = the quantity of the substance obtained from the substratum and analytically determined (ng.)

X = height of the diffusion chamber (cms.)

S = exposed surface (sq. cms.)

t = time of exposure (sec.)

D = diffusion coefficient (sq. cms sec^{-1})

X,S are characteristic parameters of the geometry of the sampler

D can result from the substance being examined and also from the

environmental conditions (temperature and pressure) in which
the work is carried out

t.Q.   are sizes which must be guaranteed, depending on the exposure
of the sampler.

The passive sampler may be considered a system for mass transport with
two resistances: the primary resistance is represented by a porous or
other kind of barrier, at the entrance of the sampler which serves as
the directional force of the molecule of the examined substance, the second
is represented by the diffusion, according to Fick's law, inside the
sampler chamber.

The general equation of a passive sampler is therefore the one which
gives due importance to the moving forces which are respectively proportio
nal to the decrease in concentration from the outside to the inside of
the sampler due to the porous, or any other kind of barrier, and to the
decrease in concentration verified in the internal diffusion chamber.

In other words from the two above mentioned resistances in series we get:

$$Q = \frac{\Delta F_I + \Delta F_2}{R_I + R_2} = \frac{\Delta F}{R_I + R_2}$$

$$\frac{R_I}{R_2} = K; \quad K = \frac{I}{I + K}$$

$$Q = \frac{\Delta F}{R_2 (I+K)} = K \frac{\Delta F}{R_2}$$

K can be found experimentally and with it the primary resistance type or
porous membrane barrier can vary, besides, it can assume the significance
of an apparent diffusion; R_2 is the resistance due to diffusion inside
the sampler chamber. From this point of view, the influence of temperature,
pressure and ventilation will be taken into consideration.

THE INFLUENCE OF TEMPERATURE AND PRESSURE

It is important to establish the influence of temperature and air pressure of the environment in order to verify the working conditions of the passive sampler.

If it is accepted that R_I is negligible respect to R_2, then the quantity of substance collected will vary according to "C" and "D" respectively, according to the concentration and the diffusion coefficient.

$$Q = f (C, D) \tag{I}$$

"C" can be obtained from the general equation for gas:

$$PV = nRT$$
$$P = n/V \ RT$$
$$C = n/V = P/T \ I/R \tag{2}$$

D can be obtained from Maxwell's equation and is a function of absolute temperature and pressure:

$$D = f \frac{T^{3/2}}{P} \tag{3}$$

by substituting (2) and (3) in (I) we get:

$$Q = f (P/T, \ T^{3/2}/P) \tag{4}$$

Granted that the dependence of D from $T^{3/2}/P$ and C from P/T is linear, and granted that there is a linear correlation between Q and the product C.D we obtain:

$$Q = f (\sqrt{T})$$

From the above mentioned points it can be noted that for the collection of a determined quantity of a required component and for the calculation of its concentration, pressure does not have any influence and temperature intervenes in the proportion of the square root of absolute temperature. In other words, the variations of "Q" to two different temperatures for a determined exposure concentration and for a determined sampler are obtained

from the relation between the square roots of the absolute temperatures.
For every 5.5°C this variation is equal to I% approximately.
Temperature only modestly influences the collected air samples; it may
even be affirmed that since the air temperature in working environments
generally ranges from I5°C to 30°C, the variation of the quantity of
polluting substance obtained by operating at the two sampled extreme
conditions can be included in the analytical errors of the method.

THE INFLUENCE OF VENTILATION

Fick's law is valid in stationary conditions, during actual operation
instead (when this law is applied to passive samplers), some not stationary conditions represented by turbulent air movement materialize, in
the case of variable laminar movements, and other errors which are difficult to identify and quantify. They are particularly important in the
determination of the value of C on the surface of the exposed sampler.
In stagnant air conditions, there is no significant transportation of
material, and the polluting substance which is on outside surface, is not
efficiently renewed (on the sampler surface), once it has entered and has
been absorbed by the sampler. This provokes a decrease of   C inside the
sampler, and therefore it does not work regularly any more. The only
process which permits the renewal of the polluting substance in stagnant
air conditions, is represented by its molecular diffusion in the outside
environment up to the external surface of the sampler. This process, in
principle, still seems to be valid, but it is increasingly slow as time
passes, as the average distance between the surface of the sampler and the
area where the polluting substance is present increases continuously.
In order to guarantee constant mass transport, so that the sampler might
work well, it is necessary instead that the exposed surface air moves;
in other words it is necessary to constantly supply, with the help of
physical forces, the presence of the polluting substance in the air, on to
the sampler's contact surface. In other words it is necessary to maintain
C constant or even variable in time, in order that the concentration will
always represent a significantly large moving force of the system. C must
always be a positive quantity.

In the case of laminar movement, perpendicular to the X axis of the sampler and at a tangent to it, the flow of air that transports the polluting substance begins to give it away when it starts touching the section of the sampler's surface and this "giving away" precess continues for the entire course of the air flow which interests the outside or exposure surface, till it comes into contact with the sampler's surface, and stops only when the air flow has passed completely through the sampler.

It is therefore obvious that as the air flow passes through the contact surface, the concentration of the polluting substance in the mentioned air flow decreases, which can be verified in the direction of its section, that is perpendicular to the direction of the flow. At the exit from the contact surface and the sampler, the flow will therefore present a variable concentration which increases with the height.

The diffusion occurs once again inside the sampler, but alsongside, a decrease in the transported mass is noted compared to the natural decrease in concentration on the outside surface of the sampler. This makes the interpretation of the transporting laws more complex, and for a practical application it is therefore necessary for the polluting substance to remain homogeneous on the entire air flow contact surface, and for the entire period of contact, and this may be obtained with rapid change of air, and that is with sufficiently rapid related movements between the air and the sampler, which therefore create a turbulent state on the outside surface of the sampler.

In turbulent state conditions it must not be excluded that one component of the air speed might be directed towards the inside of the sampler. Thus it is assumed that respect to the sampler, all the transport mechanisms act simultaneously in an addictive manner (turbulent diffusion and transport), so that the equation for the total flow of a particular component may be written as a flow outline that varies according to each mecchanism and that is:

$$J_j = -D_j \frac{dC_j}{dX} + V_x C_j$$

The terms $V_x \cdot C_j$ indicate the contribution of the nth component which moves with the additional external conventional mecchanisms :   the transported mass is added to the quota which moves along in the natural direction of the diffusion. But unfortunately, these equations can be solved only occasionally and with extreme difficulty.

The mass transportation in the sampler which occurs due to various components, therefore brings about a non-respect of the correlation of linear correspondence between transported mass and concentration. This inconvenience must be overcome because otherwise the polluting mass collected by the sampler itself will depend every time on the turbulent penetration of the air flow in the sampler. The ideal situation would be the one in which the transported mass is indipendent from the speed V also if it is known that an inertia exists in the initial phase. The above mentioned results may be outlined approximatively in figure I. There must be a constant transported mass in function of the air-speed, (curve a) and not an inclined straight line which appears when the transported mass increases along with the increase in speed (curve b).

This problem may be empirically solved by placing a porous barrier between the environment and the sampler so that it constitutes a resistance which regulates the transfer of the subject component, and impedes convective transport movements.

The single-direction flow followed by the polluting substance to reach the collection substratum must therefore cross through two resistive measures in series.

The first is represented by the anticonvective resistance called the equilibrium resistance,

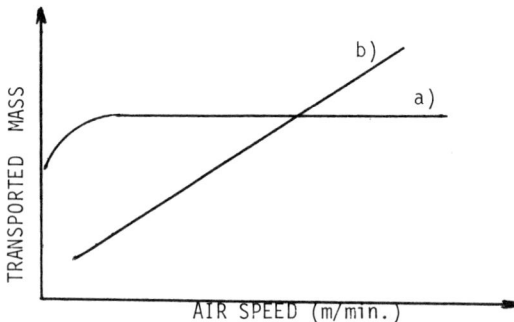

Relation between sample amounts and speed of exposure of the sampler.

and the second is represented by the diffusion chamber. The equilibrium
resistance must permit the passage of the molecules of the component to
be absorbed from the outside environment to the inside of the sampler in
a manner which is proportional to the number of molecules per unit of
volume (concentration) existing on the exposure surface and must also
this number of molecules to diffuse, in stationary conditions, in the
diffusion chamber, up to the collection substratum. In other words the
flow of molecules through the resistance must be equal (unless there are
constant factors due to the geometry of the sampler) to the diffusive
flow inside the sampler, up to the collection substratum, and thus the
collection of substances that is proportional to the outside concentration
is obtained.

The diffusing flow is controlled by the higher of the two working
resistances: the anticonvective resistance and the Fick resistance.
When the Fick resistance in the sampler chamber is higher, the fall of
concentration is practically totally governed by the resistance of the
sampler chamber itself since on contact with the collection substratum
the concentration of the examined substance, which is completely adsorbed,
is practically negligible.

The porous barrier practically acts as a directional force for ithe
molecules of the examined substance and it becomes an infinite resistance
in the case of a convective flow ($C_j V_x$) whilst it acts as a finished
resistance in the diffusive process.

The adopted attenuation mecchanisms can be recorded in the following
order of realization:

- metal netting
- slender and porous synthetic membranes
- very thick permeable membranes placed directly on the collector
- concave tubes with different ratios of length/flow
- simple increase in the heighit of the diffusion chamber.

In the practical production of passive samplers large use has been made
of the porous membranes and the concave tubes.

The phenomenon has been evidenced empirically by conceptually same passive
samplers with different dimensions of the exposed surface and of the
height of the diffusion chamber; the anticonvective resistance, embossed
polypropilene membrane, was steady.

In tables in I and 2 the experimental results are reported.
We had at our disposition two series of passive samplers with the
following dimensions:

- exposed surface 7 sq.cms. and height of the diffusion chamber 0.4 cm;
- exposed surface I2.56 sq.cms. and height of the diffusion chamber
  2.4 cm.

Both the series were exposed at different air velocities in a controlled
constant atmosphere of n - hexane of I00 mg/m^3.
It is demonstrated that the samplers with lower exposure chambers
present a diffusion resistance which is negligible compared to the
anticonvective resistance, therefore the quantity of hexane collected
is included in a wide range as regards the theoretics.
Increasing the height of the chamber the diffusion resistance increase
as to balance the convective resistance and so became the true regulator
of the diffusive system; in this situation the quantity of hexane
collected is increasingly close to the theoretical amount.
Independent from the sampler exposure or air face velocity.
Obviously to obtain a linear and proportional correspondence between
exposure concentration and quantity of substance collected from the
substratum, it is necessary to proportionally calculate the height of
the diffusion chamber and the sampler exposure surface in function of
the anticonvective resistance to be chosen according to the practical
and executive needs of the sampler itself.

THE INFLUENCE OF HUMIDITY

Relative humidity influences a sampler, be it based on active carbon,
dynamic absorption or even a passive sampler with static absorption.
Relative humidity however does not have any effect on the sampling
capacity, but intervenes on the collecting capacity of the adsorbing
material; obviously if it is sensitive to humidity.
The collection substrata of the passive samplers are generally represented
by active carbon. The various substances are adsorbed by it in different
proportions according to the law of adsorption competitivity.
Water is a polar substance and therefore does not have much similarity
with active carbon. The problem arises when it is present in a high

TABLE N° I

Experimental data obtained by passive samplers with the following characteristics:

Primary resistance: embossed polypropilene membrane

Diffusive parameters: $S = 7,065$ cm^2; h from 0,4 to 2,4 cm

Exposure: hexane I00 mg/m^3 for 60'

Total amount of hexane collected								
Exposure Velocity m/s	Height of the diffusion chamber							
	0,4cm		0,8cm		I,6cm		2,4cm	
	Theor	Exp	Theor	Exp	Theor	Exp	Theor	Exp
0,25	467	304	232	204	II6	I40	76	82
0,50		327		222		I32		9I
0,75		360		249		I35		94
I,00		383		259		I29		93
I,25		400		265		I28		94
I,50		4I0		267		I35		92
I,75		4I5		267		I35		85
2,00		425		268		I3I		84

TABLE N° 2

Experimental data obtained by passive samplers with the following characteristics:

Primary resistance: embossed polypropilene mambrane

Diffusive resistance: $S = 12,665$ cm^2; h from 0,4 to 2,4 cm

Exposure: hexane I00 mg/m^3 for 60'

Total amount of hexane collected								
Exposure Velocity m/s	Height of the diffusion chamber							
	0,4cm		0,8cm		I,6cm		2,4cm	
	Theor	Exp	Theor	Exp	Theor	Exp	Theor	Exp
0,25	828	464	4I4	298	207	I86	I38	I4I
0,50		590		346		I92		I46
0,86		704		375		222		I44
I,25		732		4I2		246		I60
I,50		744		432		244		I60
2,00		762		443		259		I80
2,50		774		446		266		I80
2,90		779		450		263		I80

percentage in the air sample; some authors place the limit of interference around 50% of relative humidity.

On the other hand industrial hygienists are well informed about the problem before using these sampling devices on site with the necessary precautions.

SAMPLER PRESERVATION

The preservation of passive samplers before use is described by the company producing the samplers. Usually an expiry date is fixed, before which the sampler must be utilized.

The preservation of the samplers after use is actually a preservation of the collection substratum and certain conditions related to these must be respected.

We have taken for example the case of active carbon, which from many points of view represents the substratum which is most commonly used.

The substance adsorbed on the active carbon after sampling and during preservation can undergo the following phenomenons:

- transformation due to a catalyzed reaction of the carbon itself
- diffusion and shifting towards another section, up to the stage in which it frees itself from the active carbon.

Transformation due to catalysis is quite improbable at room temperature since these reactions occur at high temperatures.

Diffusion and shifting are quite frequent and mostly depend on the quantity of carbon and the chemical and physical characteristics of the examined substances.

The quantity of active carbon in a passive sampler is quite limited and may be approximately between I00 and 200 mgs.

Therefore, for the solution of this problem careful attention is paid to the chemical and physical characteristics of the various substances. The diffusive phenomena are sensitive in substances with high vapour tension and low temperature boiling points.

We list hereunder the results of some of the tests carried out with a passive sampler equipped with an active carbon cloth for adsorption.

The samplers were exposed in turn to the following substances:

TETRAFLUOROETHYLENE	boiling point	$-76.3°C$
HEXAFLUOROPROPYLENE	boiling point	$-29.4°C$
CYCLOPERFLUOROBUTANE	boiling point	$- 4.0°C$

For each substance, three tests were carried out, each one composed of 6 exposures with a duration of 60-90' with concentrations of about 2 - 3 ppm. For each test, the active carbon was anasysed at different times: immediately after exposure, after 4 hours, keeping the sampler shut, after 4 hours leaving the sampler open.

The various substance were measured and it was possible to calculate the recovery percentage compared to the theoretical quantity present on the active carbon.

The results of the tests are mentioned in Table 3:

TABLE N°3 - % Recovery of substance from active charchoal.

| Substance | % Recovery of substance from active charchoal | | |
	Immediately analyzed	Closed sampler analyzed 4h after	Open sampler analyzed 4h after
$C_2 F_4$	93.6	I2.6	9.7
$C_3 F_6$	98.7	79.7	I0.8
$cC_4 F_8$	98.2	89.I	65.8

They prove: the spontaneous releasing of the adsorbed substance decreases with the increase in the boiling point of the substance itself; when the sampler is closed, a counterpressure is created when it is released and therefore this phenomenon tends to decrease; the need for an analysis, at the end of the exposure, of the collection substratum.

These data also show the possibility of a non-complete adsorption with prolonged sampling and the need to avoid errors due to loss of substances also during sampling, and to carry out short exposures, obviously for substances with low boiling points.

It is advisable in this case, to solve the problem pratically, placing the collection substratum immediately in the elution solvent of the adsorbed substance oa the end of the sampler exposure and to carry out all analyses within a short period of time. This way any errors in the results are avoided.

BIBLIOGRAFIA

BROWN R.H., HARVEY R.P., PURNELL C.J., SAUNDERS K.J.: A diffusive sampler evaluation protocol. Am.Ind.Hyg.Assoc.J. 45: 67-75, 1984.

EINFELD W.: Diffusional sampler performance under transient exposure conditions. Am.Ind.Hyg.Assoc.J. 44: 29-35, 1983.

GREGORY E.D., ELIA V.J.: Sample retention properties of passive organic vapor samplers and charcoal tubes under various conditions of sample loading, relative humidity, zero exposure level periods and a competitive solvent. Am.Ind.Hyg.Assoc.J. 44: 88, 1983.

KELLY D.W.: Dosimetry for chemical and physical agents. Ann.Am.Conf.of Gover.Ind.Hyg. volume I, 1981.

KRING E.V., DAMRELL D.J., BASILIO A.N., Mc GIBNEY P.D., DOUGLAS J.J., HENRY T.J., ANSUL G.R.: Laboratory validation and field verification of a new passive air monitoring badge for sampling ethylene oxide in air. Am. Ind.Hyg.Assoc.J. 45: 697-707, 1984.

Mc CONNAUGHEY P.W., Mc KEE E.S., PRITTS I.M.: Passive colorimetric dosimeter tubes for ammonia, carbon monoxide, carbon dioxide, hydrogen sulfide, nitrogen dioxide and sulfur dioxide. Am.Ind.Hyg.Assoc.J.46:357-362,1985.

MOORE G., STEINLE S., LEFEBRE H.: Theory and practive in the development of a multisorbent passive desimeter system. Am.Ind.Hyg.Assoc.J. 45: 145-153, 1984.

POSNER J.C., MOORE G.: A thermodynamic treatment of passive monitors. Am.Ind.Hyg.Assoc.J. 46: 277-285, 1985.

STOCKTON S.D., UNDERHILL D.W.: Field evaluation of passive organic vapor sampers. Am.Ind.Hyg.Assoc.J. 46: 526-531, 1985

UNDERHILL D.W.: Efficiency of passive sampling by adsorbents. Am. Ind.Hyg. Assoc.J. 45: 306-310, 1984.

FOWLER W.: Fundamentals of passive vapor sampling. Int. Lab. 40-48. Apr. 1983.

POZZOLI L., PEGORARO M.: I campionatori passivi da "Igiene Industriale: campionatori di gas, vapori, polveri". pag. 241-319 di L. Pozzoli, U. Maugeri. Ed. La Goliardica - Pavia - 1986.

PREDICTION OF UPTAKE RATES FOR TUBE-TYPE SAMPLERS

N. van den Hoed and Ms. M.T.H. Halmans
Koninklijke/Shell-Laboratorium, Amsterdam
(Shell Research B.V.)
Badhuisweg 3, 1031 CM Amsterdam, The Netherlands

ABSTRACT

When tube-type diffusive samplers are used for measuring air contaminants, in conjunction with thermal desorption for sample recovery, a
fall-off in the uptake rate is frequently observed. If, however, adsorbents
are selected according to criteria developed by us on the basis of the adsorption isotherm characteristics, uptake rates are constant, and in close
agreement with theoretical values that can be calculated for ideal diffusive samplers from their geometries and from the air diffusivity of
the chemical compound.

INTRODUCTION

Thermal desorption is a cost-effective means of recovering adsorbed
material from personal air-sampling devices. It has two important advantages over the commonly used solvent desorption technique: it can be used
in conjunction with on-line gas-chromatographic analysis, and the samplers
can be used repeatedly. Two different types of thermal desorption samplers
have been developed. Badge-type samplers were first introduced by SKC Inc.,
Pennsylvania (USA) in 1983, while somewhat later a sensitive dual-face
badge-type sampler was described (Lewis et al., 1985). So far these badge-
type samplers have not become very popular, presumably on account of their
limited automation potential. Tube-type diffusive samplers (Brown et al.,
1981) have proved more promising in this respect. An automated on-line
gas-chromatographic thermal desorber that can accommodate and process up
to 50 tube-type samplers has recently been introduced by Perkin Elmer. The
combination of a convenient method of sampling and cost-effective analysis
has thus now become feasible.

The relatively low mass uptake rate of tube-type diffusive samplers
has not been experienced as a serious drawback since it can largely be
compensated by efficient coupling to the gas chromatograph, so that the
whole of the sample is analysed. Often this has resulted in even higher
overall sensitivity than with solvent desorption, in which only a small
proportion of the material collected is analysed. The low uptake rate
can in fact be considered as an advantage since it means that there will
be less depletion of the air around samplers, so that the samplers will

be less susceptible to the "starvation" effect. With tube-type samplers
the air velocity need be no higher than 0.7 cm/s (Brown et al., 1981)
whereas for badge-type samplers air velocities of the order of 10 cm/s
are reported to be necessary (Tomkins et al., 1977; Van der Wal et al.,
1984).

A much more serious drawback associated with the use of thermal
desorption samplers is the non-constant uptake rate which, in particular,
is often observed in validation studies. This effect is a serious obstacle
to the official acceptance of thermal desorption samplers by governmental
institutes. One important objective of our recent work on diffusive
sampling was to find a way of counteracting or circumventing this un-
attractive characteristic of the diffusive thermal desorption samplers.

THEORY

Non-constant uptake rates by thermal desorption samplers have been

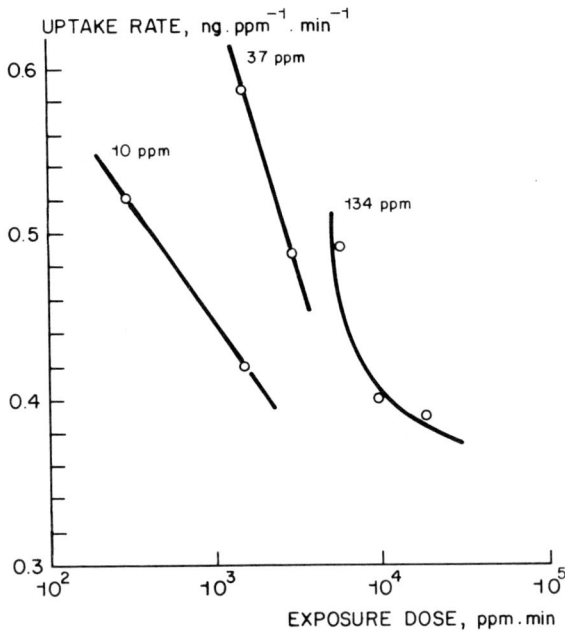

Fig. 1   Rate of 1,3-butadiene uptake by a tube-type diffusive
sampler packed with Tenax*

* Tenax is an adsorbent manufactured by Akzo Research Laboratories,
The Netherlands

reported for both tube-type and badge-type samplers. Some papers (Coutant et al., 1985; Health and Safety Executive (UK), 1985) claim that these variations are functions of time solely, so that the effective sampling rate during an actual exposure can be found from calibration graphs of up-take rate versus time. This is not in agreement with theoretical studies, which predict reduced sampling efficiency as a function of both concen-tration and time (Underhill, 1984). Figure 1 is an example of our own measurements by means of tube-type samplers: it shows clearly that con-centration does affect the uptake rate.

It is generally accepted that the phenomenon of decreasing uptake rates is associated with the use of non-optimal adsorbents. In view of the effective desorption process required for on-line GC analysis a strong adsorbent is not admissible. Quite soon after the start of the exposure the concentration of the chemical compound close to the adsorbent will deviate substantially from zero, as illustrated in Figure 2. The differential

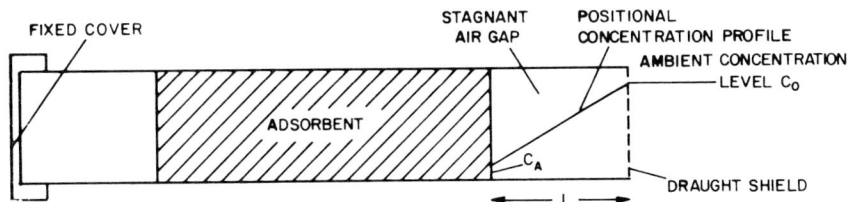

Fig. 2   Lay-out of a tube-type diffusive sampler

equation describing the uptake of adsorbate by the sampling surface reads (Underhill, 1984)

$$\frac{dM}{dt} = \frac{AD}{L} (C_O - C_A) \qquad \text{(Eq. 1)}$$

where M = mass of adsorbate sampled

A = cross-section of the sampler

D = the air diffusivity of the compound

L = length of the diffusive air zone

C_O = concentration outside the sampler

C_A = concentration adjacent to the adsorbent inside the sampler

t = exposure time

In this case equation 1 may not be simplified to the simple linear relationship

$$M = \frac{AD}{L} C_O t \qquad \text{(Eq. 2)}$$

describing the mass uptake of an ideal sampler, that is reasonably well
approximated by diffusive samplers operating on the basis of charcoal
adsorption.

Theoretical descriptions for the non-ideal situation are rather
complex since they should also account for the phase equilibrium between
adsorbent and adsorbate (the adsorption isotherm) as well as for the mass
transport in the adsorbent bed.

Solutions for the resultant differential equations have been published
in literature (Underhill, 1984, 1985; Coutant et al., 1985). Both the
studies concerned, however, have to resort to assumptions which are not
very realistic for the tube-type samplers under field conditions, viz.
instantaneous equilibrium for the complete adsorbent bed in combination
with a constant external concentration. Furthermore the assumption of
linear adsorption isotherms (Coutant et al., 1985) is questionable, for
reasons set forth subsequently.

A possible way round the theoretical problems might be to find
numerical solutions to differential equation 1 by means of computer
models, which can include all variables and parameters at any desired
level of complexity.

A modest start in this direction has been made at our institute.
Even if the adsorbed mass can be described as a function of exposure
time, and of the characteristics of the isotherm and concentration profile, it still will be uncertain whether this will provide a unique value
for the reverse relation, which would be needed in order to calculate the
time-weighted-average air concentration, independent of the actual concentration profile.

In view of the foregoing, we decided to start looking for an alternative, more experimentally orientated, solution to the non-constant
uptake rate problem. This approach is described below.

ADSORBENT SELECTION

Adsorbents for sampling/recovery by thermal desorption should show,
on the one hand, high adsorptivity to create sufficiently low "near-
sorbent" concentrations for long periods of time and, on the other hand,
sufficiently low adsorptivity to allow effective thermal desorption. The
objective of our investigation was to quantify these two qualitative
statements and to see whether there was any room for manoeuvre between
these contradictory requirements.

As already shown (Underhill, 1984) the suitability of an adsorbent with respect to its sampling efficiency is determined by the character-istics of the adsorption isotherms. In this respect Dubinin-Radushkevich- and Freundlich-type isotherms show attractive features. In such types a high adsorbed mass is in equilibrium with a very low partial pressure (hence low concentration) of the adsorbate in the gas phase. So favourable adsorbents should be sought by investigating the isotherms of adsorbent-adsorbate combinations. To measure the required isotherms we used a simple but very attractive method based on frontal chromatograpy (Gregg and Stock, 1958). It closely resembles the procedure for measuring the break-through volume of an adsorption tube for pumped sampling. The procedure is illustrated by Fig. 3, which shows an example of a frontalogram as obtained when monitoring the effluent gas of an adsorption tube after application of a stepwise concentration change at the inlet from zero to concentration C_0. At time $t = t_e$ equilibrium is attained and the amount of adsorbate in the tube is constant; the amount is represented by the shaded area. Since for commonly used adsorbents the amount of adsorbate in the gas phase is negligible in comparison with the amount adsorbed, the shaded area can be considered as the amount of material adsorbed by the adsorbent in equili-brium with the concentration C_0. By performing a series of these measure-ments over a range of concentrations, adsorption can be measured as a function of concentration.

DETECTOR SIGNAL

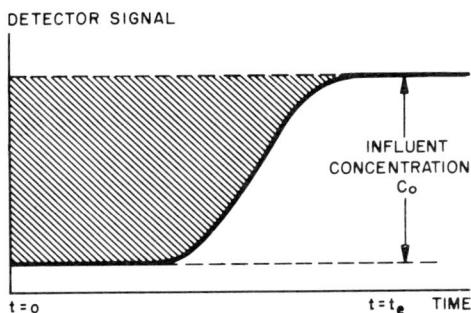

INFLUENT
CONCENTRATION
C_0

$t = o$ $t = t_e$ TIME

Fig. 3   Use of a frontalogram for the determination of adsorption isotherms

By measurements of this type we found that many adsorbents belonging to the family of porous polymers show Freundlich-type adsorptivity in combination with organic compounds. Figure 4 shows examples of such isotherms. In these isotherms the amounts adsorbed are expressed as mass per unit volume, the most representative quantity for the purposes of comparing the characteristics of tubes that are packed to the same volume. Regression analyses show that the functions can be described very accurately by the Freundlich equation, as is illustrated by the data presented in Table 1 and the lines in Fig. 4.

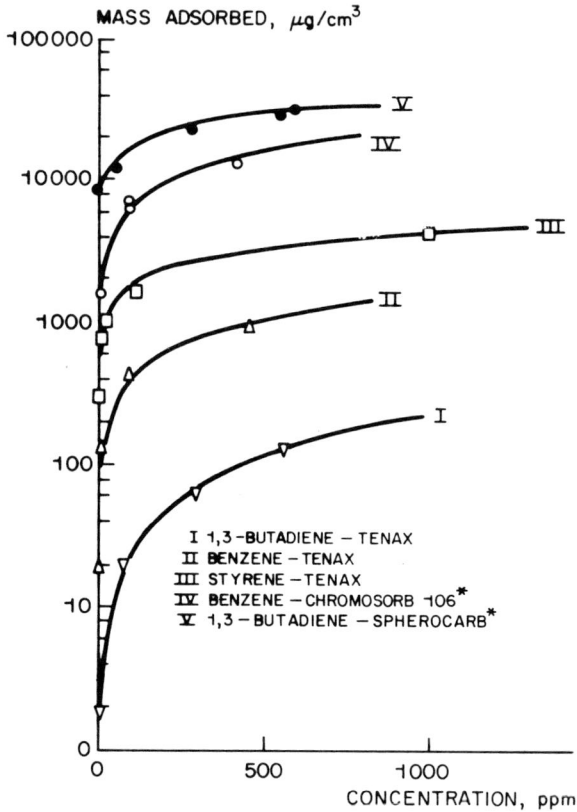

Fig. 4   Adsorption isotherms at 21 °C

* Adsorbent Chromosorb 106 ex Johns-Manville, USA; adsorbent Spherocarb ex Analabs Inc., USA.

TABLE 1   Parameters of Freundlich adsorption isotherms* found by regression analysis with C_{ref} = 1500 ppm

Adsorbate/adsorbent	a	b
	$\mu g.cm^{-3}$	(dimensionless)
Styrene/Tenax TA	5 062	2.5404
Benzene/Tenax TA	2 097	1.6062
1,3-butadiene/Tenax TA	341	1.0327
1,3-butadiene/Spherocarb	39 683	3.2962
Benzene/Chromosorb 106	30 454	1.6807

One could hope now that all these systems would show ideal diffusive sampling behaviour; unfortunately, though, this is not found in practice. For example, the lowest curve in Fig. 4, representing the combination of 1,3-butadiene with Tenax, is the system that produced the highly scattered uptake rate data shown in Fig. 1. For sampling behaviour approaching the ideal to be achieved, a very steep isotherm for low concentrations appears to be essential. The third isotherm shown in Fig. 4, which has this characteristic, turns out to represent the lower limit for an effective adsorbent. Apparently the steepness of this isotherm just meets the high adsorptivity demand needed to create a sufficiently low "near-sorbent" concentration for a sufficiently long period of time. Combinations below this level show non-constant sampling rates. The deviations become larger as the difference between an isotherm and this boundary isotherm increases.

Going upwards in Fig. 4 we indeed find a range where ideal sampling behaviour can be combined with thermal desorption. The upper boundary of this range is roughly indicated by isotherm V of Fig. 4. In the case of systems with isotherms beyond this boundary, sample recovery on thermal desorption with on-line GC is incomplete.

* According to the model W = a$(C/C_{ref})^{1/b}$, W = mass adsorbed per cm³ of adsorbent, C_{ref} = reference concentration.

Figure 5 provides an example of the performance of a system having an isotherm in between the boundaries. It illustrates the perfect constant uptake rate over at least three decades of exposure dose. These exposures were carried out in an exposure box with an air velocity of 11 cm/s with well-defined test atmospheres generated by means of dynamic air dilution (Health and Safety Executive (UK), 1981). Analyses were run on the Perkin Elmer ATD/50 automatic desorber coupled to a Hewlett Packard 5730A gas chromatograph fitted with a capillary column. The sampling tubes for these measurements were those described in literature (Brown et al., 1981), the gauze-type windscreen being used.

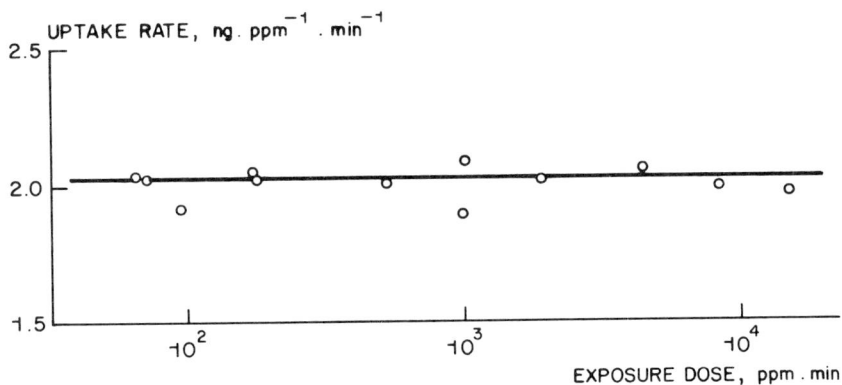

Fig. 5   Uptake rate of benzene by a tube-type diffusive sampler packed with Chromosorb 106

Another feature of the data shown in Figure 5 is that the average uptake rate, represented by the level of the horizontal line, is in very close agreement with the theoretical value that can be calculated for an ideal system from the constant AD/L in equation 2.

Table 2 gives the average uptake rates for several systems having adsorption isotherms in between the critical boundaries (curves III and V in Fig. 4) which all show constant uptake rates. As can be seen, only very small differences are found with respect to the values calculated for ideal systems. The air diffusivities of the compounds (needed for the calculation of these ideal sampling rates) were calculated by means of well-known algorithms based on molecular properties (Fuller et al., 1966).

TABLE 2   Uptake rates of tube-type diffusive samplers packed with
correctly matched adsorbents

Compound	Adsorbent	K_{ex}*	CV*	N*	K_{id}*	Relative difference $K_{ex}-K_{id}$
		$ng.ppm^{-1}.min^{-1}$	%		$ng.ppm^{-1}.min^{-1}$	%
Styrene	Tenax TA	2.39	3.7	70	2.27	5.3
Benzene	Chromosorb 106	2.03	4.0	57	2.05	-1.0
Toluene	Chromosorb 106	2.10	6.2	90	2.16	-2.8
Methyl OXITOL	Chromosorb 106	2.08	4.2	8	2.11	-1.4
n-Heptane	Chromosorb 106	1.97	6.0	70	2.05	-3.9
1,3-Butadiene	Spherocarb	1.61	5.2	20	1.59	1.3
Cyclohexanone	Tenax TA	2.48	3.2	12	2.33	6.4

* CV = coefficient of variation
 N   = number of measurements
 K_{ex} = experimental uptake rate
 K_{id} = ideal uptake rate calculated from the constant AD/L in equation 2

It should be noted that the values in Table 2 only apply to the
samplers provided with the gauze windscreens. If the silicon permeation
screens, which can be purchased optionally, are used additional resistance
to mass transfer is introduced and, accordingly, lower uptake rates can be
expected. So far, however, we have not felt any need to use these screens.
By virtue of this we are now able to predict quite accurately the uptake
rate of samplers packed with preselected adsorbents. This is a great
advantage compared with the vast amount of difficult experimental work
related to the generation of well-defined test atmospheres needed for the
experimental assessment of this quantity.

It is now quite easy to select a suitable adsorbent for a given
compound, notwithstanding the relatively narrow band within which suitable
isotherms can be found. In practice we apply the following strategy. Candidate adsorbents, preselected on the basis of data on adsorbent porosity
and surface area and on the vapour pressure (or boiling point) of the
compound, are subjected to the following procedure. One point of the isotherm of a candidate adsorbent is measured at 100 ppm. If the point is
outside the favourable range another adsorbent is tried. When a combination passes the criterion for 100 ppm an additional measurement at 500 ppm
is carried out to confirm that adsorption is Freundlich-type. If the result
is again satisfactory, the adsorbent is selected definitively for practical
application. In view of the large number of different adsorbents available

commercially nowadays, it is likely that such a search will be successful.
With some experience in this type of work a selection procedure can be completed within 8 hours.

An extensive survey of our experimental work will be published shortly
(Van den Hoed and Halmans, 1986).

The general applicability of the adsorbent selection procedure
developed has been demonstrated in a field test study involving non-
constant concentrations in the ambient air (Van den Hoed and Halmans,
in preparation).

DISCUSSION

The quantitative knowledge of the adsorption isotherms, as presented
in Table 1 in the form of the parameters of the Freundlich isotherms, can
be used to check the theoretical descriptions for diffusive samplers to
which Freundlich-type adsorption applies (Underhill, 1984, 1985).

Figure 6 shows our measurement results as dimensionless variables.
In this graph the relative mass uptake, defined as the experimental mass
uptake divided by the mass uptake for an ideal sampler (Eq. 2), is plotted

OPEN POINTS W_{ex}/W_{id} FILLED POINTS W_{th}/W_{id}

O STYRENE-TENAX TA   ■ BENZENE-TENAX TA
□ BENZENE-TENAX TA   ▲ 1,3-BUTADIENE-TENAX TA
Δ 1,3-BUTADIENE-TENAX TA
∇ BENZENE-CHROMOSORB 106
◇ 1,3-BUTADIENE-SPHEROCARB

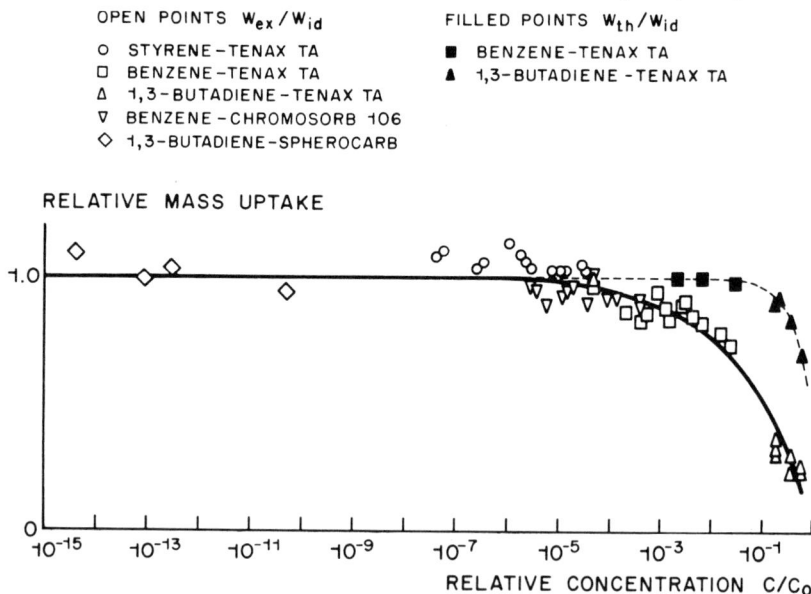

Fig. 6   Relative mass uptake by tube-type samplers. W_{ex} = experimental
mass uptake, W_{id} = ideal mass uptake according to equation 2; W_{th} =
mass uptake according to equation 1

against the relative "near-sorbent" concentration, defined as the ratio
between the concentrations adjacent to the adsorbent bed at the end of an
exposure and the concentration outside the sampler. The filled points,
connected by the dashed line, are obtained when both variables are calculated theoretically for the exposure conditions of our experiments
(Underhill, 1984, 1985). In fact that curve predicts notable deviation
from ideal behaviour when the "near-sorbent" concentration has increased
to about 10 % of the external concentration. In practice, however, these
deviations begin to occur at calculated relative concentrations of 10^{-4}, as
shown by the open points around the solid line. It can be seen that all
systems with isotherms within the critical boundaries give relative mass
uptakes around unity, whereas measurements conducted with less favourable
systems produce data points in the curved part of the solid line.

In our opinion the discrepancy between the theoretical predictions
and the experimental results is inherent to the assumption in the theory
that the whole adsorbent bed is in equilibrium with the concentration at
the adsorbent/air gap interface. Although this might still be realistic
for badge-type samplers it is apparently not applicable to the tube-type
samplers involved in our study. In these samplers the resistance to mass
transfer in the porous bed will create a much more rapid increase of the
"near-sorbent" concentration than predicted by the oversimplified model.

Notwithstanding the observed discrepancy the plot of Fig. 6 is of
practical value, since it can be used to estimate under which conditions
(in terms of concentration and exposure time) saturation effects start to
become evident, both for favourable and non-favourable adsorption isotherms.

CONCLUSION

Overall we may conclude from our experiments that tube-type diffusive
samplers, packed with adsorbents selected according to our selection procedure, are a convenient, cost-effective analytical tool for measuring
exposure to chemical hazards in the air.

REFERENCES
Brown, R.J., Charlton, J. and Saunders, K.J. 1981. The development of an
    improved diffusive sampler. Am. Ind. Hyg. Assoc. J., 42, 865-896.
Coutant, R.W., Lewis, R.G. and Mulic, J. 1985. Passive sampling devices
    with reversible adsorption. Anal. Chem., 57, 219-223.

Fuller, E.N., Schettler, P.D. and Giddings, J.D. 1966. A new method for prediction of binary gas-phase diffusion coefficients. Ind. Eng. Chem., 58, 19-27.

Gregg, S.J. and Stock, R. 1958. Sorption isotherms and chromatographic behavior of vapors. In "Proceedings of the second symposium on gas chromatography" (Ed. D.H. Desty). (Butterworth, London). pp. 90-98.

Van den Hoed, N. and Halmans, Ms. M.T.H. 1986. Sampling and thermal desorption efficiency of tube-type diffusive samplers: Selection and performance of adsorbents. Am. Ind. Hyg. Assoc. J., paper to be published.

Van den Hoed, N. and Halmans, Ms. M.T.H. A large-scale field test on the performance of tube-type diffusive samplers. Paper in preparation.

Health and Safety Executive (UK), 1981. Methods for the Determination of Hazardous Substances (MDHS): Method 3, Generation of test atmospheres of organic vapours by the syringe injection technique. Occupational Medicine and Hygiene Laboratory, London.

Health and Safety Executive (UK), 1985. MDHS: Method 50, Benzene in air. Occupational Medicine and Hygiene Laboratory, London.

Lewis, R.G., Mulic, J.D., Coutant, R.W., Wooten, G.W. and McMillin, C.R. 1985. Thermally desorbable passive sampling device for volatile organic chemicals in ambient air. Anal. Chem., 57, 214-219.

Tomkins, F.C. and Goldsmith, R.L. 1977. A new personal dosimeter for the monitoring of industrial pollutants. Am. Ind. Hyg. Assoc. J., 38, 371-377.

Underhill, D.W. 1984. Efficiency of passive sampling by adsorbents. Am. Ind. Hyg. Assoc. J., 45, 306-310.

Underhill, D.W. 1985. Correction. Am. Ind. Hyg. Assoc. J., 46, 341.

Van der Wal, J.F. and Moerkerken, A. 1984. The performance of passive diffusion monitors for organic vapours for personal sampling of painters. Ann. Occup. Hyg., 28, 39-47.

DISCUSSION - SESSION III

MOORE (UK)

Mr. Van den Hoed seems to be saying that the prediction mechanism which he has developed is doing a useful job. But there is a discrepancy between the calculated behaviour according to the isotherm and the observed behaviour of many of the samplers.

VAN DEN HOED (The Netherlands)

In tube-type samplers only the surface layers of the sorbent are immediately available to the diffusing chemical, whereas with the badge-type sampler the sorbent is all available and is used more effectively. However, the badge type sampler still suffers from the disadvantage of unsuitabily for automation.

FIRTH (UK)

An effective diffusive sampler demands that the heat of adsorption of the sorbate-analyte pair is high. If this is the case, then there will be no rise in vapour pressure at the sorbent surface and no diffusion of analyte down the tube. Therefore for both tube and badge designs, the analyte is concentrated in the top layer of sorbent and there is no real advantage in adding extra adsorbent in either case. For the tube sampler, there is some kind of frontal effect travelling down the tube. That seems to me a significant disadvantage of the tube sampler.

BERTONI (Italy)

It seems to me that the amount of absorbent placed within a passive tube sampler has a considerable influence. I think that Mr. Van den Hoed clearly illustrated the advantages of these absorbers as did Pannwitz' study. These absorbers are not affected much by air velocity and they can increase the sensitivity of the analysis by the factor of 10 in relation to systems using solvent desorption. What I find rather strange, is that Mr. Van den Hoed explored the absorption isotherms of porous polymers and Spherocarb but didn't investigate sorbent having an intermediary isotherm, such as graphitized carbon.

KENNEDY (UK)

I was wondering if it would be possible to create guidelines for the selection of a suitable sorbent based on the heat of absorption of the pollutant on the particular substrate.

VAN DEN HOED (The Netherlands)

In my opinion, adsorption isotherms are more useful than heats of adsorption in this context.

HARPER (UK)

Many authors have proposed that sampling is best achieved when the most appropriate sorbent has been chosen for each analyte; on the other hand others have expressed the opinion that hygiene requirements would be best served by the simplicity of a simple broad-range sorbent even this would result in non-ideal sampling of some species.

CRABLE (USA)

The approach that we take is that, if there is a Federal Standard or a Proposed Federal Standard for any specific analyte, we look for the best sorbent for that particular analyte.

FIRTH (UK)

We would prefer to have a single broad-range sorbent, but in reality that's not achievable. But we should move towards sorbents which are capable of acting effectively over as wide a range of different materials as possible.

COKER (UK)

I think that this highlights one of the problems at present with passive samplers. Nobody has come up with a way of using more than one sorbent with a passive sampler to take care of sampling a high volatile and fairly low volatile material simultaneously. They haven't been able to initiate one of the advantages of a pump sampler which is to use a sequence of different types of absorbents in a bed.

MOORE (UK)

The company has had multiple sorbent badge systems available for a number of years. They are available, both multiple sampling rates and multiple adsorbents; simultaneous multiple sorbents, if you wish to go that way.

COCHEO (Italy)

I'd like to ask Mr. Van den Hoed how he calibrates the tube samplers?

VAN DEN HOED (The Netherlands)

We make dilutions of the components of interest, dilutions of the components of interest, dilutions in some organic solvent, and introduce microliter amounts of these solutions into the tubes while sucking through a flow of 100 ml/min of air and continuing sucking for about 10 minutes. It's a standardized procedure.

PERKINS (UK)

In his presentation Mr. Van den Hoed said that benzene on Chromosorb 106 showed a linear rate of uptake over quite a wide concentration range. It appeared that the lower limit was about 80 ppm.minutes. I've found that, for sampling times up to about half an hour, we observed a change in rate of uptake with time. Should we not use passive sampling for short sampling periods, or for low-levels?

VAN DEN HOED (the Netherlands)

I agree: the lower limit is about 80 ppm.minutes. We haven't investigated exposure measurements below 0.5 ppm because we cannot generate standard atmospheres below this level with sufficient accuracy. We have also avoided short exposure times for the theoretical reason that the minimal sampling time should exceed the response time of the sampler, which is of the order of a few seconds, by a large factor.

FIELDS (UK)

For the past five years we've been using a method for benzene on two plants. This method uses Perkin-Elmer tubes, and is accurate down to .1 ppm. I have some field data of 25 pumped versus diffusive tubes for this range. For 20 of these results the result for the pumped tube is within two method standard deviations of the diffusive result. 15 of the 25 results were between .1 ppm and .2 ppm.

LEINSTER (UK)

The data given on the graph goes to 80 ppm.minutes. If we take the criteria of 15 minutes and .5 ppm, that gives us 7.5 ppm.minutes, i.e. an order of magnitude below that. I think that it is in this region that you can get an effect of change of rate of uptake with time with this sampling device. It all comes back to not looking at diffusive sampling or pump sampling as an absolute technique by as a means of collecting exposure data. Even if we are going to get a change of uptake rate, it is unlikely to change the decisions that we're going to make about the workplace.

VAN DER WAL (The Netherlands)

A parameter affecting the performance of diffusive samplers may be the method of determining the analytical recovery. We have found different results for styrene on charcoal when either the monitor was exposed to vapours, or the component was added as liquid with a syringe.

BROWN (UK)

This problem is usually found with substances which give very poor desorption efficiencies from charcoal. For substances, including styrene, where the desorption efficiency is close to 100%, then one wouldn't expect to see any differences.

SAMINI (USA)

I am particularly concerned about the validity of passive samplers for measuring low concentrations. In our experience with styrene, ethyl acetate and butyl acetate at concentrations below 5 ppm, a commonly used gas badge gave very erratic results. Under the same conditions, charcoal tubes gave very consistent results, although about 10% below the standard atmosphere concentration.

BROWN (UK)

This is more a comment than an answer to Mr. Samini's question. When working at low concentrations, you have to be particularly careful to avoid contamination or losses. Our experience with Perkin-Elmer tubes has shown that with proper seals, there are no significant sample losses or any pick-up of environmental contamination during storage for several months. Mr. Griepink has reached a similar conclusion with his work on the preparation of BCR certified standards. A suitable seal is a Swagelok or similar end-cap with PTFE or Teflon insert. The fitting should be hand-tight or be given a quarter turn with a spanner.

IKEDA (Japan)

Is there anybody who knows a good adsorbent for methanol and for acetone?

MOORE (UK)

We have developed and put onto the market a liquid sorbent sampler, taking the methanol into water. Water is a very good solvent for methanol.

CHOO YIN (UK)

I have been working with samplers mainly trying to get down to the low ppb levels. A major problem is actually finding an independent method which you can relate to your pasive sampling.

MILLER (UK)

We are using the Perkin-Elmer tube with a pump as the independent method.

FIELDS (UK)

Mr. Van den Hoed showed that experimental and calculated uptake rates for a number of analytes were very close for sorbents which were "good" sorbents. I wondered if this was a general phenomemon which wou could extend to other analytes; in other words if you could identify good sorbents simply from calculating a small number of uptake rates and showing that they were close to the theoretical value?

VAN DEN HOED (The Netherlands)

I think that it is general. But we cannot really prove it. We have only the statistical evidence that, for those systems which have isotherms in the operational range, the phenomenon holds. We now have some ten combinatons which follow this empirical relationship, so I'm still waiting for the first exception.

FIRTH (UK)

A point which so far hasn't been touched on concerns the analytical process. There is too much blind faith in the results coming out of analytical laboratories. We've seen samplers being compared and "wilds" being demonstrated amongst the results. When inter-laboratory assurance schemes are initially carried out, these "wilds" are exactly what come out in the analytical part of the measurement process - this is not just with diffusive samplers but with analysis generally. I suspect that some of the differences that we are seeing are coming from the analytical end of the measurement process and not just from the sampling end.

CRABLE (USA)

I think that anyone who has ever been in or operated an analytical laboratory is well aware of the need for a good quality control programme and also between lab programmes. I couldn't agree with you more that good quality control is essential to good analytical chemistry. COCHEO (Italy)

Two further comments on laboratory analysis. Generally active carbon sampling is followed by solvent desorption. The contact time between the desorption agent and the materials to be desorbed is critical and needs to be studied in depth. Secondly, carbon sulphide itself will probably disappear from the market. It's a notoriously toxic substance, so either we will have massive use of thermal desorption or we would have to use other solvents which are not so good.

KENNEDY (USA)

I know of some research that's now being done in the United States where they are looking at super-critical carbon dioxide desorption. this is a new technique which I think deserves some note.

SAMINI (USA)

Is NIOSH considering substituting the use of capillary columns for the convential packed columns for their analytical methods?

KENNEDY (USA)

We have a project which is now getting underway to look into replacement of many of the packed columns with capillary columns. But I really didn't know that you had to wait for us to do it. You're all free to try some of these things for yourself and let us know about it.

MOORE (UK)

I just wanted to put in a little optimistic note speaking on behalf of designers and producers of passive samplers. The errors that we have heard about undoubtably exist in two main areas of wind velocity effects and what one can generally call sorbent-related effects. In the case of wind velocity I don't think that the average user needs to be concerned. I think that all the manufacturers by now will provide you with a wind velocity profile of their particular dosimeter and it's largely independent of substance. The sorbent-related effects apply to the charcoal or absorption type of dosimeter. Basically they don't exist with chemically based systems in general. Adsorbent-related errors seen to me to be the main area still left to be formalized but I don't think it is an impossible task. I think that we have enough theory now to be able to put forward some fairly straight-forward guidelines.

THE EFFECTS OF AIR VELOCITY ON A LIQUID DIFFUSIVE AIR SAMPLER

Bengt-Olov Hallberg

Research Department
National Board of Occupational Safety and Health
S-171 84 Solna, Sweden

ABSTRACT

A liquid diffusive sampler has been constructed. The features of the sampler are the possibility to freely choose collection media and analytical method, the influence of face air velocity is minimized. The liquid diffusive sampler was exposed to three concentrations of ammonia with different air velocities and compared with OVM 3500 (manufactured by 3M). The result show that both samplers had a constant sampling rate and that the liquid sampler had a slightly higher recovery than the OVM 3500 at the lowest air velocity.

INTRODUCTION

The use of diffusive samplers is steadily growing. Attractive features with these samplers are their light weight and the ease of sample collection. Most commercial samplers use solid sorbents as a collection medium. By substituting the sorbent for a liquid, desorption losses and backgrounds can be eliminated. Two commercial liquid samplers, Pro-Tek^{TM} (Kring et. al, 1981) (manufactured by Dupont) and Liquid Sorbent Badge (SKC Inc.), are available. Drawbacks with the Pro-Tek system are: the sampler is not re-usable, low sampling rate and only one analytical method is possible. Our experience with the SKC liquid sampler is that the sampling rate for sulfur dioxide is only 30 % of that given in the manual. This can be due to adsorption of the sample inside the diffusive barrier.

Due to the mentioned drawbacks we decided to construct a liquid diffusive sampler of our own. Desirable features were the possibility to freely choose collection media and analytical methods. It should be possible to use the sampler for a wide number of gaseous pollutants and the influence of air velocity should be minimized. In this study our liquid sampler has been compared with the OVM 3500 (manufactured by 3M) by exposure to ammonia with different air velocities.

METHODS

The sampler is made of polytetrafluoroethylene (teflon) and the diffusive barrier consists of 88 drilled holes (diameter 1 mm, length 5mm). A teflon filter is placed between the diffusive barrier and the liquid. The liquid volume of the sampler is 4.5 ml.

The experiments were done in a chamber of polycarbonate glass. The air velocity was regulated with two speed controlled fans and monitored by a directional anemometer (TSI model 1640).

The test atmospheres were generated by the dilution of concentrated gas from a cylinder containing 3.02 % ammonia (in nitrogen) in a dynamic system (Rudling et. al, 1984). The test atmosphere was controlled by impinger sampling (six samples in each run).

Both the impinger and liquid diffusive samples were analysed with an ion chromatograph (Dionex 14). Each experiment was done with six samples. The temperature was 22.0°C and the relative humidity was 45 % in all experiments. The absorption solution used in the sampler was 0.0025 mol/l of sulphuric acid.

The sorbent in the OVM 3500 was a filter impregnated with 10 % phosphoric acid as previously described (Rudling et. al, 1984). The sampling rate of the OVM 3500 is 90 ml/min.

RESULTS AND DISCUSSION

The liquid diffusive sampler was exposed 6 hours at 6 ppm, 25 ppm and 50 ppm.The sampling rate of the liquid diffusive sampler was constant for three concentrations tested. The mean sampling rate was 18.7 ml/min and the coefficient of variation was 5 %.

The air velocities used were 0.10 m/s, 0.25 m/s and 0.80 m/s. In these experiments the liquid sampler was compared to the OVM 3500. The results of the air velocity test (table 1), show that the liquid sampler had a slightly higher recovery than the OVM 3500 at the lowest air velocity.

This work shows that it is possible to construct a simple liquid sampler. The sampler has a constant sampling rate between 36 ppm x hours to 300 ppm x hours and the influence of air velocity is small. It seems that drilled tubes as a diffusive barrier is preferable to a teflon film barrier

TABLE 1   Effect of air velocity
The samplers were exposed to 25 ppm ammonia in six hours.

Sampler	N	Air Velocity (m/s)	Recovery* (%)	Coefficient of variation (%)
Liquid sampler	6	0.10	92	4.7
OVM 3500	6	0.10	85	5.2
Liquid sampler	6	0.25	98	3.2
OVM 3500	6	0.25	100	5.7
Liquid sampler	6	0.80	99	3.3
OVM 3500	6	0.80	98	3.2

* The impinger sampling result was set to 100 %

REFERENCES

Kring, E., Lautenberger, W., Baker, B., Douglas, G. and Hoffman, R. 1981. A new passive colorimetric air monitoring badge system for ammonia, sulfur dioxide and nitrogendioxide. Am. Ind. Hyg. Assoc. J., 42, 373-381.

Rudling, J., Hallberg, B-O., Hultengren, M., Hultman, A. 1984. Development and evaluation of field methods for ammonia in air. Scand. J. Work Environ. Health, 10, 197-202.

THE EFFECTS OF TEMPERATURE AND HUMIDITY ON LENGTH-OF-STAIN DOSIMETERS

E. S. McKee, P. W. McConnaughey and C. J. Tidwell

Mine Safety Appliances Company
P. O. Box 439, Pittsburgh, PA 15230

ABSTRACT

Diffusion-type, colorimetric, length-of-stain dosimeters employ an active impregnant on an inert carrier, either a strip of paper or granular material, disposed in a glass tube. Gas to be detected diffuses from an open end of the tube to and through the indicator. A color change develops lengthwise in the indicator, the length of stain being a direct measure of concentration-time exposure.

The effect of environmental temperature and humidity on the performance of five different length-of-stain dosimeters was determined. The five dosimeters tested were those designed to detect ammonia, chlorine, hydrogen sulfide, nitrogen dioxide and sulfur dioxide. Temperatures from 40 to 90°F and absolute humidities from 3.27 to 19.5 mg/L were investigated. As a general conclusion, (1) high humidity causes a decrease in the length of stain, causing the dosimeters to read lower than they should; (2) high temperatures cause an increase in the length of stain, making the dosimeters read higher than they should. Correction factors have been derived to take care of these situations.

INTRODUCTION

There are several physical and chemical phenomena that determine the length of the stain of length-of-stain dosimeter tubes. The most important of these are (assuming concentration, time and geometry fixed): (1) diffusion rate; (2) the chemical reaction between the impregnant and the gas/vapor of interest; and (3) the physical adsorption of the gas by the inert carrier. It was the purpose of this study to determine experimentally the effects of temperature and relative humidity on the length of stain and, therefore, the indicated concentration of the various dosimeters. The results would then be analyzed to see if certain hypotheses or theories could be used to explain them.

RESULTS AND DISCUSSION

The theory of gaseous diffusion states that the rate of diffusion, or the rate of mass material transport, is independent of atmospheric pressure but is directly proportional to the square root of absolute temperature. The operating range of length-of-stain dosimeters is approximately 0° to 40°C. If the dosimeter is calibrated at 20°C, the temperature effect over the operating range would be only about ±3%. The concentration of water vapor in the air would have little effect on the diffusion rate of the gas of interest, and, therefore, very little effect on the length of stain due to diffusion. Thus, neither temperature nor humidity should cause much change in the length of stain due to diffusion.

The chemical reaction rates in these devices seem to be considerably faster than the diffusion rate, and there is so little impregnant involved that it can be assumed to be complete. Over the operating temperature range, the change in reaction rate should have little effect on the length of stain. High humidity would tend to enhance the reaction rate due to the solubility of the gases for which these dosimeters were designed; however, since the diffusion rate is still the controlling factor with respect to length of stain, even though both temperature and humidity might effect reaction rate, this would have little effect on the length of stain.

The third phenomenon, physical adsorption of the gas by the indicator and substrate, could have an effect. If physical adsorption takes place, then more gas would be taken up per unit length of indicator due to the dual effects of chemical reaction and physical adsorption, thus reducing the length of stain. With thin strips of paper impregnated with indicator, this effect should be small, but it would be greatly enhanced by water adsorption, since all of the gases are soluble in water. Thus, at high humidities it is hypothesized that the stain length should be shortened.

Temperature would interact with both the physical adsorption and the solution, in both cases causing less gas to be taken up. High temperatures cause less physical adsorption of gases and also decrease the solubility. Therefore, if all other conditions are the same, high temperature should cause an increase in the length of stain, since less gas would be adsorbed per unit length of indicator.

The results of this study are shown in Graphs 1 to 5. The Correction Factor (CF) is the factor that the concentration, read from the tube under the particular environmental conditions, must be multiplied by to give the corrected (actual) concentration under these conditions. Thus: Actual Concentration (ppm) = tube reading (ppm) x CF. If the CF is over 1.0, then the stain was too short; if the CF is less than 1.0, the stain was too long.

The Effect of Humidity

Looking at Graphs 1 through 5, it can be seen that, in all cases, the CF increases with increasing absolute humidity (except for nitrogen dioxide at 90°F, which will be discussed later). This indicates that the stain length decreases with increasing humidity, causing the dosimeter to read a lower concentration than it should, compared to the conditions under which it was calibrated (77°F, 50% relative humidity).

There are several other points of interest concerning humidity. First of all, there is plenty of water vapor present under most actual conditions to dissolve the various gases, if even part of the moisture condensed in the indicator strip. At 77°, 50% relative humidity, the concentration of water vapor is 16,000 ppm. The test concentrations of the gases at their TLVs are: NH_3 25 ppm; Cl_2 1 ppm; H_2S 10 ppm; NO_2 3 ppm; and SO_2 3 ppm. Also, the solubilities, in parts by weight per 100 parts of water, are, respectively: 44, 0.06, 0.02, 10 and decomposes. Thus, there appears to be plenty of moisture in the atmosphere to effect some solution or enhance adsorption of the gases on the indicator strip. Secondly, hydrogen sulfide is the least soluble of the gases and the H_2S dosimeter shows the least effect due to humidity. Thirdly, nitrogen dioxide apparently decomposes irreversibly when dissolved in water. There is a possibility that this is responsible for the strange effects due to humidity and temperature on the NO_2 dosimeter, compared to the other dosimeters.

The Effect of Temperature

Temperature, itself, should have little effect on the length of stain of these dosimeters, but its interaction with the physical adsorption of the gases on the indicator paper and its effect on the solubility of the gases could influence the length of stain. At higher temperatures, physical adsorption of gases on solids decreases. Also, the solubility of these gases decrease with increasing temperature. Both of these effects would tend to increase the length of stain, requiring a smaller correction factor. This is exactly what is observed as seen in Graphs 1, 2, 3 and 5. As the temperature increases, at a constant absolute humidity, the CF decreases. Again, NO_2 is an anomaly, showing little effect due to temperature up to 80°F; however, at 90°F the CR increases significantly, and does not follow the pattern seen at lower temperatures. Again, whether this is associated in some way with the decomposition of NO_2 in water is open to conjecture.

SUMMARY

The effects of temperature and humidity on length-of-stain dosimeter tubes is complex and cannot be explained quantitatively. Correction Factors must be obtained experimentally. Qualitatively, the following general conclusions can be made.

1. All of the dosimeters followed the same basic pattern, except NO_2.
2. At the same temperature, the stain length and, therefore, the indicated concentration decreased with increasing absolute humidity,

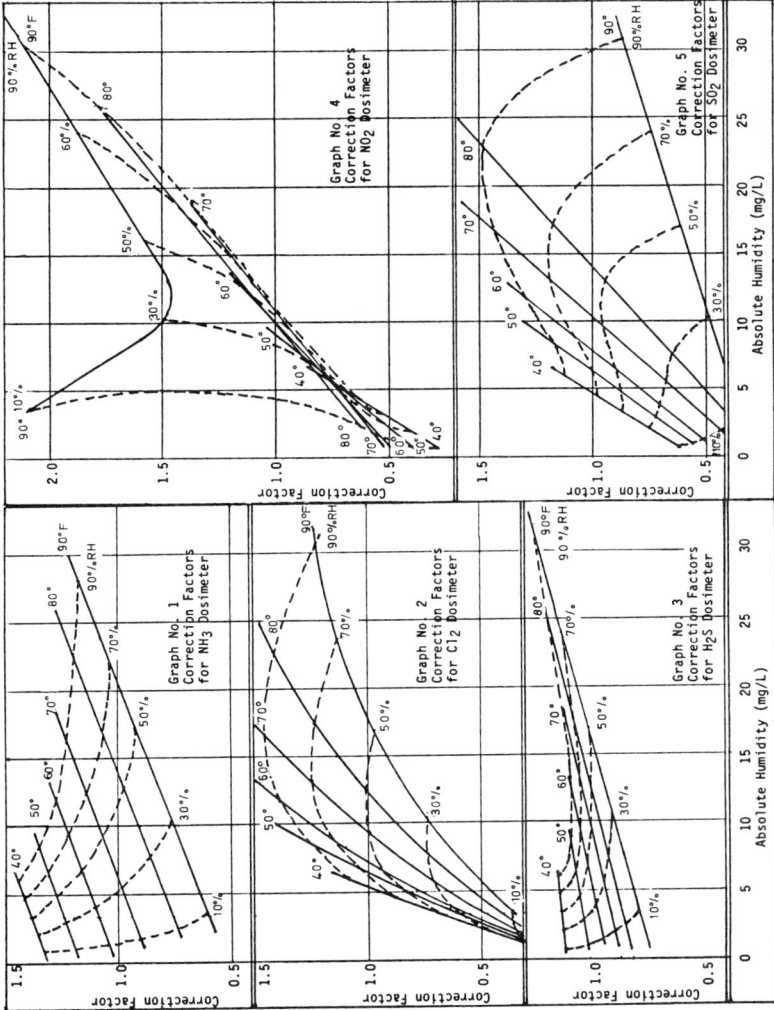

Graph No. 4 Correction Factors for NO2 Dosimeter

Graph No. 5 Correction Factors for SO2 Dosimeter

Graph No. 1 Correction Factors for NH3 Dosimeter

Graph No. 2 Correction Factors for Cl2 Dosimeter

Graph No. 3 Correction Factors for H2S Dosimeter

requiring a larger correction factor at high humidities.

3. At the same absolute humidity, the stain length and, therefore, the indicated concentration, increased with increasing temperature, requiring a smaller correction factor at high temperatures.

4. The hydrogen sulfide dosimeter showed the smallest correction factors for temperature and humidity. This could possibly be explained by hydrogen sulfide being the least water soluble gas.

5. The nitrogen dioxide dosimeter shows a different behavior with respect to the effects of temperature and humidity than the other dosimeters. This might be due to the fact that NO_2 reacts irreversibly when dissolved in water.

Finally, it appears that humidity, the solubility of the vaious gases in water, and the influence of temperature on the physical adsorption of both water vapor and the gases on the indicator paper and on the solubility of the gases, are the main factors affecting the length of stain of these dosimeters. Much more work must be done to prove these hypotheses.

INFLUENCE OF AIR CURRENTS ON THE
SAMPLING OF ORGANIC SOLVENT VAPOURS
WITH DIFFUSIVE SAMPLERS

Karl-Heinz Pannwitz

Drägerwerk AG, Lübeck, Germany F.R.

ABSTRACT

The sampling of organic solvent vapours in the air with diffusive samplers requires a certain air movement to transport the molecules of the compound to be sampled to the opening of the diffusion collector. A representative sampling can only ensue when the transport processes in the environment air are more rapid than the transport of the molecules by diffusion to the adsorption layer inside the diffusive sampler.

The minimum convection of the air which is necessary depends on the maximum possible collection rate of the sampler which essentially is a function of the diffusion cross section area and the diffusion way inside the sampler. On the other hand strong air currents can influence the analytical result by an additional not controlled mass uptake. The employment of diffusive samplers in atmospheres with strong air motions is limited by the efficiency of the diffusion barrier of the sampler.

The influence of low and strong air motions on the sampling of organic solvent vapours with the diffusive sampler ORSA 5 has been investigated. For that these samplers have been exposed to test gases of defined compositions in a wind tunnel as well as to static test atmospheres with forced convetion. Wind velocities have been generated in the range from approx. 0.25 to 250 cm/s. The results of these experiments show that the diffusive sampler ORSA 5 only needs an air motion of about 1 cm/s to achieve the maximum possible collection rate. A significant change of the collection rate has not been obtained up to air currents of up to 250 cm/s.

THEORETICAL CONSIDERATIONS

When a diffusive sampler is located with its diffusion cross section area (sampling opening) in parallel to the flow direction of a closed wind tunnel, the concentration of the compound i in a test gas flowing in this tunnel past the diffusive sampler decreases, because

a portion of the compound i diffuses into the sampling
device and is absorbed there (Fig. 1).

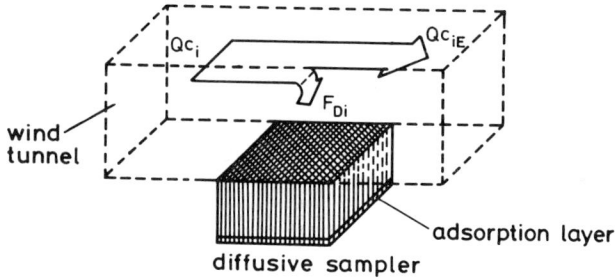

Qc_i gas flow with initial concentration of the compound i

Qc_{iE} gas flow with residual concentration of the compound i

F_{Di} mass flow directed into the diffusive sampler

Fig. 1   Diffusive sample taking out of a
passing-by gas stream

The decrease of the concentration depends on the maximum possible collection rate of the diffusive sampler
and the flow velocity of the test gas. Table I shows the
influence of the maximum possible collection rate and the
flow velocity on the actual collection rates of diffusive
samplers.
The evalutation of Table I points out that concentration determinations by means of diffusive samplers
with large diffusion cross section areas (e.g. 3 cm^2)
can lead to major measuring errors if the wind velocity in
the atmosphere is very low. The collection rates of optimal dimensioned diffusive samplers, such as ORSA 5, are
practically not influenced by low air motion.

TABLE I   Influence of the maximum possible collection
           rate (CR_{max}) of diffusive samplers and the
           flow velocity on the collection rate (CR)

Flow velocity cm/s	$CR_{max} = 10$ ng/s Collection rate in % of CR_{max}	$CR_{max} = 30$ ng/s Collection rate in % of CR_{max}
0.1	63.2	31.7
1	95.5	86.4
10	99.5	98.5
100	99.95	99.87

Note: The collection rates have been calculated for dif-
        fusive sample takings in a closed wind tunnel with
        a flow cross section area of 1 cm^2 and diffusive
        samplers with diffusion cross section areas of
        1 cm^2 and 3 cm^2.

LAB TESTS

   ORSA 5 diffusive samplers have been exposed to test
atmospheres of known Toluene concentrations in air at
defined flow conditions. 10 ORSA 5 samplers have been
opened at one end and arranged one behind the others in a
closed wind tunnel, with one of its diffusion cross section areas in parallel to the flow direction (Fig. 2).

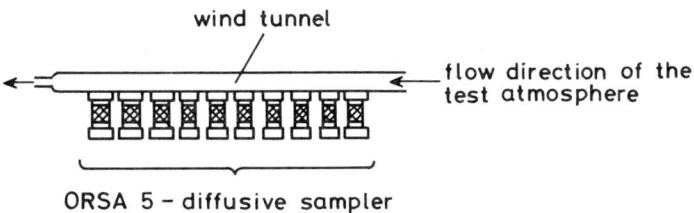

ORSA 5 - diffusive sampler

Fig. 2   Experimental set up for the loading of the
          ORSA 5 - diffusive samplers in the wind
          tunnel

Flow velocities of 0,25; 1 and 17 cm/s have been
generated. The test atmosphere has been prepared by dynamic methods. Further tests at wind velocities of up to
250 cm/s have been carried out in static test atmospheres
with forced convection. To determine the mass uptake the
charcoal of the ORSA 5-samplers has been desorbed by carbon disulphide and analysed by gaschromatography.

RESULTS AND CONCLUSIONS
        Figure 3 shows the results of the conducted measure-
ments.

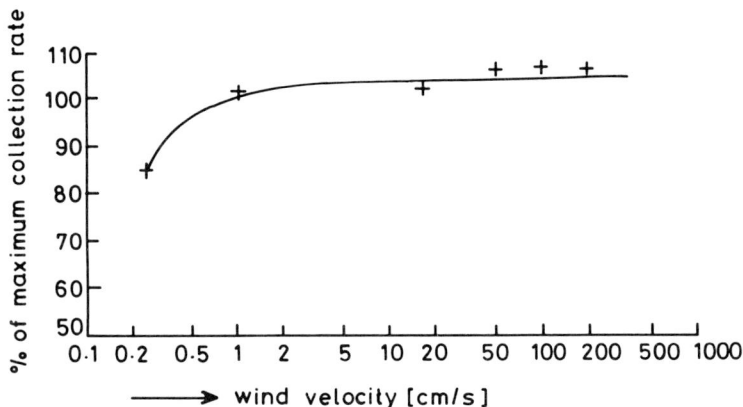

Fig. 3   Influence of the wind velocity on the
          collection rate of ORSA 5

The evaluation of the graph points out that the collection rate of the diffusion sampler ORSA 5 is independent of air motion in a wide range. The maximum possible
collection rate is already reached at a wind velocity of
1 cm/s.
Therefore this system is also suited for sampler takings
in environments with low air motions. Flow measurements
with a kata-thermometer in room without forced air
circulation have shown, that air motions between 5 and
10 cm/s can be expected.

DETERMINATION OF SOME UPTAKE RATES
PERKIN-ELMER ATD 50 / DIFFUSIVE SAMPLING

PETER ROSMANITH, JOTUN A/S, POB 400, N-3201 SANDEFJORD, NORWAY

UPTAKE RATES (Up) OF SOME SOLVENTS / MONOMERS WAS DETERMINED (at 25° C and 760 mm), AND THEIR DEPENDENCE ON CONCENTRATION AND SAMPLING TIME WAS DEMONSTRATED.

APPARATUS: MODEL 360 DYNAMIC CALIBRATOR AND DIFFUSION TUBES, AID INC..
ATD 50 AND SAMPLING TUBES CONTAINING TENAX GC (60/80 mesh), PERKIN-ELMER.
DROPPING FUNNEL, CYLINDRICAL, CAPACITY 500 mls., PYREX.

REAGENTS: SOLVENTS / MONOMERS, GR GRADE.
AIR CONTAINING LESS THAN 0.1 mg OF WATER PER CUBIC METER.

PROCEDURE: FOR EACH DETERMINATION THE DROPPING FUNNEL CONTAINING THE SAMPLING TUBE WAS ATTACHED TO THE EXIT OF THE 360 (stopcock opened, stopper of plastic with a vent opening).
USUAL ANALYTICAL PROCEDURE ON ATD 50 / SIGMA 3B GC / SIGMA 15 DATA STATION.
ATD 50 CONDITIONS:
Two stage desorption
Primary desorption temperature: 200° C
Primary desorption time: 10 mins.
Cold trap packed with Tenax
Cold trap low temperature: -30° C
Cold trap high temperature: 300° C

RESULTS: SEE TABLE 1.

CONCLUSIONS:
UPTAKE RATES (ATD 50 / DIFFUSIVE SAMPLING) DEPEND ON CONCENT-
RATION AND SAMPLING TIME, AND THEIR VARIANCE PROHIBIT EXACT
DETERMINATIONS OF AVERAGE CONCENTRATIONS, BUT PERMIT ROUGH
ESTIMATIONS OF CONCENTRATION LEVELS.

TABLE 1

SOLVENT/MONOMER	CONCENTRATION ppm	UPTAKE RATE ng/ppm/min SAMPLING TIME hours				
		1	2	4	8	16
BENZENE	2.5	2.8	2.31	1.84	1.52	1.38
BENZENE	25	2.43	2.16	1.76	1.42	1.34
BENZENE	150	2.01	1.89	1.67	1.37	1.07
TOLUENE	40	2.1	2.07	2.0	1.88	1.74
TOLUENE	250	2.52	2.12	1.84	1.53	1.14
STYRENE	10	2.88	2.56	2.49	2.4	2.38
STYRENE	25	2.95	2.66	2.53	2.35	2.16
O-XYLENE	10	2.5	2.39	2.33	2.26	2.17
M-XYLENE	10	2.34	2.29	2.22	2.17	2.11
P-XYLENE	10	2.33	2.3	2.27	2.2	2.15
ETHYLBENZENE	15	2.32	2.26	2.15	2.09	2.0
N-OCTANE	15	2.23	2.13	1.95	1.8	1.64
N-NONANE	5	3.12	2.57	2.15	2.07	1.95
METHYLENE CHLORIDE	200	1.95	1.68	1.43	1.09	0.79
1,1,1-TRICHLOROETHANE	150	8.04	5.6	3.93	2.79	1.7
ETHANOL	150	0.95	0.84	0.69	0.53	0.34
ISOPROPANOL	100	1.57	1.45	1.07	0.88	0.7
1-BUTANOL	10	1.92	1.7	1.62	1.55	1.4
ISOBUTANOL	20	2.19	1.86	1.45	1.32	1.15
ACETONE	100	1.37	1.15	1.03	0.86	0.67
METHYL ETHYL KETONE	150	1.85	1.58	1.5	1.28	1.06
METHYL ISOBUTYL KETONE	25	2.16	2.05	1.96	1.85	1.49
CYCLOHEXANONE	10	2.64	2.62	2.51	2.41	2.25
2-ETHOXYETHANOL	10	2.5	2.29	2.21	2.14	2.03
ETHYL ACETATE	150	1.96	1.82	1.65	1.46	1.22
VINYL ACETATE	175	2.01	1.76	1.59	1.45	1.15
BUTYL ACETATE	20	2.43	2.41	2.29	2.26	2.14
2-ETHOXYETHYL ACETATE	2.5	3.26	3.12	2.75	2.63	2.47
BUTYL ACRYLATE	5	2.74	2.72	2.69	2.6	2.43

COLLECTION EFFICIENCIES IN A STATIC SYSTEM

Lauri Saarinen, Maire Rothberg, Beatrice Bäck

Uusimaa Regional Institute of Occupational Health
Arinatie 3 A, SF-00370 Helsinki, Finland

ABSTRACT

A static system was used to evaluate the accuracy of diffusive samplers. The collection efficiencies (CE) were determined for different loading levels and humidities, to explore the origin of the errors. CE values deviated in some cases from desorption efficiencies (DE), which were obtained by phase equilibrium method for analytical purposes. CE values varied due to loading level of adsorbates and water, which was injected to the static adsorption chamber to maintain different humidity levels. The recoveries of the static system were also compared with samples spiked in a dynamic test chamber.

MATERIAL AND METHODS

Granular activated carbon and adsorption plates from a 3M organic vapour monitor 3500 were examined. The adsorbents were placed in gas-tight glass vessels (0.55 dm^3). The solvent mixtures of ten compounds were injected in amounts of 1 to 200 µl. Water was added in portions from 3 to 30 µl to maintain a relative humidity ranging of 10 to 100 %. The vessels were left standing overnight, and subsequently opened, and the relative humidity, temperature and head space concentration of solvent vapours were registrated. For comparison purposes, similar samples were prepared in a dynamic vapour generator chamber with known vapour concentrations. The adsorbents were desorbed in 5 ml of carbon disulfide. The test temperature was 20 $\pm$ 2°C. The adsorbents were preconditioned by oven drying for two hours. DE were determined by the phase equilibrium method.

RESULTS

The differences between CE and DE values are presented in table 1. In most measurements the difference is zero except for acetone and alcohols. The phase equilibrium method gives erraneous DE values for polar compounds.

Low loading levels (0.1-1 µl) resulted in higher CE than DE values when RH exeeded 50 %. The desorption phenomena is then favoured by high water content in the adsorbent bed.

Collection efficiencies of the ten solvents are compiled in table 2. The decreased CE values of granular carbon in table 2 are mainly caused by the high loading levels. The effect of humidity was negligible only for nonpolar compounds when RH was under 50 %.

CONCLUSIONS

Our static loading system is easy to manipulate, as well as inexpensive. Differences between CE and DE values revealed some systematic errors in the recovery of alcohols and polar compounds. The CE values determined in a static system agree with the corresponding values from a dynamic vapour generator as seen in table 3.

TABLE 1   The difference between percentages of
DE and CE values (Δ = DE-CE).

Adsorbate	Total loading (µl)				
	10	30	50	100	200
Ethanol	10	9	8	3	0
Acetone	2	3	3	1	0
i-Propanol	9	8	6	1	0
Hexane	0	0	0	0	0
Ethylacetate	0	0	0	0	0
i-Butanol	5	5	3	1	1
1,1,1-Trichloroethane	0	0	0	0	0
MiBK	0	0	0	0	0
Toluene	0	0	0	0	0
Xylene	0	0	0	0	0

TABLE 2   Recovery of adsorbates at different loading levels
and relative humidity as a percentage

Type of adsorbent	Plate						Granular			
Total loading (μl)	1		10		50		10	50	100	200
Relative humidity (RH %)	8	50	8	50	8	50	40	50	90	100
Ethanol	-	-	-	-	67	65	72	73	42	22
Acetone	96	94	95	92	69	68	94	83	28	10
i-Propanol	90	81	92	81	86	83	86	88	60	30
Hexane	102	99	99	98	98	92	101	101	43	7
Ethylacetate	101	100	99	97	94	90	97	99	62	20
i-Butanol	92	81	92	88	93	90	88	95	91	67
1,1,1-Trichloro-ethane	99	99	99	97	96	89	101	99	59	19
MiBK	99	99	98	95	98	97	99	101	97	65
Toluene	100	100	98	96	99	97	100	101	99	69
Xylene	98	98	98	94	100	98	100	101	103	91

TABLE 3   Comparison of collection efficiencies in static and dynamic
sampling systems. In both cases total loadings were 50 μl of
solvent mixture, RH % ~ 30 and temperature 22 $\pm$ 1 °C

Adsorbates	Sampling system	
	dynamic	static
Ethanol	66	70
Acetone	97	72
i-Propanol	86	87
Hexane	-	98
Ethyl Acetate	96	94
i-Butanol	89	92
1,1,1-Tricholoroethane	-	94
MiBK	93	98
Toluene	92	98
Xylene	-	97

THE EFFECT OF FACE AIR VELOCITY ON THE RATE OF SAMPLING OF AIR

CONTAMINANTS BY A DIFFUSIVE SAMPLER

Behzad S. Samimi
Graduate School of Public Health
San Diego State University
San Diego, CA 92182-0405, USA

ABSTRACT

The influence of concentrations and air currents on the rate of sampling of organic vapors was studied on NMS Gasbadge. The samplers were exposed within controlled atmosphere of an inhalation chamber, to predetermined air velocities and concentrations of styrene and acrylate monomers. Sampling rate did not change significantly with change in concentration of organic vapors. However, increased air velocities, particularly those perpendicular to the diffusional path, caused significant increase in sampling rate of air contaminants, which was independent of concentration. Compared to the actual concentration of organic vapors within the chamber and those measured by means of standard charcoal tubes, the NMS Gasbadges underestimated the concentrations significantly, under 22 cm/sec air velocity; whereas, doubling this velocity resulted in overestimation of concentrations by Gasbadges. Recommendations included necessity of improvement in design of diffusional chamber, providing more efficient windshield and establishment of correction factors for air velocity.

INTRODUCTION:

Among several environmental factors in the workplace that may influence the rate of sampling of air contaminants by a diffusive dosimeter, air movement seems to be a prime factor. Temperature and barometric pressure have negligible influence on the rate of sampling (Hearl and Manning, 1980; Jonas, et al., 1981; Lautenberger et al., 1980; NIOSH, 1977). However, air currents seem to influence the sampling rate to a great extent. A minimum air velocity of 7.62 cm/sec has been described as the boundry condition to overcome the external resistance to mass transfer associated with convection increase (Hearl and Manning, 1980; NIOSH, 1977). However, the influence of air currents, particularly those perpendicular to the diffusional path, have not been adequately studied (Samimi, 1983; Samimi, 1985). In this study, the influence of air velocity on the rate of sampling of organic vapors was investigated using NMS Gasbadge.

METHODS:

Known concentrations of organic vapors, i.e. ethyl acrylate, butyl acrylate and styrene were produced within an inhalation chamber by means of a dynamic system. Metered air was introduced into a solvent evaporator containing the desired solvent. The organic vapor was then introduced into the chamber where it was thoroughly mixed with the incoming air stream. The concentration of organic vapor was controlled by (1) rate vapor generation, (2) temperature of solvent, and (3) rate of dilution air. The concentrations of organic vapors were monitored continuously by means of a recording infrared gas analyzer (Samimi, 1983).

Dosimeters and standard charcoal tube samples were collected simultaneously within the chamber. Sampling time varied from 30 minutes to 9 hours. Due to limited adsorption capacity of charcoal tubes, for extended sampling period, more several charcoal tube samples were collected consecutively throughout a single exposure period. Two air velocities were experimented: 22 cm/s and 44 cm/s. Air streams were perpendicular to the dosimeters surface. The concentration of organic vapors varied from 50-120 ppm for styrene, 5-25 ppm for ethyl acrylate and 10 ppm for butyl acrylate. Samples were desorbed by carbon disulfite and analyzed by gas chromatography according to methods recommended by NIOSH (Tomkins and Goldsmith, 1977). The concentration of organic vapors within the chamber were maintained with minimal fluctuation throughout the exposure periods of up to 9 hours, i.e. the maximum standard deviation of concentrations did not exceed 5% of the mean concentration.

The dosimeter diffusive constant (Dc) was calculated for each exposure trial using the following equation: $Dc = (Wr)(Kc)/(C)(M)(t)(Ed)$; where: Dc = Diffusive constant for the dosimeter (cm^2/s); Wr = Weight of organic vapor recovered (μg); C = Mean concentration of organic vapor within chamber($\mu g/g=ppm$); M = Molecular weight of organic vapor (g-mole); t = sampling time (s); Ed = Efficiency of desorption from carbon felt (unitless); Kc = Exposure condition constant derived from: (24,450) (1.31)/(9.54); where: 24,450 = molar volume of gas under $25^{o}C$ and 760 mm Hg in milliliters; 1.31 = depth of diffusion path in Cm; 9.54 = cross-sectional area of diffusion chamber in square centimeter.

The sampling rates (Rs) were also calculated in micrograms of organic vapor adsorbed per ppm of organic vapor per minute of sampling time. Both Dc and Rs values were compared to examine the change in sampling rate under various conditions. Relatively low standard deviations (0.5 to 6.5% of the mean) among Dc and/or Rs values for 12 to 23 dosimeters used for each batch of experimental condition, indicated the fair precision and consistency of sampling rates by the NMS dosimeters under same sampling conditions.

RESULTS:

Minimal variations were found among DC and/or Rs values, for all three organic vapors, when air velocity was maintained constant even when concentration or sampling time changed. For example, under air velocity of 22 cm/s, there was an insignificant ($F=0.1; p>0.1$) increase in mean value of Dc for styrene when concentration was increased from 50 ppm (mean Dc $=0.05588$ cm^2/s) to 100 ppm (mean Dc $=0.05768$ cm^2/s); mean Rs values were 0.105 μg/ppm,min. and 0.106 μg/ppm,min. for 50 and 100 ppm, respectively. Similar results were found for ethyl acrylate at air velocity of 44 cm/s i.e. there was an infignificant increase ($F=0.5$; $P>0.1$) in mean Dc value when concentration was increased from 5 ppm (mean Dc = 0.07305 cm^2/s) to 25 ppm (mean Dc = 0.07397 cm^2/s); the corresponding mean values for Rs were 0.126 and 0.128 μg/ppm,min.

Significant increase in Dc and Rs were observed when concentrations of organic vapors were maintained constant but air velocity was increased. For example, at concentration of 100 ppm styrene, mean Dc value increased from 0.05768 to 0.07959 cm^2/s ($F=81; P<0.005$) when air velocity was increased from 22 to 44 cm/s; the corresponding mean Rs values were 0.106 and 0.134 μg/ppm,min. Similar results were found for the other two organic vapors. For example, at concentration of 25 ppm of ethyl acrylate, mean Dc value increased significantly ($F=23.7$; $P<0.005$) from 0.06018 to 0.07397 cm^2/s (Rs increased from 0.106 to 0.128 μg/ppm, min, respectively) when air velocity was doubled from 22 to 44 cm/s. For butyl acrylate at concentration of 10 ppm, the corresponding increase was from 0.04845 to 0.05710 cm^2/s ($F=27.5; P<0.005$) for mean Dc and from 0.110 to 0.128 ug/ppm,min for mean Rs.

Compared to the actual concentration of organic vapors within the chamber, under air velocity of 22 cm/s, NMS dosimeters consistently under estimated the concentration of all three organic vapors by an average of 12% (range: 8-16%) of the actual concentrations. Whereas, under 44 cm/s air velocity, the dosimeters showed consistent overestimation of concentration by an average of 11%(range: 9-15%) of the actual concentrations. Standard charcoal tube samples were not affected by the change in air velocity; however, the concentration of organic vapors assessed by charcoal tubes were consistently lower than the actual concentrations within the chamber by an average of 9% (range: 3-18%).

It was concluded that increased air velocity, particularly perpendicular to the dosimeter surface, is likely to increase the sampling rate of a diffusive dosimeter. Air velocities higher than those currently experimented, may even more increase the sampling rate by pushing more contaminated air into the diffusional cavity. Whereas, air currents with lower velocities, seem to be unable to overcome the external resistance to mass transfer associated with convection increase and result in reduced sampling rates. More work on improving design of future diffusive samplers combined with vigorous testing under known atmospheres, providing more efficient windshield, and last but not the least, establishment of correction factors for wind velocity for each brand name of diffusive dosimeter by the manufacturer, seem to be the right steps towards improving the sampling accuracy of these popular and convenient air sampling devices.

REFERENCES:

Hearl, F., and M. P. Manning, 1980: Transient Response of Diffusion Dosimeters. Amer. Ind. Hyg. Assoc. J. 41:778-783.
Jonas, L. C., C. E. Billings, and C. Lilis, 1981: Laboratory Performance of Passive Personal Samplers for Waste Anesthetic Gas (Enflurane) Concentrations. Am. Ind. Hyg. Assoc. J. 42:104-111.
Lautenberger, W. J., et al, 1980: A New Personal Badge Monitor for Organic Vapors. Am. Ind. Hyg. Assoc. J. 41:737-747.
NIOSH, 1977: Manual of Analytical Methods. 2nd Ed.
Samimi, B., and L. Falbo, 1983: Validation of Abcor Organic Vapor Dosimeter Under Various Concentration and Air Velocity Conditions. Am. Ind. Hyg. Assoc. J. 44:402-408.
Samimi, B., and L. Falbo, 1985: Comparison of Standard Charcoal Tubes with Abcor Gasbadges Within Controlled Atmosphere. Am. Ind. Hyg. Assoc. J. 46:49-52.
Samimi, B., 1983: Calibration of MIRAN Gas Analyzers; Extent of Vapor Loss Within A Closed Loop Calibration System. Am. Ind. Hyg. Assoc. J. 4:40-45.
Tomkins, F. Jr. and R. Goldsmith, 1977: A New Personal Dosimeter for the Monitoring of Industrial Pollutants. Am. Ind. Hyg. Assoc. J. 38:371-377.

VALIDITA' DEL CAMPIONAMENTO PER DIFFUSIONE ANCHE

CON CONCENTRAZIONE ATMOSFERICA VARIABILE

N. Zurlo, F. Andreoletti[°]

[°]Clinica del Lavoro "L. Devoto"
Università degli Studi, Milano, Italia

ESTRATTO

Viene valutata teoricamente l'incidenza della variazione della
concentrazione atmosferica C_a dell'analita sulla precisione del prelievo per diffusione. Risulta che il deficit accumulato nell'assorbimento durante l'incremento ΔC_a, pari a $0.18\,\Delta C_a$ H A, dove H e A sono
rispettivamente profondità e sezione della camera di diffusione, viene recuperato durante la fase opposta quando la concentrazione diminuisce di ΔC_a. Alla fine del prelievo si ha un eccesso nell'assorbimento pari a $0.32\,C_a$ H A.

INTRODUZIONE

In condizioni d'equilibrio con concentrazione atmosferica C_a e
coefficiente di diffusione D, nella camera di diffusione di profondità H e sezione unitaria $(A=1)$ si ha:

$$(1) \quad C = C_a\left(1 - \frac{x}{H}\right) \quad \frac{dC}{dx} = -\frac{C_a}{H} \quad \frac{dq}{dt} = -C_a\frac{D}{H} \quad q_a = C_a\frac{H}{2}$$

Il "volano" q_a contenuto nella camera pari al flusso di equilibrio
F_e per il tempo $t_a = H^2/2D$ è indispensabile per creare il gradiente
dC/dx e segue C_a nelle sue variazioni assorbendo o restituendo parte
del flusso con conseguente ritardo o anticipo nel prelievo.

FORMAZIONE DI q_a

Il flusso di q è molto rapido (camera di profondità L=0) all'ini-
zio del prelievo o quando la concentrazione passa da 0 a C_a, in seguito diminuisce verso il valore F_e mentre L passa da 0 ad H.

La grandezza L è funzione di D (dimensionalmente pari al quadrato
di una lunghezza diviso per un tempo) ed indicando con E e k due coef-
ficienti si può porre:

(2) $\qquad\qquad L^2 = E\ D\ t \qquad\qquad q_a = k\ C_a\ L$

Sulla superficie di contatto aria–camera si ha:

(3) $\quad (\dfrac{dC}{dx})_{x=0} = \dfrac{C_a}{L} \qquad -\dfrac{dq}{dt} = C_a\ (\dfrac{D}{Et})^{\frac{1}{2}} \qquad q_a = 2\ C_a\ (\dfrac{Dt}{E})^{\frac{1}{2}}$

per cui è $k = 2/E$; ma deve pure essere:

(4) $\qquad\qquad\qquad \dfrac{dC}{dt} = D\ \dfrac{d^2C}{dx^2}$

Queste condizioni sono assolte ponendo:

(5) $\quad C = C_a\ \exp\left[-(x/L) - \dfrac{1}{2}(x/L)^2 - \dfrac{4-E}{12}(x/L)^3 - \dfrac{3-E}{12}(x/L)^4 \ldots\right]$

dove:

(6) $\qquad\qquad E = \pi \qquad\qquad L = (\pi D\ t)^{\frac{1}{2}}$

L'equilibrio (L=H) viene raggiunto nel tempo $t_e = H^2/\pi\ D$ quando è entrato $q = C_a\ 2H/\pi$; detraendo q_a si ha che l'assorbitore ha ricevuto $0.137\ q_a$ mentre avrebbe dovuto ricevere $F_e \cdot t_e = Ca\ H/\pi$. Si è pertanto accumulato un deficit, o ritardo, di $0.363\ q_a$, mentre il restante $0.637\ q_a$ rappresenta il surplus rispetto al flusso F_e entrato nella camera.

FINE PRELIEVO

A prelievo ultimato, l'assorbimento prosegue fino all'esaurimento del volano q_a, con recupero del deficit e del surplus che equivale ad un prolungamento del prelievo per $t = H^2/3D$. Si hanno le condizioni:

(7) $\quad (\dfrac{dC}{dx})_{x=0} = 0 \qquad \dfrac{dC}{dt} = D\ \dfrac{d^2C}{dx^2} \qquad \dfrac{1}{C}\ \dfrac{d^2C}{dx^2} = \text{costante}$

Queste condizioni sono assolte ponendo:

(8) $\qquad\qquad C = \dfrac{\pi}{4}\ C_a\ \exp\left[-(\dfrac{\pi}{2H})^2\ D\ t\right]\ \cos \dfrac{\pi}{2H}\ x$

L'andamento della concentrazione, inizialmente dato dalla curva a) della Figura 1 prosegue secondo le curve b), c), ecc.

EVACUAZIONE DI q_a DURANTE IL PRELIEVO

Se durante il prelievo C_a scende a zero si crea immediatamente un

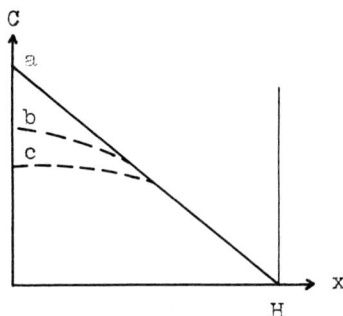

FIGURA 1   Variazione nel tempo dell'andamento della concentrazione all'interno della camera di diffusione durante l'esaurimento a fine prelievo

flusso in uscita mentre continua quello verso l'assorbitore. Si formano così due camere di diffusione (Figura 2) una verso l'esterno con H crescente da 0 ad $H/2$ e l'altra verso l'assorbitore con H decrescente da H ad $H/2$.

La concentrazione massima C_M all'interno della camera, inizialmente pari a C_a e posta sulla superficie di contatto con l'aria, si sposta verso $H/2$ diminuendo gradatamente. Quando C_M è in $H/2$ le due camere, che contengono ciascuna $0.25\ q_a$, sono simmetriche ed il loro esaurimento prosegue in parallelo.

Mentre C_M si sposta verso $H/2$ a valle di $H/2$, dove $C < C_M$, defluisce $0.117\ q_a$ che si somma allo $0.25\ q_a$ già preesistente; complessivamente l'assorbitore riceve $0.367\ q_a$ con recupero del deficit accumulato durante la costituzione del volano (in effetti l'evacuazione avviene secondo la (8) ed è praticamente completa in t_e).

CONCENTRAZIONE ATMOSFERICA VARIABILE

Se la concentrazione atmosferica oscilla intorno al valor medio per ogni variazione ΔC_a si ha la corrispondente variazione Δq_a, che avviene con le modalità già viste, con deficit nell'assorbimento pari a $0.363\ q_a$ quando ΔC_a è positivo e corrispondente recupero quando ΔC_a è negativo.

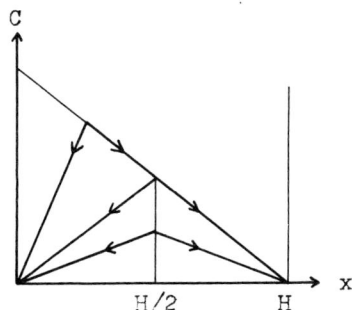

FIGURA 2 Variazione nel tempo dell'andamento della concentrazione all'interno della camera di diffusione durante l'esaurimento in seguito a riduzione a zero della concentrazione atmosferica

VARIAZIONI ISTANTANEE

Si dimostra che il deficit nell'assorbimento ed il relativo recupero si verificano anche con variazioni di durata $t_i < t_e/64$.

Con $t_i \leq t_e/2$ con incremento di concentrazione ΔC_a entra nella camera la quantità:

$$q_i = \frac{2H}{\pi} \left(-\frac{t_i}{t_e}\right)^{\frac{1}{2}} \Delta C_a$$

con accumulo del deficit:

$$F_e \cdot t_i = \frac{H}{\pi} \frac{t_i}{t_e} \Delta C_a = q_i \left(-\frac{t_i}{4t_e}\right)^{\frac{1}{2}}$$

CONCLUSIONI

Lo studio analitico mette in evidenza che il "volano" q_a, direttamente proporzionale alla concentrazione atmosferica C_a e alla profondità della camera di diffusione H, non incide significativamente sul prelievo in quanto il flusso verso l'assorbitore segue fedelmente le variazioni della concentrazione atmosferica, con leggero ritardo quando
essa aumenta, prontamente recuperato quando essa diminuisce.

EFFECT OF AIR TURBULENCE ON DIFFUSIVE SAMPLING

N. Zurlo, F. Andreoletti (1)

ABSTRACT

The extension h of the diffusive passage of a passive sampler caused by the inadequate velocity of ambient is calculated theoretically. The theory, confirmed by experiment, indicates that h depends on velocity u, diffusion coefficient D and length L of the sampler in the direction of u on the basis of $h = f(DL/u) 1/2$ and is not affected by depth H.

----

(1) Clinica del Lavoro " Luigi Devoto" , University of Milan.

INTRODUCTION

In the diffusion chamber of a passive sampler in equilibrium with the atmospheric concentration Ca of the analyte, the concentration gradient dC/dx and the flow through the chamber at the level of the absorption layer is zero, then:

$$C = Ca \left(1 - \frac{x}{H} \right) \qquad dC/dx = -\frac{Ca}{H} \qquad dq/dt = Ca \frac{D}{H} A$$

where D is the diffusion coefficient of the analyte and A is the section of the chamber perpendicular to the direction of diffusion x.

This is true only if the air is renewed instantly on the air-chamber contact surface. In fact, the air with an initial concentration Ca is depleted as it flows over the contact surface and the level at which the concentration is still Ca moves steadily farther from the contact surface and diffusion occurs through a chamber of depth Hc =H+h, where h is the distance from the contact surface of the level at which the concentration is still equal to Ca.

The purpose of this research is to determine a theoretical value for h and to verify it experimentally.

THEORETICAL CALCULATIONS

In the area outside the diffusion chamber near the air-chamber contact surface the concentration Ce and the gradient, in relation to the distance x from the surface itself, may be expressed:

$$Ce = Ca \left(1 - \frac{h - x}{H + h} \right) \qquad dCe/dx = \frac{Ca}{H + h}$$

Consider the element, outside the chamber, of section dxdz perpendicular to the direction y of the air velocity u, at distance Ly. If dV = u, dz dx dt gives the volume of air which passes in the period of time dt through this section.

If $\Delta C = Ca - Ce = Ca \dfrac{h - x}{H + h}$

gives the concentration difference in respect of Ca in relation to distance x, the quantity of analyte dq which during
the time dt passes through the layer of thickness dx and width
dz is given as:

$$dq = \Delta C \cdot dV = Ca \frac{h - x}{H + h} u \, dz \, dx \, dt$$

and by integration: $\quad (1) \quad q = \int Ca \dfrac{h - x}{H + h} u \, dz \, dx \, dt$

This quantity should equal the
quantity q' which was absorbed by diffusion during the period
of time dt by the area upstream of the distance Ly for which h
is considered and which is given by:

$$(2) \quad q' = \int D \frac{Ca}{H + hy} \, dz \, dy \, dt$$

where hy represents the distance from the air-chamber contact
surface of the level of concentration Ca in position y.

By taking (1) and (2) as equal, we have:

$$(3) \quad \int \frac{h - x}{H + h} u \, dx = \int D \frac{dy}{H + hy}$$

If we assume for u the value given by Prandtl (1952) expressed
as a function of the distance from the surface, of the "form"
of the surface itself and of the Renolds number, we can work
out the integral of the first side of equation (3) for x which
varies from 0 to h.

If we take as a first approximation hy to equal the average
value h which is assumed in the integration interval and which
proves to be 2/3 of h at less that 1%, we find a value for the
integral on the second side of equation (3) for y which varies
from 0 to L, where L is the length of the contact surface in
the direction of u.

Solving equation (3) with respect to h, we arrive at a first
approximation of the average value $\bar{h}$:

$$\bar{h} = f \, (\, D \, L/u \,)^{1/2}$$

where f is a function of the "form" of the sampler and of the
Reynolds number. The value of h therefore depends on the
diffusion coefficient D, length L of the contact surface and
air velocity u, whereas it is not affected by the depth H of
the diffusion chamber.

Experimental confirmation

In order to confirm experimentally the average extension h as
a function of L and u, an experiment was conducted using
cylindrical diffusion chambers of diameter L between 1 and 5
cm and depth H between 0.5 and 5 cm, for air velocity u
(parallel to the contact surface) varying between 0 and 200 cm
s -1 and with diffusion coefficient D ranging between 0.07 and
0.26 cm2 s -1. Experiments were also carried out using
diffusers with a conical chamber (or at least truncated) with

depths up to 15cm.

The results of the experiments gave an initial approximation:

$$\bar{h} = (2 D \frac{L}{3 + u})^{1/2}$$

The constant 3 in the denominator must be attributed to convective movement in the room where the experiments were conducted; the room was approximately 40 m3 in size, in a state of rest and without sources of heat, and the temperature variation was less than 0.5oC in the course of 24 hours.

Table 1 shows the values of $\bar{h}$ as a function of L and u for D = 0.1 cm2 s -1.

Table 1 - Values of average extension $\bar{h}$ as a function of length L of the air-chamber contact surface and air velocity u, for D = 0.1 cm2 s -1

| L, cm | $\bar{h}$, cm | | | |
	u=0 cm s^{-1}	u=3 cm s^{-1}	u=9 cm s^{-1}	u=45 cm s^{-1}
1	0.25	0.18	0.12	0.06
3	0.45	0.32	0.22	0.11
5	0.58	0.41	0.29	0.14

CONCLUSIONS

On the basis of the theory outlined above, by increasing the depth of the diffusion chamber of a passive sampler it is possible to reduce the effect of extension h, which in an atmosphere in a state of rest is practically negligible with H of about 5 cm.

BIBLIOGRAPHY

Prandtl, L., 1952 The essentials of fluid dynamics (Blackie and Son Ltd., Glasgow).

HSE PROTOCOL FOR ASSESSING PERFORMANCE

J G Firth

Health & Safety Executive
Occupational Medicine and Hygiene Laboratories
403 Edgware Road
London, NW2 6LN
England

ABSTRACT

The ease of use of diffusive samplers and the relatively low unit costs of occupational exposure measurements taken with them, makes HSE wish to promote their widespread use in industry. However, it is important that users are confident in the results obtained with these devices. Therefore, in 1983, HSE published a protocol for assessing the performance of diffusive samplers in the series "Methods for the Determination of Hazardous Substances" (MDHS) as MDHS 27. The protocol identifies the fundamental performance parameter as the "uptake rate" and sets out how it should be evaluated against variables which can affect its value. These variables include airborne concentration levels of the vapour to be measured, exposure time and air movement, as well as environmental variables such as temperature, pressure and humidity. Their effects are measured in laboratory tests and in field tests which use both personal sampling and static sampling comparisons with pumped sampling systems. In addition, tests for the stability of the sample after adsorption and storage tests are required.

INTRODUCTION

In 1983, the Health & Safety Executive's Committee of Analytical Requirements (CAR), published a protocol for assessing the performance of diffusive samplers as publication No 27 in the series "Methods for the Determination of Hazardous Substances". The Committee has overall responsibility for the publication of methods of analyses recommended by HSE as well as for quality assurance schemes and regards the establishment of protocols as an essential base for the development and implementation of such standard procedures.

The protocol for diffusive samplers was developed by Working Group 5 of CAR, which has prime responsibility for the development and evaluation of occupational hygiene measuring systems based on diffusive sampling. The majority of its members are drawn from industrial organisations and have first-hand experience with diffusive samplers. The ease of use of diffusive samplers and the low costs of measurements obtained with them, particularly when they are combined with automated systems of analyses,

make them devices which HSE would like to see in widespread use, for carrying out surveys and for enforcement purposes and also by industry itself
for self-monitoring. However, reservations by certain potential users and
apparent differences in experience obtained by some actual users, made the
production of a standard protocol for assessing the performance of them,
an important factor in gaining their widespread acceptance.

    It was essential that the protocol applied to all types of diffusive
samplers which are likely to be developed. Such samplers are available in
a variety of shapes and sizes and although, for most of them, the rate of
sampling is controlled by gas phase diffusion, which is essentially predictable it is also possible for the sampling rate to be controlled by diffusion through a liquid or solid phase where the effects of variables such
as temperature are greater and less predictable.

GENERAL CONSIDERATIONS

    Any sampler is basically a device which obtains a measurable amount of
the appropriate material which can then be reliably related back to the
amount of that material in the matrix which was sampled. Samplers used in
occupational hygiene monitoring, obtained this relation by sampling for a
measurable time at a known rate. A diffusive sampler controls its rate of
sampling by a diffusive process and is defined in the protocol as:-

> "A device which is capable of taking samples of gas or vapour
> pollutants from the atmospheres at a rate controlled by a
> physical process such as diffusion through a static air layer
> or permeation through a membrane, but which does not involve
> the active movement of air through the sampler".

It is thus obvious that the parameter of fundamental importance in a diffusive sampler is the sampling rate or uptake rate. This must be capable of
reliable prediction when variables which might affect its value when the
sampler is in use, are changed. The uptake rate (U) is defined as

$$U = \frac{\text{observed mass uptake of diffusive sampler } (ng)}{\text{"true"concentration } (ppm) \times \text{ exposure time } (min)} \quad \text{--- (1)}$$

Vapour concentration is generally measured in ppm (v/v) and exposure time
in minutes. Thus, convenient units for the uptake rate are ng $(ppm)^{-1}(min)^{-1}$.
These units are equivalent dimensionally to ml $(min)^{-1}$ where the volume
refers to the volume of air from which the mass of pollutant has been totally extracted by the sampler. These units are useful when comparing

sampling rates of diffusive samplers, with those of pump samplers.

Basically, a diffusive sampler consists of an adsorbent to adsorb the material being measured and between the adsorbent and the atmosphere being monitored is a rate limiting diffusion path. This diffusion path is generally a column of air which is stabilised by being enclosed within a tube, or within the pores of a membrane. Therefore, in principle, the rate of uptake is controlled solely by the geometry of the sampler and the effects of variables such as temperature and pressure on the diffusion coefficient of the substance being monitored, since it is assumed that the concentration immediately above the absorbent is zero. Thus, the concentration gradient along the diffusion path, in steady state conditions, can be taken as constant and equal to the concentration in the atmosphere being measured, divided by the length of the diffusive path. However, adsorption onto the absorbent material, will obey the various types of adsorption isotherms depending on the strength of adsorption and hence once material has started to adsorb onto the absorbent, the vapour pressure of it immediately above the absorbent, will start to rise. If the substance is not strongly adsorbed, or the ratio of available surface area to the amount of material adsorbed, is not high, then the vapour pressure above the absorbent, can rise to levels which effectively reduce the rate of diffusion in the sample. In addition, adsorbed vapour can permeate through the adsorbent not exposed directly to the gas phase by surface migration and desorption readsorption processes and so effecting the vapour pressure immediately above the exposed surface of the adsorbent.

The above considerations mean that the protocol must take into account the effects of loading and exposure time on the uptake rate as well as environmental variables such as temperature, pressure and humidity, which affect the diffusion coefficient of the vapour, the vapour pressure above the adsorbent and the migration of material across the adsorbent. Similarly, air flow over the samplers which could affect the diffusive path, must also be taken into account as must interferents, which would reduce the amount of available surface area on the absorbent for the substance being measured. This could mean that a large number of tests would need to be specified to assess the effect of these variables on the sampling rate. The effects of these variables will then be statistically evaluated to determine the confidence with which the sampling rate can be predicted in various circumstances.

It is essentially true that the greater the number of tests carried out, the more reliable will be the conclusions drawn from them.   However, a protocol must be realistic and attempt to reduce the number of tests which are carried out to the minimum necessary to obtain data of sufficient reliability.   The HSE Protocol allows a two stage process with an initial evaluation of the effects of the variables with a minimum number of tests and if those variables are then shown to have significant effect, a larger number of tests to define these effects, is then specified.

In principle, all the tests which are necessary to define the performance of a diffusive sampler, could be carried out within the laboratory.   However, it was felt that field testing would be an advantage because for some designs of sampler the physico-chemical processes defining the rate of sampling, were not clearly apparent and because diffusive samplers are essentially devices operating in a steady state condition. Sudden changes in concentration levels in the atmosphere around the sampler can upset such steady state conditions and affect the rate of sampling. Field trials of the devices in the situations in which they are likely to be used, are one way in which it is possible to determine whether or not such effects encountered in practice are of any significance.

LABORATORY EXPERIMENTS

The basic laboratory experimental procedure consists of exposing the diffusive samplers to a standard atmosphere of the appropriate gas or vapour, for a known period of time.   Standard atmospheres are generated using either the syringe (1) or permeation tube (2), or an equivalent technique.   The "true" concentration of the vapour in the test atmosphere, is calculated from the syringe injection rate or the rate of loss of vapour from the permeation tube and the volumetric flow of diluent air.   This flow is determined with a calibrated test meter in the usual way.   The temperature, pressure and humidity are measured and the air flow within the standard atmosphere apparatus is at a level where the effects of air flow on uptake rate are negligible.

The concentration of vapour in the apparatus is verified with a secondary method, which is usually a pump adsorbant tube, or a pump absorbant bubbler method.   This method will have been validated over the ranges of pollutant concentration, sampling time, etc, used in the evaluation of the diffusive sampler.   The true concentration is regarded as verified if the results of the secondary reference method are within 10% of the calculated concentration. If this is not found to be the case, then a fresh standard atmosphere is set up or the reference method is changed. This procedure is only used to check the concentration of the standard atmosphere and measurements by the reference method are not taken as part of the validation of a diffusive sampler. The effects of exposure variables are measured using six diffusive samplers to investigate each variable.

The first set of tests involves the assessment of external air movement on the sampler uptake rate. In stagnant air conditions, the atmosphere around the entrance to the sampler, can be depleted as vapour is removed by the adsorbant, effectively increasing the diffusion path and reducing the rate of uptake. Generally, above a minimum air speed, the uptake rate remains constant. The uptake rate is measured at different air speeds and the air speed at which 95% of the maximum value is achieved, is noted. All subsequent laboratory experiments are carried out at air speeds above this value.

The effect of concentration and time of exposure are determined with each experiment being carried out using six samplers. These experiments are carried out according to the two factor designs shown in table 1. The full design would involve 9 separate experiments but a step-wise approach can be used in which the experiments indicated by four shaded areas of table 1 are carried out first. Only if the statistical analysis of these experiments show a variation of uptake rate are the remainder of the experiments carried through.

TABLE 1

Concentration	Exposure time		
	30 min	120 min	480 min
0.1 E.L.	///////		///////
1.0 E.L.			
2.0 E.L.	///////		///////

The investigation of the effects of temperature, pressure and humidity are separately carried out using a 2^3 factorial experiment in which the effects of each of the factors is measured at two levels and combined as shown in table 2. This design requires 8 experiments but these can be reduced if it is known that one of the variables has no effect on the uptake rate of a particular pollutant.

TABLE 2

Vapour loading ppm x hrs	Air temperature			
	5°C Relative humidity		30°C Relative humidity	
	20%	90%	20%	90%
0.5 x (EL)				
2 x (EL)				

The effect of potential interference is measured in the manner outlined above at a given set of environmental factors, by exposing the 6 samplers to the vapour of interest and any co-pollutants that are likely to be encountered in the practical use of the sampler and are likely to interfere.

In addition to the above performance tests, the effects of exposure to zero concentrations of the vapour are measured. This is to ensure that desorption of the adsorbed material during or after use does not occur to any significant extent. The test involves exposing the sampler to clean air for a period of at least 4 hours after a known exposure. In practice, two extra samplers are used in each of the tests where the effects of concentration and time of exposure are being investigated.

The effects of storage are examined by using two further samplers in each of the 480 min tests on the effects of concentration and time. These extra samplers are stored at normal laboratory temperatures for up to two weeks, followed by analysis, when the results are compared with the unstored samplers. If the amounts of material on the stored sample falls rapidly then the time at which results between stored and unstored samplers differs by more than 20%, is determined.

From the laboratory experiments, a standard uptake rate, defined by equation 2 is generally determined.

$$\text{standard uptake rate} = \frac{\text{observed mass uptake of sampler (ng)}}{\text{concentration (ppm) x exposure time (min)}} \quad --(2)$$

The standard uptake rate refers normally to standard conditions of 20°C, 101 K Pa and 50% relative humidity. The random error in the uptake rate is expressed either as a coefficient of variation or as repeatability.

FIELD EXPERIMENTS

Field tests are carried out to investigate the effects of random or systematic variations in concentrations which might be encountered in the field on uptake rate.   Two types of comparisons are made; personal sampling and static sampling.

For personal sampling trials, a minimum of 20 comparisons are made between a diffusive sampler and the reference sampling system, which is normally a pump absorbant system.   The two samplers are worn simultaneously and as close together as possible on one lapel of a person carrying out normal work practices.   Comparisons are chosen to cover a wide range of field conditions and exposure times.   The results between the two methods are compared by a linear regression analysis.   However, the reference method has, in general, inaccuracies at least as large as the diffusive method and so the regression comparison is only a crude approximation to the true functional relationship between the two methods.   The estimated value of each coefficient in the regression analysis and the correlation coefficient, are reported.

Multiple comparisons using static sampling are made wherever possible to obtain estimates of the random error measurement of each method and to permit a more detailed study of the differences between them.   A set of at least 6 diffusive samplers and an equal number of reference samplers are mounted on a grid, keeping the distances between them as small as

possible and the overall sampling area as small as possible.   The grids are arranged so that nominally a diffusive sampler is paired with a pump sampler.   The grids are then exposed at different locations in a workplace environment over a wide range of pollutant levels, to give at least three blocks of data.

CONCLUSIONS

The protocol is a minimal procedure which enables the performance of one type of diffusive sampler to be determined for one type of vapour. It generates sufficient data for an evaluation of the performance to be made with a minimum amount of experimental work.

At the present time, HSE has not imposed minimum performance standards for diffusive samplers.   This is essentially because workplace monitoring is carried out for a number of purposes, ranging from compliance testing to a rough estimate of exposure levels. Qualitative evaluations of exposure are generally made using techniques which have a coefficient of variation of about 25%, whilst compliance testing is carried out using techniques with an overall coefficient of variation of less than 10%. This may well be the way in which acceptance of diffusive samplers will move but at the present time the rate of development of samplers is sufficiently rapid for the imposition of acceptance standards to be delayed.

REFERENCES

Health & Safety Executive. Methods for the Determination of Hazardous
    Substances. Generation of test atmosphere by the syringe injection
    method. MDHS 3. HSE London 1981.
Ibid. Generation of test atmospheres by the permeation tube technique
    MDHS 4. HSE London 1981.

APPLICATIONS OF THE HSE DIFFUSIVE SAMPLER EVALUATION PROTOCOL

R.H. Brown

Health and Safety Executive
Occupational Medicine and Hygiene Laboratory
403 Edgware Road, London NW2 6LN, England

ABSTRACT

The HSE diffusive sampler evaluation protocol (HSE, 1983) has been used to evaluate the performance of several different diffusive samplers for a variety of pollutants. Examples of the measurements, statistical analyses and conclusions will be given for a selection of evaluations, making particular reference to the Perkin-Elmer diffusive sampler.

In laboratory experiments designed to determine the effect of exposure parameters on the sampling rate, three distinct types of behaviour were noted. Depending on the sampling capacity and nature of the adsorbent, the uptake rate was found to be either constant (e.g. for butadiene on Molecular Sieve), to vary with time of exposure only (e.g. for benzene on Tenax) or to vary with exposure dose (concentration x time; e.g. for nitrous oxide on Molecular Sieve).

In field experiments designed to compare diffusive methods with independent test methods, in all cases no significant systematic differences were found in the sets of results, provided an uptake rate appropriate to the exposure conditions was used. The observed precisions in these field tests were, on average, 12 +6% for the diffusive methods and 13+5% for the pumped methods, indicating that diffusive methods are no better and no worse than conventional, accepted methods.

INTRODUCTION

The development of novel sampling devices based on diffusion for estimating personal exposure to toxic pollutants in the workplace has led to a challenge to existing conventional sampling methods that use personal sampling pumps. It is important that before these new methods are introduced, they should be fully evaluated and shown to be at least as accurate and reliable as the methods they are intended to replace. The Health and Safety Executive (HSE) has developed a protocol for such an evaluation (HSE, 1983), which has been described in the paper previous to this in the Symposium (Firth, J.G.). This paper gives some examples of the use of the protocol in practice.

LABORATORY EXPERIMENTS

The main features of the HSE protocol are the determination of the dependence of sampler performance on exposure variables when sampling from standard atmospheres in the laboratory and a comparison of the diffusive sampler with an independent test method, usually a pumped method, in determining exposures to toxic pollutants in the workplace.

The most important exposure variables are exposure time and exposure concentration and the laboratory tests in the HSE protocol can be illustrated by reference to these parameters in particular.

Ideally, the sampling rate of a diffusive device should be constant over the range of exposure conditions likely to be encountered in the workplace. This may usually be achieved if an appropriate adsorbent is available. For example, butadiene may b sampled on Molecular Sieve 13X in the Perkin-Elmer diffusive tube. An analysis of variance on six replicates in a 3-level, 2-way exposure test for this example is presented in Table 1.

TABLE 1   Butadiene laboratory exposures.

Variable	F-value	Significance
between concentrations	0.87	NS
between times	1.07	NS
concentration x time interaction	1.74	NS

NS = not significant (5% level)

For many organic vapours sampled on the Perkin-Elmer tube however, adsorbents are not available that provide both an absolute zero sink and also allow desorption at temperatures belc those which cause thermal decomposition. If porous polymers are employed in these situations, the loss of zero sink will result i a small reduction in sampling rate compared with the theoretical maximum. Empirically, the sampling rate (expressed in ng ppm-1 min-1 or cm3 min-1) is found to be dependent on the sampling time but not on the pollutant concentration (Table 2).

TABLE 2   Laboratory exposures to acrylonitrile, benzene
            and styrene.

| Variable | F-value and significance | | |
	Acrylonitrile	Benzene	Styrene
between conc.	1.0 NS	3.4 NS	2.0 NS
between times	40.6 **	49.7 **	17.2 **
interaction	2.9 NS	9.2 *	0.25 NS

NS = not significant (5% level)
* = significant (5% level)
** = significant (1% level)

In situations of this type, it is necessary to determine the
sampling rate for the range of exposure times likely to be
encountered in the workplace, typically 10 min to 8 hours, so that
an appropriate factor may be applied (Table 3).

TABLE 3   Sampling rates for acrylonitrile, benzene
            and styrene on Tenax.

	Acrylonitrile	Benzene	Styrene
theoretical rate	100%	100%	100%
10 min sample	80%	89%	93%
8 hour sample	38%	55%	84%

As might be expected, the reduction in sampling rate is
smallest for the least volatile adsorbate. It can also be
minimised by choosing the strongest suitable adsorbent; thus
acrylonitrile on Porapak N behaves in the same way as benzene on
Tenax.
With other adsorbents, for example Molecular Sieve or carbon,
and weakly retained adsorbates, the sampling rate may be dependent
on both sampling time and pollutant concentration, i.e. exposure
dose (Table 4).

TABLE 4   Nitrous oxide laboratory exposures.

Variable	F-value	Significance
between concentrations	7.0	**
between times	5.6	**
concentration x time		
interaction	0.4	NS

NS = not significant (5% level); ** = significant (1% level)

In this case, it is necessary to use an iterative process with a plot of uptake rate against exposure dose in order to get the best estimate of the exposure concentration.

FIELD TRIALS

The two types of field trial in the HSE protocol are designed to test for the presence of any systematic difference between the diffusive method and an independent test method in field use.

Personal field comparisons

In the first of these trials, comparisons are made between single samplers of each type simultaneously used as personal monitors. One appropriate statistical analysis method for such data is linear regression analysis (y = log of diffusive result; x = log of pumped result) and some typical results are summarised in Table 5.

TABLE 5   1 + 1 Personal field comparisons.

Pollutant	Sampler	n	r	Slope	Intercept
acrylonitrile	P/E-Porapak N	60	0.98	0.95	0.01
benzene	P/E-Porapak Q	103	0.98	0.89	0.04
butadiene	P/E-Mol Sieve	24	0.94	0.94	-0.07
carbon disulfide	P/E-Spherocarb	14	0.95	1.03	-0.15
halothane	3M-carbon	27	0.94	0.88	0.03
nitrous oxide	P/E-Mol Sieve	24	0.93	1.02	-0.11
styrene	P/E-Tenax	25	0.94	0.99	0.01
trichloroethane	P/E-Tenax	40	0.91	1.01	0.04
average			0.95	0.91	-0.02

n = no of comparisons
r = correlation coefficient
P/E = Perkin-Elmer

In each example in Table 5, the 95% confidence intervals of the slopes and intercepts respectively embrace 0 and 1, indicating the absence of any systematic difference between the methods. Th slopes are on average slightly below 1.0, but this is a consequence of using monovariate regression analysis in these examples. Thus if the plot for butadiene is reversed (i.e. y = log of pumped result), the slope is still below 1.0 (0.91).

Static field comparisons

The second type of field trial involves multiple replicates of diffusive and test sampler and is usually conducted in a static position because of the impracticability of multiple personal sampling. This type of experiment checks for any systematic difference in the results given by the two methods, and also gives an independent estimate of the precision of each method under field conditions. Table 6 gives some typical results; it also includes, for comparison, diffusive sampler precision data derived from laboratory experiments.

TABLE 6   Multiple static field comparisons.

Analyte	No. of sets	No. in set	Precision (CV, %) Diffusive Lab	Diffusive Field	Pumped Field
acrylonitrile	2	4 + 4			
	2	5 + 5	9	11	15
benzene	4	3 + 3			
	11	2 + 2	9	11	13
butadiene	3	7 + 7	15	16	20
carbon disulfide	3	6 + 6	7	20	5
styrene (P/E)	3	4 + 4	12	8	10
styrene (3M)*	41	2 + 2	9	4	15
average			10	12	13

CV = coefficient of variation
* = personal trial
P/E = Perkin-Elmer

One particular difficulty with the multiple comparison trial is that for the recommended 6 + 6 replicates a personal exposure test is not practical. Badge-type samplers require a minimum localised air velocity to function optimally and this air movement cannot be guaranteed in static positions. They are normally only used in personal exposure estimations. A compromise is therefore to expose a smaller number of replicates (say two of each type of sampler) and have a larger number of sets, as in the example for styrene (3M) in Table 6.

REFERENCE
Health and Safety Executive. 1983. Protocol for assessing the performance of a diffusive sampler. Methods for the Determination of Hazardous Substances. MDHS 27 (HSE, London).

PROTOCOL FOR THE EVALUATION OF PASSIVE MONITORS

Mary Ellen Cassinelli,* R. DeLon Hull,** John V. Crable,*Alexander W. Teass*

National Institute for Occupational Safety and Health
*Division of Physical Sciences and Engineering
Methods Research Branch
**Division of Biomedical and Behavioral Science
Applied Biology and Physics Branch
4676 Columbia Parkway, Cincinnati, Ohio 45226, USA

ABSTRACT

An evaluation protocol was devised specifically for passive monitors. It tests those aspects intrinsic to diffusive sampling, as well as those aspects common to both dynamic and passive sampling. The following characteristics were identified as critical to the performance of passive monitors and studied: recovery; capacity; reverse diffusion; storage stability; accuracy and precision; shelf life; behavior in the field; and the effects of concentration, exposure time, face velocity, humidity, interferences, orientation, and temperature. The experimental design obtained a maximum of information from a minimum of experimentation. Data interpretation was based on statistical treatment using generally accepted criteria. This protocol is intended as a set of guidelines for the evaluation of passive monitors.

INTRODUCTION

Since the late 1970's, when passive sampling devices impacted the market in great numbers, the industrial hygiene community has been concerned over the reliability of these devices for monitoring the concentration of gases and vapors in air. A comprehensive laboratory evaluation project for passive organic vapor monitors was carried out by Utah Biomedical and Test Laboratory (UBTL) Division of University of Utah Research Institute under a contract with the National Institute for Occupational Safety and Health (NIOSH) (Perkins et al., 1981). The results of this study, along with those from other laboratory and field comparisons of passive monitors with conventional sampling methods (Jones et al., 1979; McCammon et al., 1980; Hickey and Bishop, 1981; Rinehart, 1981; Stricoff and Summers, 1981; Cassinelli and Hull, 1983), have demonstrated the need for characterizing the performance of passive monitors.

Passive monitors rely on the principle of diffusion to effect analyte collection. Because passive (diffusive) and dynamic (pumped) sampling systems operate by different mechanisms, different tests are required to characterize their performance. The evaluation protocol was devised to test those aspects intrinsic to diffusive sampling, along with those aspects common to both dynamic and passive sampling. The approach taken differed from that of others (Coulson, 1981; Brown et al., 1984) in that not only were characteristics critical to the performance of passive monitors identified and studied, but also bases for the interpertation of the results were stated. The experiments were designed to characterize under controlled conditions the following critical aspects: recovery; sampling rate and capacity; reverse diffusion; storage stability; temperature; the effects of varying analyte concentration, exposure time, face velocity, humidity, interferences and sampler orientation; accuracy and precision; shelf life; and behavior in the field. The order of the experiments was arranged for the most efficient testing of the monitors, with six factors tested simultaneously by means of a sixteen–run fractional factorial experimental design (du Pont, 1975). Replication of the passive samplers and independent measurements was sufficient for statistically valid comparisons. Interpretation of the data was based on statistical tests using generally accepted criteria, e.g., $\pm25\%$ accuracy, 10% differences at the 95% confidence level, etc. (Taylor et al., 1977; Busch and Taylor, 1981).

The protocol for the evaluation of passive monitors was subjected to laboratory use by applying it to commercially–available passive monitors for sulfur dioxide (Cassinelli et al., 1985), formaldehyde (Kennedy and Hull, 1986) and ammonia (Hull, 1986). Minor modifications made in the protocol as a result of the laboratory work are incorporated into the protocol presented in this paper.

EVALUATION OF PERFORMANCE CHARACTERISTICS

The protocol for the evaluation of passive monitors is summarized in Table 1, which presents each performance characteristic, the corresponding evaluation experiment, and the basis for interpreting the data.

Analytical recovery

The experiment for analytical recovery serves a twofold purpose, the determination of the desorption efficiency of solid sorbents, and the determination of recovery as a means of checking sample workup and instrumental response. This characteristic applies only to passive monitors which can be spiked without destroying the integrity of the monitor, i.e., monitors with solid sorbent pads and those liquid–containing monitors from which the liquid is removed for analysis. Recovery was determined by spiking the sorbent with known quantities of analyte at levels representative of exposures at 0.1, 0.5, 1.0, and 2.0 times the appropriate

TABLE 1   Protocol for the evaluation of performance characteristics of passive monitors.[a]

Characteristic	Experimental Design	Interpretation of Results
1. ANALYTICAL RECOVERY	Spike 16 monitors, 4 at each of 4 concentration levels (0.1, 0.5, 1.0 & 2.0 x STD). Equilibrate about 12 h, and analyze.	For the 3 higher levels require $\geq$75% recoveries with $\bar{s}_r \leq$0.1.
2. SAMPLING RATE AND CAPACITY	Expose monitors (4 per time period) for 1/8, 1/4, 1/2, 1, 2, 4, 6, 8, 10 & 12 h to 2 x STD, 80% RH and 20 cm/s face velocity. Plot concentration vs. time exposed. Determine MRST and SRST.	Verify sampling rate. State useful range at 80% RH & 2 x STD. Capacity= sample loading corresponding to the downward break in conc. vs. time curve from constant concentration. SRST=time linear uptake rate achieved. MRST=0.67 x capacity (1 analyte) MRST=0.33 x capacity (Multi-analyte)
3. REVERSE DIFFUSION	Expose 20 monitors to 2 x STD, 80% RH for 0.5 x MRST. Remove and analyze 10 monitors. Expose others to 80% RH and no analyte for remainder of MRST.	Require $\leq$10% difference between means of the two monitor sets at the 95% CL.
4. STORAGE STABILITY Laboratory Analyzed:	Expose 3 sets of monitors (10 per set) at 80% RH, 1 x STD, and 0.5 x MRST. Analyze first set within 1 day, second set after 2 weeks storage at about 25 °C, third set after 2 weeks storage at about 5 °C.	Require $\leq$10% difference at the 95% CL between means of stored monitor sets and set analyzed within 1 day.
Direct-Reading:	Expose 10 monitors to storage test conditions. Take readings within 5 min, then at 15, 30, 60 & 120 min.	State maximum time after exposure when readings are $\geq$90% of initial readings.
5. FACTOR EFFECTS	Test the following factors at the levels shown. Use a 16-run fractional factorial design (4 monitors per exposure) to determine significant factors.	Indicate any factor that causes a statistically significant difference in recovery at the 95% CL. Investigate further to characterize its effect.
	Factor Test Levels analyte concentration 0.1 & 2 X STD exposure time SRST & MRST face velocity 10 & 150 cm/s relative humidity 10 & 80% RH interferent 0 & 1 X STD monitor orientation parallel & perpendicular (to air flow)	
6. TEMPERATURE EFFECTS	Expose monitors (10 per temp.) to 0.5 x STD at 10, 25 & 40 °C for 0.5 x MRST.	Define temperature effect and verify correction factor, if provided.
7. ACCURACY AND PRECISION	Calculate precision and bias for monitors ($\geq$10 per conc. level) exposed to 0.1, 0.5, 1 & 2 x STD at 80% RH for $\geq$0.5 x MRST. Use data from previous experiments.	Require bias within $\pm$25% of true value at 95% CL with precision ($\bar{s}_r$) $\leq$10.5% for 0.5, 1 & 2 x STD levels.
8. SHELF LIFE	Observe monitors throughout evaluation for changes in blank values, physical appearance, etc. Test monitors from more than one lot, if possible.	Note shelf storage time at which changes begin to occur. Indicate whether correctable or not.
9. BEHAVIOR IN THE FIELD	Consider problems not predictable from laboratory experiments.	Record temperature, humidity, air velocity, other contaminants, etc.
Area Sampling:	Expose passive monitors and independent method samplers (13 each) to the same environment.	Calculate precision and bias. Compare with laboratory results.
Personal sampling:	Conduct personal sampling with $\geq$25 sampler pairs. Place pairs of passive monitors and independent samplers on the same lapel of each worker.	Calculate bias. Compare with area sampling and laboratory results.

[a] The abbreviations are: STD = the appropriate exposure standard [OSHA permissible exposure limit, ACGIH threshold limit value or NIOSH recommended standard]; $\bar{s}_r$ = Pooled relative standard deviation; RH = Relative humidity; MRST = Maximum recommended sampling time; SRST = Shortest recommended sampling time; and CL = Confidence level

exposure standard (x STD), allowing the analyte and sorbent to equilibrate at room temperature, and analyzing the sample. Spiking is normally done by pipetting a solution of the analyte (preferably 10 uL or less) directly onto the sorbent. For monitors containing solid sorbents, the acceptability criterion was similar to that used previously for solid sorbent methods (Taylor et al., 1977).

Sampling rate and capacity

Although approximate sampling rates may be calculated from the steady-state form of Fick's First Law (Rose and Perkins, 1982), the actual sampling rate must be verified in the laboratory for several reasons: (a) the calculated sampling rate contains some error, particularly in the diffusion coefficient; (b) assumptions made in developing Fick's equation may not always be true; (c) the use of draft shields or other mechanisms to prevent face velocity effects may alter the sampling rate; (d) physical effects, such as affinity of the analyte for sampler housing, may perturb the actual sampling rate; and (e) the diffusion coefficient of the analyte gas or its solubility in a permeable membrane may not be known.

Ideally, at constant concentration passive monitors should have a constant uptake rate , i.e., mass of analyte collected per unit time. For monitors with a fixed diffusion zone, the uptake rate, or the effective sampling rate, is assumed to be constant. However, for monitors with packed sorbent beds or length of stain color change strips, the diffusion path lengthens with time; therefore, the sampling rate decreases with time. Monitors of this type require a family of calibration curves to determine the concentration of the analyte. The investigation of sampling rate and capacity verified the calibration of monitors with variable sampling rates, as well as the sampling rate of monitors with fixed diffusion zones and monitors controlled by permeation.

To determine sampling rate and capacity, four passive monitors per time period were exposed to an analyte concentration of 2 x STD, 80% relative humidity, and a face velocity greater than 20 cm/s for ten time periods ranging from 7.5 min to 12 h. Because the actual mass collected by some passive monitors cannot be determined from the information provided by the manufacturers, the concentration as determined by each group of monitors was plotted against time (Figure 1). The linear horizontal portion of the curve indicates a constant uptake rate. As the sorbent in the monitor approaches its capacity, the uptake rate decreases. This decrease ultimately becomes significant enough to cause the concentration measured with the passive monitor to decrease with time. For practicality, the capacity of the passive monitor is defined as the loading, in micrograms or in

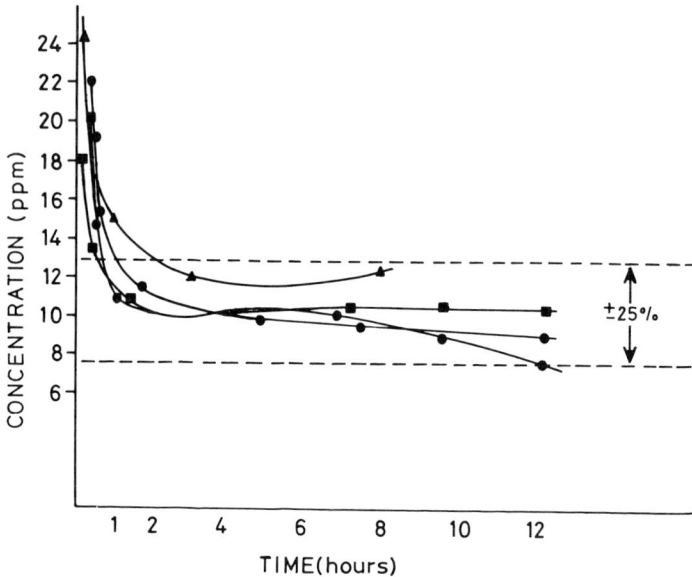

Figure 1   Passive monitor response with respect to time

parts–per–million–hours, corresponding to the beginning of this decrease in concentration. To compensate for any reduction in capacity caused by the sorption of interfering gases or water vapor, a maximum recommended sampling time (MRST) was specified. For single–substance passive monitors, the MRST was set at two–thirds the time required to reach the capacity. If the curve of concentration vs. time remained horizontal for 12 h, the MRST was set at 8 h. A MRST of two–thirds may not be sufficient for multiple substance monitors, such as organic vapor monitors with charcoal sorbents, exposed to an atmosphere containing several contaminants. In such cases a MRST of one–third seems appropriate. Most of the passive monitors in this study did not reach a steady sampling rate for at least 30 to 60 minutes (Cassinelli et al., 1985; Kennedy and Hull, 1986; Hull, 1986); therefore, a minimum sampling time was recommended. The shortest recommended sampling time (SRST) was defined as the time at which the concentration falls within $\pm 25\%$ of the reference concentration, determined by the independent methods.

Reverse diffusion

Reverse diffusion may be a significant problem for weakly–retained species (a) when the analyte is competing for sorbent sites with other species including water vapor, (b) when a high peak exposure is followed by very low or no analyte exposure for a significant period of time, and (c) when monitors are stored for long time periods after exposure (Posner, 1981; Moore, 1981). To verify that reverse diffusion did not occur, 20 monitors were exposed to an analyte concentration of 2 x STD and 80% relative humidity for one–half the MRST (about 4 h in most cases). One set of 10 monitors was removed and analyzed. The second set of 10 monitors was exposed to 80% relative humidity but no detectable analyte concentration for the remainder of the MRST. To determine if the data met the criterion of not greater than 10% difference at the 95% confidence level, the results were tested through the use of a statistical equation which compares means ($\overline{X}$) and variances (s^2) for a stated difference between two sets of data containing n_1 and n_2 samples (Walpole and Myers, 1978):

$$T' = \frac{(\overline{X}_1 - \overline{X}_2) - d_o}{(s_1^2/n_1 + s_2^2/n_2)^{0.5}} \tag{1}$$

The T' distribution above has the following degrees of freedom:

$$df = \frac{(s_1^2/n_1 + s_2^2/n_2)^2}{\dfrac{(s_1^2/n_1)^2}{n_1 - 1} + \dfrac{(s_2^2/n_2)^2}{n_2 - 1}} \tag{2}$$

The null hypothesis states that the difference between the means of the two sets of data ($\overline{X}_1 - \overline{X}_2$) is equal to a stated difference (d_o). The alternate hypothesis states $\overline{X}_1 - \overline{X}_2$ is greater than d_o. In this work the largest acceptable difference was 10%; thus, $d_o = 0.1 \overline{X}_1$.

Storage stability

Since a time lag occurs between the exposure of the sampling devices and analysis in the laboratory, the exposed monitors must be stable. To determine stability, three sets of 10 monitors were exposed to the test conditions specified in Table 1. The first set was analyzed within one day. The second set was stored at room temperature (about 25 $^{\circ}$C) for at least two weeks, and the third set refrigerated (about 5 $^{\circ}$C) for at least two weeks. The air samples were considered stable if the difference in average concentration between the stored sample set and the set analyzed within one day of exposure was 10% or less at the 95% confidence level. This difference was tested as described above. Although storage at room temperature is preferable, reacted sorbents may not be stable at ambient temperatures for extended periods of time; hence, both ambient and refrigerated storage experiments were conducted simultaneously.

For direct–reading passive monitors, long storage periods are unnecessary. However, continued diffusion within the sampler or fading of a reacted sorbent may limit the time during which accurate readings may be taken. To test these devices, a set of 10 monitors was exposed to the test conditions and readings were taken within 5 min. Readings were taken again on this same set of monitors at 15, 30, 60, and 120 min after exposure. The maximum time during which the average reading of the samplers was at least 90% of the average initial readings was noted.

Factor effects

Six factors were grouped into a 16–run fractional factorial experimental design, illustrated in Table 2 (du Pont, 1975; Box et al., 1978). The design is capable of revealing any factor having a significant effect on the performance of the passive monitor. In some cases, significant two–factor interactions, where two factors together give an effect different from that predicted by their separate behaviors, can be observed. Test factors are assigned to each of the X columns. A positive sign (+) in the X column indicates the high test level and a negative sign (−) indicates the low test level for that factor is used in the particular run.

The six factors were evaluated at levels which were chosen to bracket conditions found in the field. The factors with their recommended test levels are stated in Table 1. Exposure times were the SRST and MRST determined in the sampling rate and capacity experiment. Interferences tested were either present at 1 x STD or absent (0 concentration). The range of 10–150 cm/s brackets the range of face velocities encountered in a typical workplace. Convection induced currents have minimum velocities of 5 to 10 cm/s (Hemeon, 1963). In an indoor facility the air velocity is probably less than 150 cm/s, the face velocity created by briskly walking in still air. For routine outdoor use, samplers should be verified for higher face velocities (about 900 cm/s). Low face velocities may lead to undersampling, causing starvation (Rose and Perkins, 1982), and high face velocities to oversampling, where turbulence in the diffusion zone of the monitor may change the sampling rate. The orientation of a passive monitor may result in poor contact between a liquid absorber and the diffuser or permeation membrane, or result in air flow into the diffusion zone causing a change in the sampling rate of the device.

TABLE 2   Sixteen–run fractional factorial.

Run No.	Concen-tration	Humidity	Interfer-ence	Time	Face Velocity	Orienta-tion	Result
	X 1	X 2	X 3	X 4	X 5	X 6	
1	−	−	−	+	+	+	
2	+	−	−	−	−	+	
3	−	+	−	−	+	−	
4	+	+	−	+	−	−	
5	−	−	+	+	−	−	
6	+	−	+	−	+	−	
7	−	+	+	−	−	+	
8	+	+	+	+	+	+	
9	+	+	+	−	−	−	
10	−	+	+	+	+	−	
11	+	−	+	+	−	+	
12	−	−	+	−	+	+	
13	+	+	−	−	+	+	
14	−	+	−	+	−	+	
15	+	−	−	+	+	−	
16	−	−	−	−	−	−	

The design provides an estimate of the main factor effects free of two factor interactions, an estimate of two or three factor interactions, as well as an estimate of experimental error. Procedures for data workup are described elsewhere (du Pont, 1975; Cuendet et al., 1983; Abell, 1984). Factors that were shown to be significant at the 95% confidence level were further tested to characterize the effect and to determine the range over which accurate results may be obtained.

Temperature effects

Diffusion coefficients of gases and vapors vary proportionally with the absolute temperature raised to the 1.5 power. [There are slight variations in force constants and collision integrals, but these can be disregarded (Palmes et al., 1976).] Since the gaseous volume also varies with temperature, the analyte concentration on a mass per unit volume basis varies inversely with temperature (Rose and Perkins, 1982; Palmes et al., 1976; Tomkins and Goldsmith, 1977). Consequently, the quantity of material sampled varies as a function of absolute temperature to the 0.5 power (i.e., $T^{1.5}/T$). Since the quantity of material sampled by a diffusional sampler is proportional to the square root of the absolute temperature, the ratio of the concentration found at low temperature, C_L, to the concentration found at high temperature, C_H, is equal to the ratio of the square roots of the corresponding absolute temperatures:

$$C_L / C_H = T_L^{0.5} / T_H^{0.5} \qquad (3)$$

This relationship may be perturbed by the effect of temperature on the material used for draft shields and on the sorbent/analyte affinity. The temperature experiment was included to characterize the effect of temperature on the performance of the passive monitor and to determine the adequacy of the manufacturer's correction factor (when provided). Ten passive monitors per temperature were exposed to low, ambient, and high temperatures (10, 25, and 40 $^{\circ}$C). The data were normalized prior to making the calculations. The high and low temperature data were compared to the theoretical ratio (Cassinelli et al., 1985; Kennedy and Hull, 1986), or compared to the reference method data (Hull, 1986). The latter comparison may be made if the reference methods are unaffected by temperature or are temperature correctable. For permeation samplers, temperature effects are due to the nature of the permeable membrane and may be quite large or small depending on the membrane material used (Mullins and Anders, 1981). If a temperature correction is provided by the manufacturer, it should be verified.

Accuracy and precision

The accuracy and precision of passive monitors were calculated over the concentration range of 0.1 to 2 x STD. The acceptability criterion for bias was $\pm 25\%$ of the reference value at the 95% confidence level with a pooled relative standard deviation ($\bar{s}_r$) less than or equal to 10.5% for the range of 0.5 to 2 x STD. While the data for the 0.1–x–STD level were not included in the comparison with the criterion, the monitors were tested at that level, because the 0.1–x–STD level is of importance when a worker is exposed to several toxic substances, the effects of which are additive for the purpose of evaluating overall exposure (ACGIH, 1985). The data for this characteristic were obtained from previously run experiments: the 2–x–STD concentration level data from the 4– to 8–h samples from the sampling rate and capacity experiment, the reverse diffusion experiment, and runs 2, 4, 13, and 15 from the factorial experiment; the 1–x–STD level from the storage stability experiment; the 0.5–x–STD level from the temperature experiment; and the 0.1–x–STD level from runs 1, 3, 14, and 16 of the factorial experiment.

Shelf life

Observations made throughout the evaluation of commercial passive monitors revealed some effects of aging, although the monitors were stored according to the manufacturers' recommendations. Most noticeable were variations in blank values and in relative recoveries. Losses of liquid sorbent and effects later attributed to changes in moisture content of solid sorbents also were observed. In at least one case (Kennedy and Hull, 1986) the monitor lost liquid reagent during storage prior to exposure. To accomodate for this loss, the sample workup was modified to include adjustment of the liquid volume to the original level. Since the aforementioned conditions affect the results obtained for the passive sampler, the shelf life characteristic was added to the protocol. The experiment involves observance throughout the monitor evaluation of any changes in blank values, physical appearance, etc., noting the shelf storage time at which changes begin to occur, and indicating whether or not the changes are correctable. If possible, monitors from more than one lot should be tested for indications of aging effects.

Behavior in the field

When evaluating passive monitors in field environments, it is important to be aware of conditions which prevail in the field that are not predictable from the laboratory experiments. Passive monitors fully characterized in the laboratory prior to field testing will have known operating ranges and limitations; therefore, any changes in performance may be attributed to field conditions. A walk-through survey of the test area prior to conducting a field evaluation will provide an opportunity for selecting acceptable sampling locations and for determining if the environmental factors are within the operational ranges of the monitors. Company records, if available, may also assist in the selection of sampling locations. The relative humidity, temperature, velocity of air movement, etc. which prevail during sampling and the presence of other contaminants in the atmosphere should be documented. Human factors, such as worker movement which would restrict exposure of the monitor, tampering, etc., should be noted during sampling.

For area sampling, passive monitors and independent method samplers should be physically located such that both sample from the same air mass at a face velocity high enough to prevent passive monitor starvation. This was accomplished with a field sampling chamber capable of holding both independent and passive samplers and of maintaining a face velocity greater than 20 cm/s (Cassinelli et al., 1985; Hull, 1986). To detect a 15% difference at the 95% confidence level between passive and independent sampling methods when a relative standard deviation (s_r) of 10% for each method is assumed, thirteen samples for each method were required (Busch and Taylor, 1981). A direct–reading instrument may serve as an additional independent method, as well as a monitor for concentration variation during the exposure.

To determine the ability of a passive sampler to monitor worker exposure, personal sampling was conducted with each passive sampler paired with an independent–method sampler. Because of potential concentration gradient problems, both samplers were placed on the same lapel of the worker. Since the movement of workers from place to place within a work area may result in lesser exposures than at stationary locations, and since usable data are obtained only from samplers exposed to concentrations within their working ranges, at least 25 sampler pairs were used.

For field samples, the bias of the passive monitors was based on the reference concentration determined with the independent–method samplers. Because the true analyte concentration in the field is not known, and the results obtained for both passive monitors and independent–method samplers are subject to variations caused by field conditions, the field data were compared with the laboratory results. Data analysis was conducted using a paired t–test (duPont, 1975; Natrella, 1963). Although the concentrations may be plotted to indicate correlation between the methods, traditional least squares analysis may not be appropriate to fit a line to these data, because both methods (variables) are subject to errors of measurement (Tuggle and Hawkinson, 1981).

NOTES ON IMPLEMENTATION OF THE PROTOCOL

For effective implementation of the evaluation protocol, test atmospheres at the appropriate analyte concentration, relative humidity, and face velocity must be produced. The exposure chamber must be constructed of material inert to the subject analyte, be capable of generating atmospheres of acceptably constant composition, and have the capacity to hold the required numbers of passive monitors and independent–method samplers. In our work, recirculating chambers were employed in order to limit the amount of analyte gas needed. For the sulfur dioxide work a chamber was constructed of plexiglas[R] and acrylic–coated galvanized sheet metal ductwork (Cassinelli et al., 1985). A Teflon[R]–lined plexiglas chamber was constructed for the formaldehyde (Kennedy and Hull, 1986) and ammonia (Hull, 1986) studies.

Reference data are determined from the average of at least two independent monitoring methods with at least four samples per method. Independent methods include sampling and measurement methods of demonstrated accuracy, calibrated instrumentation with demonstrated accuracy and stability of calibration, and the calculated exposure chamber concentration. If permeation tubes or certified gas cylinders are used as analyte sources, the permeation rate of the tube or the concentration in the cylinder should be verified routinely.

Test conditions during the evaluation should be as close as possible to actual field use conditions. Throughout the laboratory evaluations the passive monitors were handled in the same manner as an industrial hygienist would handle them in the field, except that the monitors were exposed to known, controlled atmospheres. When provided, manufacturers' instructions for handling and storing the monitors were followed.

Good laboratory practice should prevail throughout the study, from the handling of the monitors and analysis of the samples to the calibration of instrumentation and documentation of all observed conditions and results.

SUMMARY AND CONCLUSIONS

The protocol, presented in this paper evaluates passive monitors over their useful operating ranges. It is not intended as a strict pass or fail evaluation. The criteria presented were those used in our evaluations and are intended merely as guidelines for others. The protocol was used in the evaluation of commercial passive monitors for three analytes (sulfur dioxide, formaldehyde, and ammonia) and was found to have the flexibility to effectively evaluate monitors of various designs—solid and liquid sorbents, diffusion- and permeation-controlled, direct-reading and laboratory analyzed. Thus, the protocol seems generally applicable to passive monitors and should serve as a guide for others in the evaluation of passive monitors.

REFERENCES

Abell, M.T. 1984. Using an 'electronic spreadsheet' for Plackett–Burman calculations. Trends in Analytical Chemistry, 3, VII–X.
ACGIH. 1985. Threshold Limit Values and Biological Indices for 1985–86, Appendix C. (American Conference of Governmental Industrial Hygienists, Cincinnati, OH).
Box, G. E. P., Hunter, W.G., Hunter, J.S. 1978. Statistics for Experimentors. (Wiley and Sons, New York)
Brown, R. H., Harvey, R.P., Purnell, C.J. and Saunders, K.J. 1984. A diffusive sampler evaluation protocol. Am. Ind. Hyg. Assoc. J. 45, 67–75.
Busch, K. A. and Taylor, D.G. 1981. Statistical Protocol for the NIOSH Validation Tests. In "ACS Symposium Series 149" (Ed. G. Choudhary). (ACS, Washington, DC). pp. 503–517.
Cassinelli, M.E. and Hull, R.D. 1983. Passive monitoring for nitrous oxide. NTIS No. PB85–221323. (National Technical Information Service, Springfield, VA).
Cassinelli, M. E., Hull, R.D. and Cuendet, P.A. 1985. Performance of sulfur dioxide passive monitors. Am. Ind. Hyg. Assoc. J. 46, 599–608.

Coulson, D. M. 1981. Method Validation. In "Annals of the American Conference of Governmental Industrial Hygienists, Dosimetry for Chemical and Physical Agents, Vol.1" (Ed. W.D. Kelley). (ACGIH, Cincinnati, OH) pp 83-90.

Cuendet, P., Cassinelli, M.E. and Hull, R.D. 1983. Enchantillonneurs passifs pour le dosage de l'anhydride sulfurex: Evaluation en atmosphere controlee, essais factoriels. Arch. Mal. Prof., 44, 509-511.

E. I. duPont de Nemours and Co. 1975. Strategy of Experimentation, 2nd Ed. (Wilmington, Delaware)

Hemeon, W. C. L. 1963. Plant and Process Ventilation, 2nd Ed. (Industrial Press, New York)

Hickey, J. L. S. and Bishop, C.C. 1981. Field comparison of charcoal tubes and passive vapor monitors with mixed organic vapors. Am. Ind. Hyg. Assoc. J., 42, 264-267.

Hull, R. D. 1986. Performance of passive monitors for ammonia, Submitted to Applied Industrial Hygiene.

Jones, W., Palmes, E.D., Tomczyk, C. and Millson, M. 1979. Field comparison of two methods for the determination of NO_2 concentration in air. Am. Ind. Hyg. Assoc. J., 40, 437-438.

Kennedy, E. R. and Hull, R.D. 1986. Evaluation of the DuPont Pro-Tek[R] Formaldehyde badge and the 3M Formaldehyde monitor. Am. Ind. Hyg. Assoc. J., 47, 94-105.

McCammon, C.S., Edwards, S.L., Hull, R.D. and Woodfin, W.J. 1980. A comparison of four personal sampling methods for the determination of mercury vapor. Am. Ind. Hyg. Assoc. J., 41, 528-531.

Moore, G. 1981. Letter to the Editor. Am. Ind. Hyg. Assoc. J., 42, A-26.

Mullins, H. E. and Anders, L.W. 1981. A New Innovative Diffusional Monitor for Sampling Ethylene Oxide in Air. 3M Technical Report, R-AIHAI (71.1)R.

Natrella, M. G. 1963. Experimental Statistics, National Bureau of Standards Handbook 91. (NBS, Washington, D. C.)

Palmes, E. D., Gunnison, A.F., DiMattio, J. and Tomczyk, C. 1976, Personal sampler for nitrogen dioxide. Am. Ind. Hyg. Assoc.J., 37, 570-577.

Perkins, J.B., Price, N.H., Eggenberger, L. and Burkart, J.A. 1981. An evaluation of passive organic vapor monitors. NIOSH Contract No. 210-78-0115.

Posner, J.C. 1981, Letter to the Editor. Am. Ind. Hyg. Assoc. J., 42, A-28.

Rinehart, D.S. 1981. Field Evaluation Experiences. In "Annals of the American Conference of Governmental Industrial Hygienists, Dosimetry for Chemical and Physical Agents, Vol.1." (Ed. W.D. Kelley). (ACGIH, Cincinnati, OH) pp. 117-124.

Rose, V.E. and Perkins J.L., 1982. Passive dosimetry – State of the art review. Am. Ind. Hyg. Assoc. J., 43, 605-621.

Stricoff, R.S. and Summers, C. 1981. An evaluation of organic vapor passive dosimeters under field use conditions. In "ACS Symposium Series 149" (Ed. G. Choudhary). (ACS, Washington, DC). pp. 209-221.

Taylor, D. G., Kupel, R.E. and Bryant, J.M. 1977. Documentation of the NIOSH Validation Tests, DHEW (NIOSH) Publication No. 77-185.

Tomkins, F. C. and Goldsmith, R.L. 1977. A new personal dosimeter for the monitoring of industrial pollutants. Am. Ind. Hyg. Assoc. J., 38, 371-377.

Tuggle, R. M. and Hawkinson, R.W. 1981. Incorrect use of t-tests. Am. Ind. Hyg. Assoc. J., 42, 325-326.

Walpole, R. E. and Myers, R.H. 1978. Probability and Statistics for Engineers and Scientists, 2nd Ed. (MacMillan, New York) pp. 249-255.

VERIFICATION OF PASSIVE MONITOR PERFORMANCE: APPLICATIONS

Eugene R. Kennedy[*], Mary Ellen Cassinelli[*], R. DeLon Hull[**]

National Institute for Occupational Safety and Health
[*]Division of Physical Sciences and Engineering
Methods Research Branch
[**]Division of Biomedical and Behavioral Science
Applied Biology and Physics Branch
4676 Columbia Parkway, Cincinnati, Ohio 45226

ABSTRACT

The performance of several commercially-available passive monitors was studied using an experimental protocol designed to serve as a general reference for such investigations. Compounds used for this verification were formaldehyde, sulfur dioxide and ammonia. Problems revealed by the protocol included variable sampling rate, variable blank values, humidity and interference effects, high bias for short sampling periods and limited capacity. The results obtained with this protocol demonstrated its applicability for the verification of passive monitor performance.

INTRODUCTION

Passive monitors represent an attractive alternative to conventional pump and sorbent tube or bubbler sampling methods. However, most passive monitors currently available are not supplied with all of the information needed by the industrial hygienist to determine which device is most suitable for a particular monitoring situation. Since past experience with passive samplers indicated potential problems with their performance (Perkins et al., 1981), Cassinelli and co-workers developed an experimental protocol for the verification of passive monitor performance. The details of this protocol have been presented in another paper (Cassinelli et al., 1986). To test the ability of this protocol to assess passive monitor performance, the protocol experiments were followed using commercially-available passive monitors for formaldehyde, sulfur dioxide and ammonia. Detailed results of these studies are reported elsewhere (Cassinelli et al., 1985; Kennedy and Hull, 1986; Hull, 1986). The purpose of this paper is to show the utility of the protocol for verification of the performance of passive monitors.

PROTOCOL RESULTS

A brief summary of the types of passive monitors used is included in Table 1. Included in this study were monitors which were either solid-sorbent based, liquid-absorbent based, or length-of-stain direct-reading. Also one of the

TABLE 1   Summary of passive samplers used in this study.

Badge Type	Sampling and Analysis Summary
3M Monitor	
Model 3750/3751 Formaldehyde Monitor	Reagent coated sorbent pad, chromotropic acid method
Sulfur Dioxide Monitor	Direct reading badge
DuPont Pro-Tek[R] Badge	
Formaldehyde, Type C-60	Liquid absorbent, chromotropic acid method
Sulfur Dioxide, Type C-20	Liquid absorbent, self-contained reagent system spectrophotometric method
Ammonia, Type C-10	Liquid absorbent, self-contained reagent system spectrophotometric method
MSA Vaporgard Dosimeter Tube	
Sulfur Dioxide	Reagent-impregnated length-of-stain strip
Ammonia	Reagent-impregnated length-of-stain strip
REAL BioBadge[R]	
Sulfur Dioxide	
(REAL-TCM)	Tetrachloromercurate solution, West-Gaeke analysis
(REAL-IC)[a]	Water, ion chromatography
Ammonia	
(REAL-BA)	Boric acid solution, Nessler's reagent method
(Permeation-IC)[a]	0.01N Sulfuric acid, ion chromatography
SKC Liquid Dosimeter	
Ammonia	0.01N Sulfuric acid, ion chromatography

[a]   Device was used with an absorbent and analysis method not recommended by the manufacturer.

monitor types (REAL BioBadge[R]) used a permeation sampling mechanism instead of the diffusion mechanism used in the other devices. The variety of monitor types provided a diversified test of the protocol experiments, which are detailed in Table 1 of Cassinelli et al., 1986. Table 2 summarizes the performance of the monitors in each of the protocol experiments.

Experiment 1 – analytical recovery

Only one of the monitors used a solid sorbent which required a desorption step in its analysis. Recovery was found to be acceptable over the range of interest.

TABLE 2 Summary of passive sampler experiments used in this study.

Monitor Number	Badge Type	1	2	3	4	5	6	7
	3M							
1.	Model 3750/3751 Formaldehyde Monitor	O	X	X	–	X	–	–
2.	Sulfur Dioxide Monitor	–	X	–	–	–	–	–
	DuPont Pro–Tek[R] Badge							
3.	Formaldehyde, Type C–60	–	O	O	O	P	O	O
4.	Sulfur Dioxide, Type C–20	–	O	O	O	O	O	O
5.	Ammonia, Type C–10	–	O	O	O	O	O	O
	MSA Vaporgard Dosimeter Tube							
6.	Sulfur Dioxide	–	P	X	O	P	P	X
7.	Ammonia	–	P	O	O	X	P	O
	REAL BioBadge[R] Sulfur Dioxide							
8.	(REAL–TCM)	–	O	O	X	O	X	O
9.	(REAL–IC)	–	P	O	O	P	O	O
	Ammonia							
10.	(REAL–BA)	–	O	O	O	O	X	O
11.	(Permeation–IC)	–	O	O	O	P	O	P
	SKC Liquid Dosimeter							
12.	Ammonia	–	O	O	O	P	O	O

O Monitor met evaluation criteria or had no sampling limitations.
P Monitor partially met evaluation criteria or had defined sampling limitations.
X Monitor failed evaluation criteria or had major sampling limitations.
– Monitor was not evaluated with this experiment.

Experiment 2 – sampling rate and capacity

Of the 12 monitors which were used for this experiment, 5 monitors did not meet protocol evaluation criteria for the full term of the exposure. These results were not necessarily unexpected, since monitors 6 and 7 had sampling capacity limitations placed on them by the manufacturer. Although monitor 9 did not have sufficient capacity to sample for the full 12 h of exposure, a maximum exposure time could be determined from the results of this experiment. This monitor used a sorbent solution which had not been specified by the manufacturer. Monitor 1 performed so poorly that it was unusable for determining concentration.

One unexpected finding of this experiment was a high initial bias which was observed with several of the monitors, irrespective of analyte. After 1 to 2 h of exposure, the bias was usually negligible. This finding did point out the need to specify a minimum sampling time to ensure that the monitor would give an accurate result.

Experiment 3 – reverse diffusion

In this experiment all but two of the samplers met the protocol performance criteria. Monitor 6 was a direct reading monitor which exhibited diffusion along its color indicating strip. Monitor 1, while not exhibiting reverse diffusion of the analyte from the sorbent, had extremely poor accuracy. This device was later removed from the market by the manufacturer for redesign.

Experiment 4 – storage stability

Only monitor 8 did not meet performance criteria. The manufacturer of this device has changed the recommended absorbent solution to be used in the monitor, which should solve the instability problem.

Experiment 5 – fractional factorial

Of the 11 monitors which had been evaluated with this experiment, only 4 monitors did not exhibit significant effects for the six factors and six interactions studied. Monitor 9, which was known to have limited capacity, and monitor 1, which was known to have a relative humidity sensitivity, were included in the experimentation to test the power of the experiment to detect these factors. The results of the experiment revealed exposure time to be significant with monitor 9 at the 95% confidence level and relative humidity to be significant at the 99% confidence level for the monitor 1. Monitor 1 was also found to have a significant effect due to face velocity. For monitor 3, effects of interference and face velocity were significant at the 99% confidence level. None of the two–way and three–way interactions were significant at the 95% confidence level. However, for some of the monitors, significance at the 90% confidence level was observed. While findings of significance at the 90% confidence level for a particular factor or interaction still met the protocol performance criteria, this information may be useful to the end user of the monitor for the explanation of sampling results.

One disadvantage noted with this experiment was that when a device had a large amount of imprecision associated with it, the imprecision would obscure any factor which might have had a significant effect on that device's performance. This problem was noted with monitor 6.

Experiment 6 – temperature effects.

Of the 10 monitors studied with this experiment, only four did not perform acceptably. With monitors 6 and 7, the manufacturer did not recommend use of the devices at the higher temperature. With monitors 8 and 10, the chemistry used in the sampling solutions in the monitors may have been affected by the elevated temperatures.

Experiment 7 – precision and accuracy

Results from Experiments 2, 3 and 4 were used for the calculation of precision and accuracy data. Other exposures were made at lower concentrations to cover the 0.1 and 0.5 times the recommended exposure standard. Of the 10 monitors studied, only two monitors did not meet the protocol precision and accuracy criteria. Later the manufacturer of monitor 6 found the particular lot of badges used in this study was defective, thus explaining the poor precision and accuracy observed with this monitor in many of the other experiments. Monitor 11 used an absorbing solution not recommended by the manufacturer.

SUMMARY

From the 12 monitors which were studied, monitors 1 and 2 were dropped from the study due to either poor performance or unavailability of monitors from the manufacturer. Of the remaining 10 monitors, only monitors 4 and 5 met the protocol performance criteria for all of the experiments. For the remaining monitors, monitors 3, 10 and 12 met performance criteria for 5 experiments, monitors 8, 9 and 11 met performance criteria for 4 experiments, monitor 7 met performance criteria for 3 experiments and monitor 6 met performance criteria for 1 of the experiments.

Except for Experiment 1, every experiment demonstrated at least one performance problem with at least one of the monitor types studied. Experiment 5 was the most time consuming to perform since it involves 16 separate exposure sub–experiments. This experiment did yield the most information about each passive monitor tested. The additional power of this experiment lies in the fact that it can detect more than one significant factor or interaction of factors and can serve as an estimate of precision of the monitors. Experiment 2 provided a good estimation of monitor performance over a 12-h period. It detected sampling rate or capacity limitations on 5 of 12 devices. This experiment also allowed the sampling rate for ammonia to be calculated for one of the experimental devices.

Some of the problems observed with passive monitors were found only after devices which had been aged under storage conditions were analyzed and these results interpreted. If all of this testing had been done with fresh sampling devices, many of the problems, such as humidity sensitivity, blank variability and liquid sorbent volume variation, would have gone unnoticed. Based on the results of this study, this factor of diffusive monitor aging was addressed in the most recent revision of the protocol (Cassinelli et al., 1986).

Another problem experienced during the course of the protocol experiments was the proper method for the study of direct reading monitors in the storage stability experiment. The solution to this problem was to include specific instructions for the evaluation of direct reading monitors in the most recent revision of the protocol (Cassinelli et al., 1986).

The basic concept of passive monitoring is very attractive, but it has some limitations which often are not recognized. The series of experiments discussed in this paper and a previous paper (Cassinelli et al., 1986) have helped define the operational limitations of passive samplers for formaldehyde, sulfur dioxide and ammonia. The evaluation experiments have exhibited the flexibility to evaluate both diffusion and permeation controlled monitors, and provided an estimate of the experimental error for each monitor. These passive monitor evaluations (Kennedy and Hull, 1986; Cassinelli et al., 1985; Hull, 1986) have helped to refine the basic experimental evaluation protocol used in these studies to a point approaching general applicability to the field of passive monitoring.

REFERENCES

Cassinelli, M.E., Hull, R.D., Crable, J.V. and Teass, A.W. 1986. Protocol for the Evaluation of Passive Monitors. Presented at the International Symposium Workplace Air Monitoring, Luxembourg, Sept., 22–26.

Cassinelli, M.E., Hull, R.D. and Cuendet, P.A. 1985. Performance of Sulfur Dioxide Passive Monitors. Am. Ind. Hyg. Assoc. J., 46, 599–608.

Hull, R.D. 1986. Performance of Passive Samplers for Ammonia. In preparation.

Kennedy, E.R. and Hull, R.D. 1986. Evaluation of the DuPont Pro-Tek[R] Formaldehyde Badge and the 3M Formaldehyde Monitor. Am. Ind. Hyg. Assoc. J., 47, 94–105.

Perkins, J.B., Price, N.H., Eggenberger, L. and Burkart, J.A. 1981. An Evaluation of Passive Organic Vapor Monitors. NIOSH Contract No. 210–78–0115.

PROCEDURES FOR COMPARING THE PERFORMANCE OF PUMPED AND
DIFFUSIVE SAMPLERS

Dr P. Leinster, Mr D. Irvine and Mr M.J. Evans

Group Occupational Health Centre
BP International Limited

ABSTRACT

Methods for assessing the relative performance of
pumped and diffusive sampling techniques for the
determination of airborne contaminants are discussed.
Correlation has often been used to analyse such data; the
reasons why this is an incorrect approach are given. As
both methods measure with error, and there is no direct
means of determining the absolute airborne concentration,
standard linear regression techniques were considered to
be inappropriate, and more sophisticated analyses not
worthwhile.

The statistical approach adopted investigates the
differences between the results obtained using the two
sampling techniques. It provides a simple estimate of the
lack of agreement, or bias, and a measure of the
repeatability, both graphically and by the use of analysis
of variance.

The criteria for determining the equivalence of the
two techniques depend on the ultimate use of the data; if
the same decisions are taken from both sets of results
then the two methods are considered to be equivalent.

The results of field trials are used to illustrate
the approach adopted; they emphasise the requirement for
field trials in addition to laboratory evaluations when
determining whether a new sampling technique yields
comparable results to an established method.

INTRODUCTION

To date, there has been a mixed reception amongst occupational hygienists as to the acceptability and applicability of commercially available diffusive monitors. Some types have been withdrawn from the market, or modified, due to poor field performance, and the instructions provided with others have been altered by the manufacturers as field data have become available.

Before any diffusive monitors are used routinely laboratory and field trials should be conducted, preferably by the user or other independent workers. Such evaluations should provide information regarding the following parameters:

- Rate of uptake of contaminant(s) as a function of concentration and time
- Effect of external air flow
- Capacity of the sorbent for the compound being sampled
- Interferences by other contaminants
- Effect of changes in ambient conditions (temperature, humidity, pressure)
- Sample storage stability prior to analysis
- Precision and accuracy of the overall measurement

METHODS

Methods of comparing the performance of pumped and diffusive samplers have been investigated. The performance of Perkin Elmer diffusive monitors was compared with pumped samplers, both containing Tenax as the sorbent, in determining airborne benzene concentrations during motor spirit handling operations. In view of the mixture of organic vapours encountered and the nature of the operations, involving rapid fluctuations in airborne concentrations, this was considered to be a 'worst case' situation. If good agreement could be obtained between the two techniques in this application it was thought that they were likely to be comparable in many other 'workplace' atmospheres.

The Perkin Elmer diffusive monitors were calibrated as a function of time to determine the appropriate rate of contaminant uptake. A standard atmosphere, generated using a syringe injection technique (Health and Safety Executive), was passed through a large cylindrical vessel (600 mm x 50 mm OD) in which the diffusive monitors were placed. The concentrations of the generated atmospheres were checked by pumped sampler and by infra-red techniques.

The pumped samplers and the Perkin Elmer diffusive monitors were analysed by a thermal desorption/gas chromatography technique which has been described by Baxter et al (1980) for the determination of benzene using pumped samplers.

In the field trials reported the two techniques were compared by paired sampling (attaching a diffusive and a pumped sampler next to each other on the same lapel of the individual being monitored) and by grouped sampling (placing five diffusive and five pumped samplers side by side at fixed positions).

STATISTICAL APPROACHES

It is evident from a study of the literature that there is no consensus regarding a protocol for testing and assessing whether one sampling technique is equivalent to, better than, or worse than another.

Typically a scatter diagram of the results is plotted to determine whether there is any obvious divergence between the two techniques. For example, Figures 1A and 1B give paired benzene results for sampling periods of <60 minutes (typically 10-25 minutes) and >60 minutes (typically 200-500 minutes) respectively; the qualitative impression gained is that there is less agreement between the two monitoring techniques for the shorter monitoring periods. In the majority of cases a correlation coefficient is then calculated and tested for statistical significance.

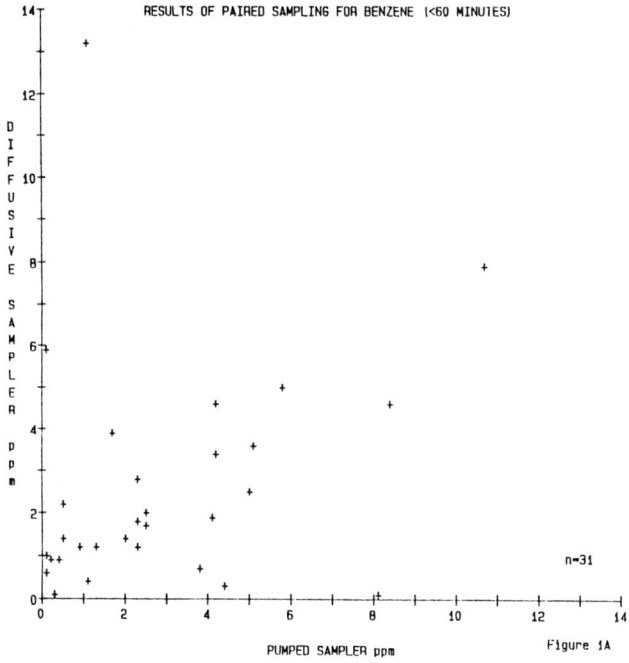

RESULTS OF PAIRED SAMPLING FOR BENZENE (<60 MINUTES)

n=31

PUMPED SAMPLER ppm

Figure 1A

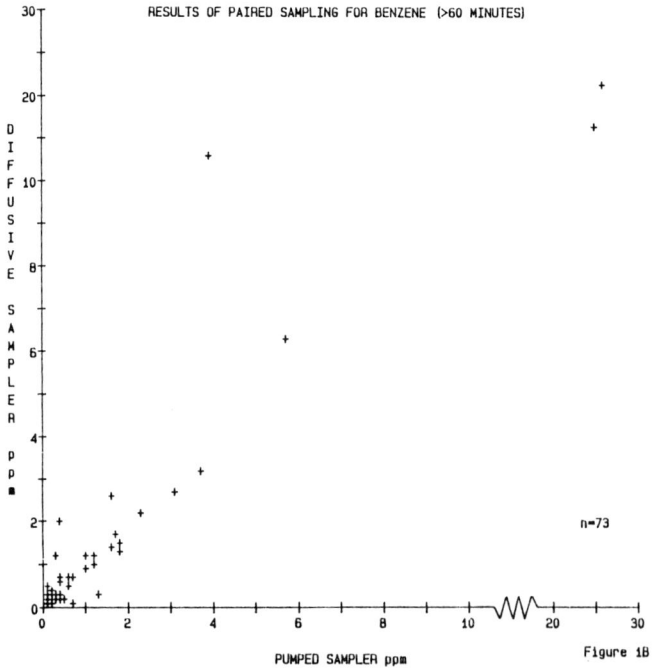

RESULTS OF PAIRED SAMPLING FOR BENZENE (>60 MINUTES)

n=73

PUMPED SAMPLER ppm

Figure 1B

However, as Altman and Bland (1986) point out, the use of the correlation coefficient (r) in method comparison is totally inappropriate. Their reasons for this conclusion are as follows:

o   r does not measure agreement, it measures the strength of the association.

o   Changes in the scale of measurement affect agreement but not r.

o   r increases as the range of measurement increases.

o   The statistical significance test referred to above is a test for non zero correlation; one would anticipate the methods are correlated since they are theoretically measuring the same quantity.

o   One can produce examples with high correlation and poor agreement.

Despite these pitfalls the next step in the majority of published work is linear regression.

In simple regression analysis it is assumed that the independent variable is measured without error. This is not the case in the comparison of two sampling techniques as there is measurement error associated with both and there is no absolute means of determining the airborne concentration.

Functional analysis, which attempts to define the true underlying relationship between the two methods and takes account of the errors in both techniques, can be applied. However the approach is complicated and is more specific to the problem of calibration rather than to method comparison. For practical purposes what is needed is a method that will estimate the magnitude of the difference between the two techniques. Judgement about the acceptability of the difference can then be made by incorporating practical/operating considerations. In this context agreement and precision are considered to be the important qualities to examine in the assessment of the comparability of two techniques.

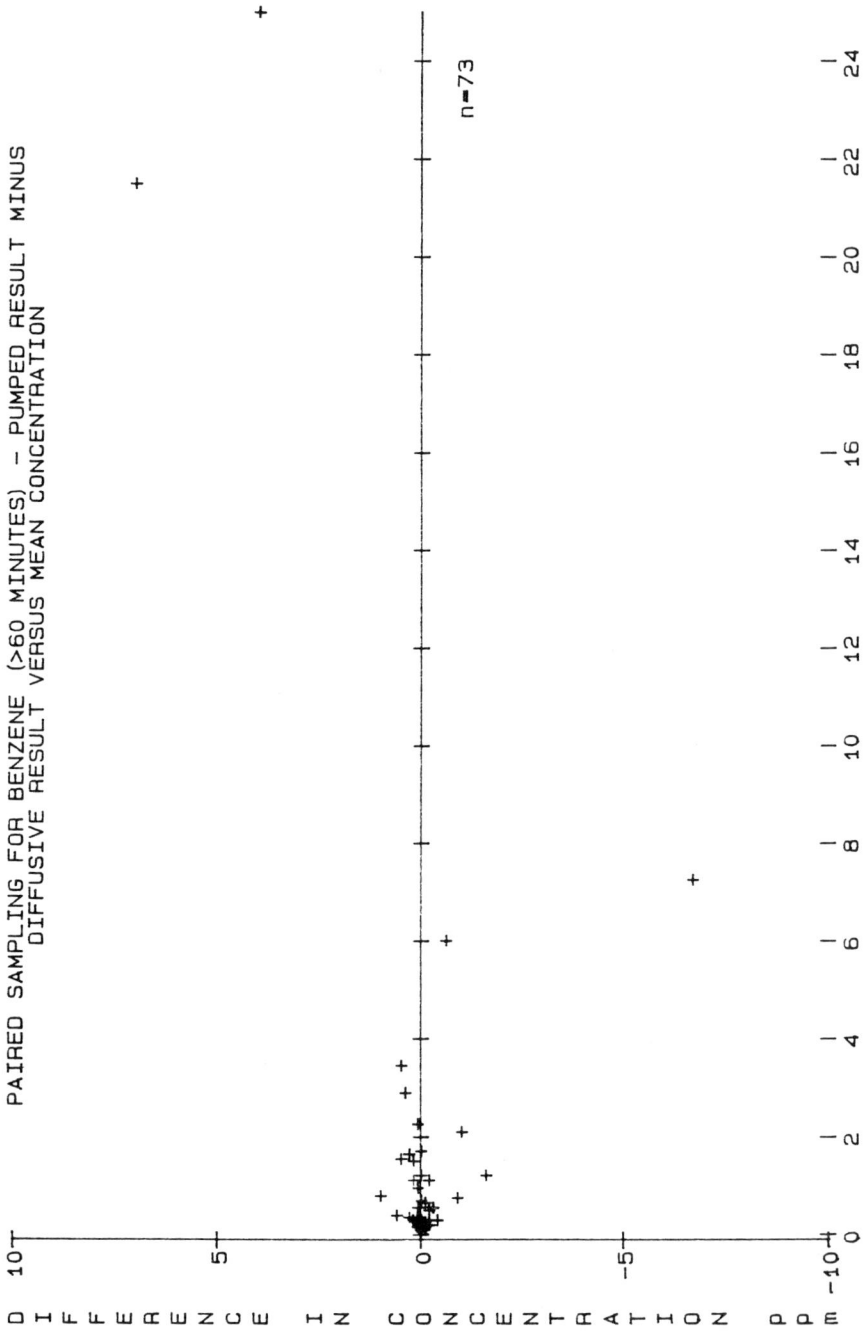

Figure 2

AGREEMENT

A simple estimate of the agreement can be obtained by calculating the mean and standard deviation of the differences (Altman and Bland 1986, Armitage 1971). If the differences demonstrate no dependence on concentration then the mean difference is an estimate of the bias between the techniques, the 95% confidence limits on the bias being referred to as the limits of agreement. If the differences do indicate a dependence on concentration, which can be formally tested by calculating a correlation coefficient, a log transformation of the data may be required.

Figure 2 illustrates the differences between the pumped and diffusive results plotted against mean concentration. The correlation coefficient r = .54 is significant at the 1% level indicating a dependence upon concentration. A logarithmic transformation of the data, Figure 3, removes the dependence on concentration (r = .05, nonsignificant). The mean of the differences (log) is - .06 with 95% confidence interval (- 1.242, 1.122). By transforming these logarithms back to the original scale we obtain an estimate of the bias as .94 with associated limits of agreement (.29, 3.07). This indicates that at the extreme the pump technique could be registering as little as one third of the diffusive value or three times as much.

PRECISION

An estimate of the precision is provided by the residual mean square derived from a one way analysis of variance for each technique.

An initial plot of the standard deviation of each group of 5 samples was carried out against the mean concentration to examine whether there was any dependence upon concentration. The estimate of the standard deviation at the lower concentrations was different to that at the higher levels. A logarithmic transformation

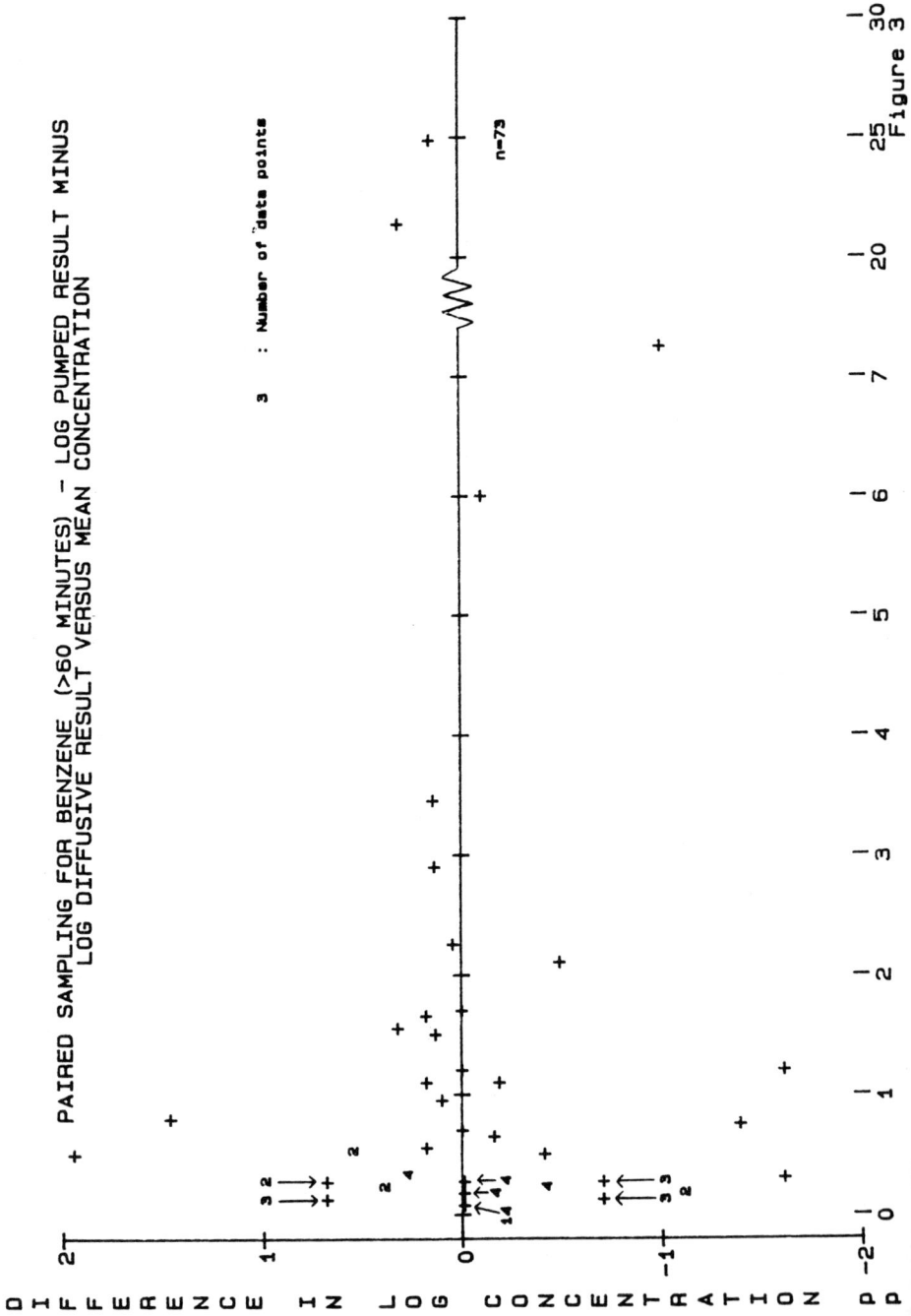

Figure 3

of the data removed this dependence and an analysis of variance was performed. Separate estimates of variance provided by the residual mean square of (.038 (log), GSD 1.22) for the pumped technique and (.100 (log), GSD 1.37) for the diffusive indicate that the diffusive monitor is in general more variable than the pumped, (the F statistic, the ratio of the larger to the smaller variance, was 2.56, significant at the 1% level). However sole reliance upon a single summary statistic from these data must be viewed with caution. Examination of the individual data showed that the estimate of residual mean square was highly influenced by one aberrant set of data for both pumped and diffusive monitors.

A further way of comparing the grouped data would be to conduct a two way analysis of variance which provides an estimate of the variance of the difference between the two techniques. However the hypothesis of a composite estimate of variance which applies to both pumped and diffusive monitors seems untenable in the light of the significant difference in the estimates obtained in the one way analysis. For this reason we have adopted an alternative approach, plotting the ratio of the respective coefficients of variation against concentration (Figure 4) which gives a qualitative impression of the relative precision in different concentration ranges. By assuming that the two techniques are proportionally related this approach is a special case of the sensitivity analysis proposed by Mandel (1964).

For low concentrations (< 2 ppm) the pump results appear more precise (ratio < 1.0). Thereafter the concentration is not adequately represented, with only an indication, maybe surprisingly, that the position is reversed and the diffusive precision is at least as good as the pumped. This illustrates an important problem with field trials; it is difficult to obtain an even spread of data over the concentration range.

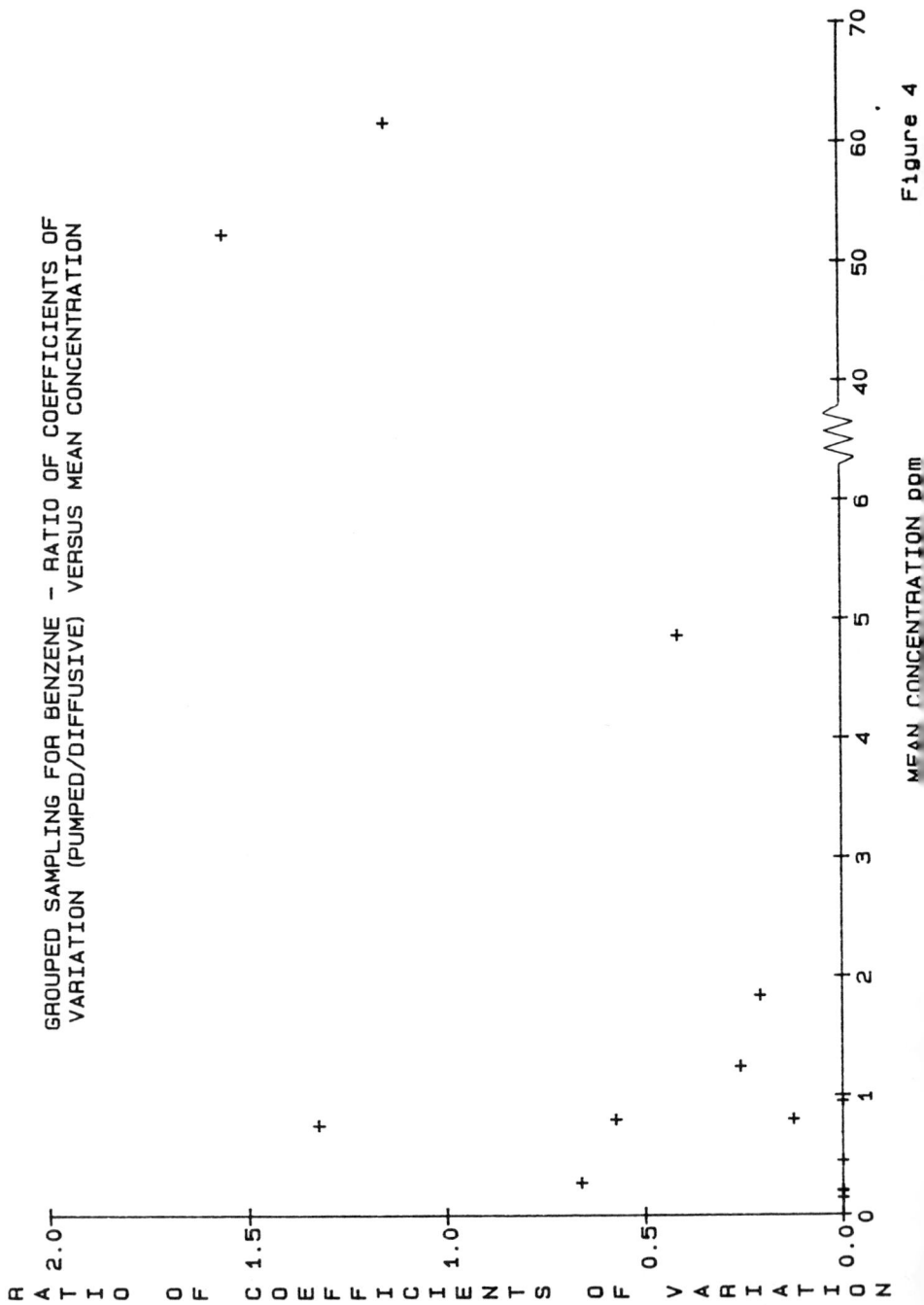

GROUPED SAMPLING FOR BENZENE — RATIO OF COEFFICIENTS OF VARIATION (PUMPED/DIFFUSIVE) VERSUS MEAN CONCENTRATION

Figure 4

DISCUSSION

Outliers

In assessing the acceptability of an alternative
technique or, for that matter, examining the performance
of an existing technique, the frequency of occurrence and
nature of 'outlier' observations is of great importance.
In the analysis presented outliers have only been excluded
if there is an obvious explanation for the discrepancy,
e.g. pump failure, proof of contamination. The inclusion
of remaining aberrant values may still affect summary
statistics when simple transformations of the data do not
remove their influence. However the recognition that
outliers can occur is important in the comparison of the
techniques.

Bias

There is frequently a bias between the results
obtained using the two techniques. If there is a large
discrepancy the laboratory calibration should be checked;
if there continues to be a difference between the
laboratory and field data a scaling factor, based on the
bias, can be applied.

The limits of agreement (the 95% confidence limits on
the bias) provide an indication of the likely
discrepancies that will occur. Recognising that an
increase in sample size will reduce the limits, the
acceptability of the limits obtained must be judged on the
basis of the operational decisions to be taken.

Precision

There is an indication, from the data for both paired
and grouped samples, that the diffusive monitors were less
precise at low concentrations (< 2 ppm). This could be
due partly to the smaller amount of material collected on
the diffusive monitors resulting in an increased
analytical error. This has important implications
concerning the suitability of a monitoring technique; for
example, the precision required if the hygiene standard is

100 ppm and all the results are less than 1.0 ppm, is different to that if the limit is 2.0 ppm and the results are in the range 1.0-2.0 ppm.

Operational decisions

Another more qualitative approach in comparing the two monitoring methods is to consider whether the same occupational hygiene decisions would arise from the data obtained using either of the two techniques. A possible sampling strategy is to assign exposure categories according to the relationship between the 90th percentile value of the distribution (assuming it to approximate to log normal) of at least ten results, obtained for a particular job function, and the appropriate hygiene standard. The results from the two techniques, when sampling for benzene, presented in Table 1, are in good agreement and the interpretation of the data from a health protection viewpoint is identical.

CRITERIA FOR ACCEPTANCE OF A DIFFUSIVE MONITORING TECHNIQUE

It is not possible to set precise criteria in the general case as factors such as the reasons for data collection and the magnitude of exposure in relation to the hygiene standard must be considered. The most important factor in determining the acceptability of a diffusive monitor is that the precision should be as good as the pumped sampler. If a bias is indicated a correction can be applied although the reason for the discrepancy between the laboratory and field comparisons should be investigated.

The relevant question when assessing the acceptability of diffusive monitors is not whether they are generally satisfactory for sampling organic vapours, but whether a given type of diffusive monitor is an appropriate sampling technique for a particular situation. Well designed field trials and appropriate statistical analysis of the results are vital in demonstrating this.

TABLE 1 Comparison of data obtained using pumped and diffusive samplers.

Number of Samples	Sampling Duration (minutes)	Diffusive Sampler			Pumped Sampler		
		Geometric Mean (ppm)	Geometric Standard Deviation (ppm)	90th Percentile Value (ppm)	Geometric Mean (ppm)	Geometric Standard Deviation (ppm)	90th Percentile Value (ppm)
73	>60	0.8	4.8	5.8	0.7	5.6	6.3
31	<60	1.5	3.3	6.9	1.4	3.8	7.8

REFERENCES

Altman, D.G. and Bland, J.M. (1986). Statistical Methods
    for Assessing Agreement Between Two Methods of Clinical
    Measurement. Lancet No. 8476, 307-310.
Armitage, P. (1971). Statistical Methods in Medical
    Research. Blackwell Scientific Pulications.
Baxter, H.G., Blakemore R., Moore, J.P., Coker, D.T. and
    McCambley, W.H. (1980). Ann. Occup. Hyg. 23, 117-132.
Health and Safety Executive. Generation of Test Atmospheres
    of Organic Vapours by the Permeation Tube Method.
    Methods for the Determination of Hazardous Substances
    Number 4.
Mandel, J. (1964). The Statistical Analysis of Experimental
    Data. Interscience Publishers.

DISCUSSION - SESSION IV

NORBACK (Sweden)

It's not a question but a comment on short-time sampling. In Sweden we recommended short-time values of 15 minute periods for almost every component besides the ordinary 8 hour TLV. My experience from field evaluations of badge type diffusive samplers using active carbon is that minute sampling is as good as 8 hours.

KRISTENSSON (Sweden)

Dr. Kennedy, have you tried the NIOSH protocol on some common solvents, for example: toluene or styrene?

KENNEDY (USA)

We did not do that work because some of that work was originally done under contract by a separate laboratory in the United States.

CRABLE (USA)

There is a contract report which you can get out of NTIS (The National Technical Information Service).

SAMINI (USA)

The NIOSH Protocol examines the effect of air velocity, both perpendicular and parallel to the direction of the dosimeters. Have you found any significant difference between the two?

KENNEDY (USA)

We did not find orientation to be a significant factor but I think that most of the badges which we tested had a wind screen on them. This might not be the case if the badge did not have a wind screen.

BROWN (UK)

We do have a section in the HSE Protocol which is concerned with determining the effects of air-velocity on sampling uptake rate as Dr. Firth described. We don't make any specific comment about the orientation of the sampler and this is primarily because in our experience up to now we have never seen any effect of orientation. But obviously if one felt like doing thism it would be perfectly feasible.

CHOO YIN (UK)

What are the costs involved in setting up the validation procedures?

FIRTH (UK) (representing HFC)

We have not lookled at the costs of setting up a validation protocol for diffusive samplers because it's going to be absorbed into the running of the laboratories such as ours without any difficulty. I would point out that you should also be validating sampling procedures which you are going to use. One of the main cost factors are in field trials. We attach great importance to the field work particularly the use of samplers on individuals but I expect this will be something that you will be able to carry out within your normal procedures for monitoring workers.

CRABLE (USA)

We have not made an estimate of the cost, but it's not insignificant. On the other hand, anybody who goes into the field to do an evaluation without having properly evaluated equipment is wasting a great deal of money.

BERLIN (CEC)

Should the field trial results be part of the information that the manufacturer provides with the equipment?

FIRTH (UK)

The field trials should be carried out by a regulatory body or by the manufacturer. But if the user should choose to do it then we wouldn't want to stop him from doing so.

There are methods published in the MDHS series on the generation of standard atmospheres. These are normally in the ppm range since it is the concentration range that most exposure limits are in.

LEHMANN (FRG)

Can our current knowledge concerning the use of the different diffusive samplers be summarized in a way which would allow us to make recommendations for the use of the diffusive sampler in practice?

LEINSTER (UK)

We are not talking about a generic sampling technique. We are
talking about individual types of monitors. You cannot come
up with a recommendation which says diffusive monitoring is
the best thing and that everybody should use it. We have to
have specific data on a specific monitor for a specific
application before we can come up with any recommendations.

CRABLE (USA)

I can't really respond to the situation in the FRG but in the
USA, there is a general misconception that any method we have
developed is automatically accepted by NIOSH and therefore
becomes the legal procedure for measuring in the workplace.
This is legally not the case. The Standards are just a
numerical value and however you arrive at that value must have
back-up data to prove its precision and accuracy. There is no
requirement that you use an accepted OCEA procedure or an
accepted NIOSH procedure. I can foresee the day when passive
monitors which have been thoroughly evaluated will be used in
regulatory compliance evaluations.

BROWN (UK)

I personally feel that it is entirely proper for us to
consider diffusive sampling in general. Pump samplers are
accepted as a general recipe in spite of the fact that they
have limitations in terms of saturation capacity and so on.
It's equally undeniable that diffusive samplers have their
limitations. It's obvious that some specific designs have
faults and it is equally obvious that in many cases it can be
demonstrated quite conclusively that the results given by
diffusive samplers are just as good as the results given by
methods which are accepted. We have to decide whether or not
in general diffusive sampling is an acceptable way of doing
things. With regard to the question of national standards, I
think that all we can do at this stage is to publish and
disseminate the information we have, as best we can, through
the open literature and through methods such as the MDHS
series which carry an element of official approval about them.
It's been mentioned many times that the cost of evaluations is
very high and I think that when it comes down to it the cost
of the evaluation of a pump sampler will be almost as high if
it were done to the same degree as you would a diffusive
sampler. It's of the utmost importance therefore that we
should share the information that we have and not repeat what
has already been done by someone else at vast expense.

CRABLE (USA)

I was around at the beginning of the charcoal tube, and you
may be surprised to know that it wasn't readily accepted. As
a matter of fact, we could not get anyone to manufacture them

for us because it was said that there was no demand. So, for the first few years we made our own and we had to create a market by ordering them. When we finally convinced some people to make them for us, they were faulty and eventually we had to set up a quality control scheme for these people. So I think that the fact that there are some problems early on with some of these devices doesn't upset me. I think that the future is very bright.

GUILD (USA)

Both the HSE and the NIOSH protocols talked about interfering compounds but no-one has mentioned actual data showing interfering compounds.

BROWN (UK)

In our experience, particularly with the tube-type sampler, we have never in practice observed any interferences, and hence our lack of comment. this is not to say that interferences cannot occur; it is a well known phenomenon with active carbon that you can get displacement of one sorbed material by another.

KETTRUPP (FRG)

Are there any particular difficulties associated with generating standard atmospheres?

BROWN (UK)

One of the big problems that we have encountered in the evaluation of diffusive samplers is that it is actually very difficult to generate standard atmospheres. The accuracy with which you can generate a standard atmosphere is of the same order of magnitude, actually almost exactly the same numerically as the accuracy of the diffusive sampling methods themselves. in my experience it is difficult to generate a standard atmosphere of better than plus or minus 5% coefficient of variation and certainly if you go down to parts per billion, then the accuracy is going to be down to plus or minus 30% or worse. Most of the diffusive error may in fact be coming from our original generation of the standard atmosphere.

CRABLE (USA)

I'd like to comment. Obviously as it's been stated now and I'll repeat, the generating of a standard atmosphere is not very simple. We do recommend two estimations for the independent determination. In the case of formaldehyde we had three independent sampling techniques plus a continuous

monitor.   You   can also calculate from the air-flow   and   the
analyte input.   So there are ways of homing in on   what   the
standard atmosphere is.

FIELD COMPARISON OF CHARCOAL TUBES AND PASSIVE DOSIMETERS

FOR THE DETERMINATION OF SOLVENTS IN AIR

G.B. Bartolucci*, L. Perbellini**, G.P. Gori*, F. Brugnone**, E. De Rosa*

*Istituto di Medicina del Lavoro, Universita' di Padova, Italy.
**Istituto di Medicina del Lavoro, Universita' di Verona, Italy.

ABSTRACT
    Results from contemporaneous samples of solvents in air performed with personal sampling pumps connected to charcoal tubes and passive dosimeters were compared. Excellent correlations were found between data obtained with the two sampling systems for styrene, toluene and n-hexane. Less good but statistically significant correlations were obtained for other 6 solvents found in mixture with n-hexane. The data showed that passive dosimeters are reliable and represent a valid alternative to traditional active samplers.

INTRODUCTION

    The use of passive dosimeters for determining the solvent concentrations in air has increased considerably in the last years. Comparative studies on the reliability of passive dosimeters with respect to traditional systems with sampling pumps and activated charcoal tubes, both in the laboratory and in the field, have generally given good results (Rose and Perkins, 1982).

    The aim of the present study is to compare field data of solvent concentrations obtained with active and passive samplers, in the latter case using a sampler constructed in Italy.

MATERIALS AND METHODS

    Our study consisted in the determination of the concentrations of various solvents, used both alone and in mixture in different industrial sectors.

    Fifty-two styrene concentration measurements were carried out in two plants where polyester resins were used, 44 toluene concentration measurements in a printing industry, and 56 measurements of 7 solvents in mixture (n-hexane, 2-methylpentane, 3-methylpentane, methylcyclopentane, cyclohexane, acetone and ethyl acetate) in a shoe manufacturing plant. Jumbo-type activated charcoal tubes connected to personal Dupont P 4000 samplers (flow rate 1 l/min) and TK-200 passive dosimeters from Zambelli (Milan, Italy) (whose tecnical characteristics were described by Pozzoli et al., 1981) were contemporaneously used (each sampling lasting from 2 to 3 1/2 h). The two adsorbing systems were placed in the breathing area of the exposed workers, at a distance of 10-20 cm from each other.

Analyses of the charcoal tubes and passive dosimeters were carried out by gas chromatography after desorption with CS_2 according to the NIOSH method. Desorption efficiency for all solvents was higher than 90%.

RESULTS

The mean concentrations of the various solvents measured with the two sampling systems were quite similar and far lower than the respective TLVs-ACGIH. As regards single samplings, the TLV for styrene was exceeded in 2 cases, that for n-hexane in 5 cases, and that for cyclohexane in one (both with active and passive samplers). The solvents were detected over a sufficiently wide range of concentrations to permit a linear regression analysis, whose results are shown in table 1.

Table 1: Relationships [slope, intercept and 95% confidence limits (CL), correlation coefficient (r)] between values of solvent concentrations obtained using charcoal tubes (x) and passive dosimeters (y).

Solvents	pairs of samples	range of data(mg/m^3)	slope	+ 95%CL	intcp	+ 95%CL	r
Styrene	52	6- 278	1.01	0.09	1.03	8.84	0.96
Toluene	44	21- 297	0.96	0.05	4.00	8.25	0.99
N-Hexane	56	7- 352	0.97	0.05	0.39	4.58	0.98
2-Methylpentane	56	2- 126	0.83	0.10	6.41	4.66	0.92
3-Methylpentane	56	2- 126	0.91	0.08	-1.05	3.28	0.95
Methylcyclopentane	56	1- 107	0.84	0.07	-0.45	2.42	0.96
Cyclohexane	56	24-1288	0.95	0.10	26.14	38.97	0.93
Acetone	56	2- 123	0.89	0.12	-0.85	2.58	0.90
Ethyl Acetate	56	10- 413	0.80	0.13	24.60	20.60	0.86

Regarding the three more important solvents, i.e., styrene, toluene and n-hexane, we found excellent correlations between the data obtained with the two sampling systems. For the other 6 solvents (toxicologically less interesting) found together with n-hexane, the data were in agreement for cyclohexane, while the correlations were less good particularly for the solvents occurring in low concentrations. The more discordant data were those for ethyl acetate, probably because this solvent is more polar than others (Hagberg and Sällsten, 1984). An additional error that might justify some discrepancies between pairs of values is the positioning of the two sampling systems with respect to the polluting source, as reported recently (Van Der Wal and Moerkerken, 1984).

CONCLUSIONS

The values obtained with passive dosimeters turned out to be substancially similar, in most cases, to those measured using active samplers. These results would seem to be particularly convincing, in view to the fact that they were obtained in field investigations. It must be noted here that also using the traditional system with pump and charcoal tube it is possible to make an evaluation error of the order of 5-10% (above all due to problems connected with calibration and constant flow rate). In conclusion, our results show that passive dosimeters are reliable and represent a valid alternative to the active samplers for the determination of solvents in air.

REFERENCES
Hagberg S, Sällsten G: Field comparison of charcoal tubes and passive vapor monitors for organic vapors. Abstracts of International Conference on Organic Solvent Toxicity, Stockholm, 15-17 October 1984, p. 52.
Pozzoli L, Cottica D, Ghittori S: A new passive organic vapor device. Proceedings of American Industrial Hygiene Conference, Portland, 25-29 May 1981, pp. 48-49.
Rose VE, Perkins JL: Passive dosimetry - state of art review. Am Ind Hyg Assoc J, 43, 605-621, 1982.
Van Der Wal JF, Moerkerken A: The performance of passive diffusion monitors for organic vapours for personal sampling of painters. Ann Occup Hyg, 28, 39-47, 1984.

DEVELOPMENT AND EVALUATION
OF A FORMALDEHYDE VAPOR GENERATION SYSTEM

L.M. Blade - NIOSH, Cincinnati, USA
A. Gold - University of North Carolina, Chapel Hill, USA
R. Hornung - NIOSH, Cincinnati, USA

ABSTRACT

Formaldehyde, a colorless, strongly odiferous gas, is regarded by NIOSH as a human carcinogen based on the results of animal bioassays (indicating the induction of nasal carcinomas in laboratory rats and mice) plus the consistency of evidence from epidemiologic data. These findings, along with the commercial importance of formaldehyde, have generated much interest in various methods for the measurement of airborne concentrations of this gas. One of the most widely used of these methods involves the collection of formaldehyde with a midget impinger containing an aqueous solution, and subsequent analysis using the chromotropic acid procedure (NIOSH 3500); however, this method has limitations on the sampling parameters which may be used. In order to increase the flexibility and usefulness of this method, a laboratory evaluation of the collection efficiency of the midget impinger for formaldehyde under a wide range of sampling conditions (flow rate, time, total air volume) was undertaken. This required development and evaluation of an apparatus to generate known concentrations of formaldehyde in air. A generation system design based upon the direct injection of formalin, a commercial preparation of formaldehyde in aqueous solution, into the carrier gas stream was adopted. Injection from a gas-tight syringe through a septum port was controlled by a syringe pump, and thorough mixing occurred in a glass condenser. Evaluation of the system was accomplished by sampling the output concentration, and determining the percent recovery (defined as the ratio of the sampled concentration to the concentration calculated based on the generation system parameters). When injection was performed at room temperature, low recovery was found, along with a buildup of solid, white residue at the injection point. This problem was eliminated by incorporating into the system a heated injection block, operated at 170 C. Generator efficiencies at concentrations ranging from approximately 0.4 to 6.4 ppm averaged over 87%. At 0.1 ppm, these averaged only 80%, and were significantly different from those at the higher levels. Therefore, the system was judged to be suitable for reliably generating concentrations ranging from 0.4 to 6.4 ppm Further work is recommended to determine the source of the apparent formaldehyde loss in the 0.1 ppm range.

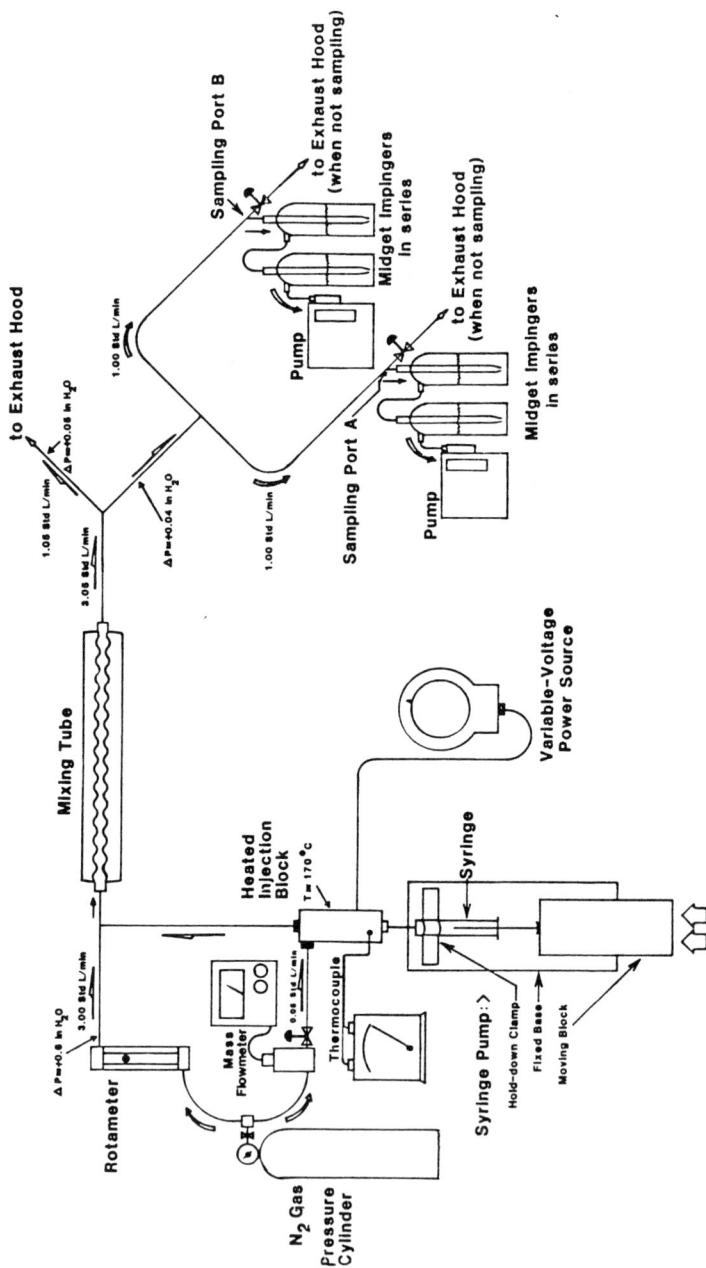

FIGURE.   SCHEMATIC DIAGRAM OF DYNAMIC
FORMALDEHYDE VAPOR – GENERATION SYSTEM
AND SAMPLING SYSTEM WITH
TYPICAL OPERATING PARAMETERS

SUMMARY OF EVALUATION

STATISTICAL ANALYSIS

NUMBER OF VALUES	MEAN	% GENERATOR EFFICIENCY SAMPLE STANDARD DEVIATION	
ALL DATA	24	85%	4.8%
ALL DATA EXCLUDING			
0.1-PPM LEVEL	18	87.3%	3.4%
0.1-PPM LEVEL	6	80%	3.5%
0.4-PPM LEVEL	6	87.5%	2.6%
1.6-PPM LEVEL	6	85.9%	3.3%
6.4-PPM LEVEL	6	88.6%	4.0%

* "% Generator Efficiency" actually represents the combined generator and collection efficiency. The collection efficiency at 1 Lpm for 1 hr is approximately 95 to 98% (well characterized in the literature). Remaining losses due to generator efficiency rate.

ANALYSIS OF VARIANCE (ANOVA): NESTED ANOVA WITH MULTIPLE COMPARISONS
          (SAS GENERAL LINEAR MODELS PROCEDURE)

1. Concentration level was determined to be a source of significant variation (P=0.013) from a constant generator efficiency rate.

2. This effect was determined to be due to a low average generator efficiency at the 0.1-ppm level.

INTERPRETATION

Generator efficiency is significantly reduced at the 0.1-ppm level.

CONCLUSIONS

1. Precise, reproducible concentrations of formaldehyde vapor, equalling 87.3% of the calculated concentration, can be generated using this generation system.

2.   An   exception   to the above statement must be   made   when
levels around 0.1 ppm are desired, as low generator efficiency
occurs in this range.

3.   Losses   of formaldehyde vapor in the system, due to   wall
effects   becoming   quantitatively   important   at   the   lowest
levels, must be suspected.

RECOMMENDATIONS

1.   The generation system may be used to produce formaldehyde
vapor concentrations of 0.4 to 6.4 ppm.

2.   Further   work   is needed to determine the source   of   the
lowered generator efficiency in the 0.1-ppm range.

TRACE GAS TESTING IN PIGGERIES
COMPARATIVE TESTS WITH DRAGER TUBES
FOR SHORT-TERM MONITORING AND
DRAGER DIRECT-READING DIFFUSION TUBES

Dr. F-W Busse and U.W. Schwarz

Weser-Ems Chamber of Agriculture, Animal Health Department, Osnabruck Branch, Neuer Graben 19/21, D - 4500 Osnabruck Dr U.W. Schwarz, Dragerwerk AG, Sales Department for Gas Measuring Technology, Moislinger Allee 53/55, D-2400 Lubeck

SUMMARY

The best possible air quality is needed in rearing sheds if the health of the animals is to be protected and their performance maintained. The quality of the air depends on the concentration of the trace gases ammonia, carbon dioxide and hydrogen sulphide, as well as on temperature and water vapour content. Short-term monitoring with Drager tubes is suitable for establishing instantaneous values. Long-term monitoring with direct-reading diffusion tubes provides new scope for determining trace gases in the air on a long-term basis.

INTRODUCTION

In animal rearing establishments, physical, chemical and biological factors may reduce the quality of the ambient air to such an extent that the animals may become diseased or their performance reduced (1-10). A number of practical tests have been carried out on this (6,8,9 and 10). Guidelines have also been laid down, such as the DIN standard 18910 "Climate in enclosed animal housing" (11), TRgA 900 (12) or the BMELF report on humane conditions for keeping domestic animals (13), as well as optimum values for the performance and health of pigs (von Mickwitz and Kalich - 14). Since there can be considerable fluctuations in ambient air quality both over the day and over the year, comparative short-term and long-term monitoring of certain trace gases was carried out in the present study (15). The relation was established between the values determined and the health of the animals in the individual production areas.

EQUIPMENT AND METHODS

Parallel tests were conducted in four establishments over a period of 8 weeks during the winter, both in the weaned piglet shed and in the gilt rearing shed. Short-term monitoring was carried out to determine the trace gases ammonia, carbon dioxide and hydrogen sulphide using the Drager

gas-detection pump and Drager tubes ammonia 5/a, carbon
dioxide 0.1%/a and hydrogen sulphide 0.5/a. Continuous
monitoring was carried out over a period of 48 hours using
the Drager direct-reading diffusion tubes ammonia 10/a-D,
carbon dioxide 1%/a-D and hydrogen sulphide 10/a-D.
Short-term monitoring involved determining the gas levels
after the morning feed on two days a week in the inhalation
area of the animals in both types of shed. Long-term monitoring was continuous. In addition to the analysis of the
gases, the water vapour content and the temperature inside
the shed were determined.

RESULTS AND DISCUSSION

There were considerable differences between the establishments in some of the ammonia and carbon dioxide levels
determined in the individual sheds. In individual cases the
levels greatly exceeded the limit values given in the literature (2,6,8 and 9). Continuous monitoring also revealed that
limit values were exceeded at times. The results of the
short-term monitoring tallied with the values determined by
Busse (10) at various visits to different sheds. There was a
close correlation between the air quality in the sheds and
the health and/or performance of the animals. Respiratory
diseases, cannibalism and reduced performance were observed
among the animals when the optimum values of 10 ppm for
ammonia and 0.15% by volume for carbon dioxide were exceeded.
The trace gas hydrogen sulphide was present in relatively
high concentrations only during certain operations (mucking
out etc.).

BIBLIOGRAPHY

1. PFEIFFER, H.: Einfluss der Umgebungstemperatur auf die
Leistungsfahigkeit von Mastschweinen, Mo. Heft Vet. Med.
Bd. 22 (1967) 980-984

2. MOTHES, E.: Stallklima, Umweltanforderungen,
Warme-Wasserdampf - und Gasabgabe der Tiere, Deutsche Bauin-
formation Berlin 1969

3. BLENDL, H.M., M. Suss, G. Koller und J. Hartl: Ein
Beitrag zum Problem Kannibalismus beim Schwein. Zuchtungskunde Bd. 43 (1971) Heft 4, 268-282

4. GOBULEW, G.W.: Einfluss des Mkiroklimas auf die Leistungsfahigkeit der Schweine. Internat. Z. Landwisrtschaft
(1971) 61-65

5. JERICHO, K.W.F. und T.L. Church: Cannibalism in Pigs.
Can. Vet. J. 13 (1972) 156-159

6. ADAM, T. and U. Andreae: Toleranzgrenzen fur gasfor-
mige Umweltfaktoren bei landwirtschaftlichen Nutztieren.
Zuchtungskunde Bd. 3 (1973) 45.Jg. 162-177

7.   PASCHERTZ, H. Gesundheitliche Probleme in Mastbetrieb.
Mitt. d.DLG 18 (1975) 1009-1011

8. MATTHES, S.: Art und Zusammensetzung der Luftverunreini-
gungen in der Nutztierhaltung und ihre Wirkung in der
Stallumgebung. Dtsch. tierarztl. Wschr. 86 (1979) 154-155

9. DONHAM, K.J. and Popendorf, W.J.: Ambient Levels of
Selected   Gases   Inside   Swine   Confinement   Buildings.
Am.Ind.Hyg.Assoc.J. 46 (11) 1985, 658-661

10.   BUSSE,F.-W.: International Pig Veterinary Society 9th
Congress, Barcelona (1986), Proceeding 345

11. DIN 18910: Klima in geschlossenen Stallen, Beuth-Verlag
(1974), Berlin 30

12.   TRgA 900: MAK-Werte, Maximale Arbeitsplatzkonzentrationen und Biiologische Arbeitsstofftoleranzwerte, Bundesarbeitsblatt 10 (1984)

13.   GUTACHTEN DES BMELF vom 19.4.1974: Zur tierschutzge-
rechten Haltung der Nutztiere

14.   V.   MICKWITZ und J. Kalich zitiert in: Klinik der
Schweinekrankheiten, Schoper-Verlag, Hannover (1980)

15.   WEMERY, U. und H.G. Hilliger: Zum Umfang und zur
Veranderlichkeit der in einem einstreulosen Mastschweinestall
gebildeten Wasserdampf- und Kohlendioxidmengen, Zuchtungskunde Bd. 46 (1974) 293-301

EUROPEAN INTERLABORATORY COMPARISON OF PASSIVE SAMPLERS

FOR ORGANIC VAPOUR MONITORING IN INDOOR AIR

Maurizio DE BORTOLI
Commission of the European Communities,
Joint Research Centre, Ispra (Italy)

Lars MØLHAVE
University of Aarhus, Hygiene Institute,
Aarhus (Denmark)

Detlef ULLRICH
Federal Health Office, Institute for Water,
Soil and Air Hygiene
Berlin (Federal Republic of Germany)

ABSTRACT

An interlaboratory experiment was made with the aim of testing passive
samplers for volatile organic compounds in indoor air and, at the same
time, evaluating the agreement among different laboratories in such determinations. 3M OVM-3500 passive samplers were exposed simultaneously in a
large test chamber to the vapours of 1-butanol, pentanal, 1.1.2-trichloro-
ethane, 1-octene, butylacetate, 3-heptanone, 1.2-xylene, a-pinene and n-
decane. Lots of three samplers were mailed to the participating laboratories, three of which exposed also their active samplers for thermal desorption. Comparing the results obtained with passive and active samplers,
pentanal is largely underestimated with passive samplers, probably because
of irreversible adsorption. Also for the other polar compounds tested passive samplers give significantly lower results (up to 50%), whereas the
agreement is excellent for the apolar compounds. Individual passive samplers exposed under the same conditions seem to give different results.
The agreement among six laboratories may be quantified with a relative
standard deviation ranging between 9 and 20% for the different compounds
excluding pentanal.

INTRODUCTION

   In the framework of a European co-operation in the field of Indoor

Air Quality sponsored by the Commission of the EC and its Joint Research

Centre (JRC) an important task is that of assessing or developing approp-

riate investigation techniques, which can be adopted by laboratories in different countries to build up a common data base on indoor air pollution in the Community. This paper describes an interlaboratory experiment which had the objective of evaluating a commercial passive sampler and at the same time estimating the agreement among several european laboratories in the determination of different organic compounds at the relatively low concentrations found in indoor air.

EXPERIMENTAL SECTION

Many passive samplers model OVM-3500 (3M-Company) were exposed in a large test chamber at the Hygiene Institute of the University in Aarhus (Denmark) during four days. The atmosphere contained 9 volatile organic compounds (see Table 1) with concentrations reaching a total of 5 mg/m^3. Eight laboratories were involved in the experiment (those of the authors and those listed in the Acknowledgement section): only six of these, however, analysed the OVM 3500 samplers (three each). Some of the laboratories mailed to Aarhus their own active samplers (Tenax or Porapak, thermal desorption) for parallel sampling. For the extraction of the compounds from the samplers the procedure fixed by the manufacturer was adopted with one exception: one of the laboratories used dimethylformamide instead of carbon disulfide as solvent. The analyses were carried out by gas-chromatography and triplicate determinations on each sampler extract were asked of the participants, but only two complied with this requirement. The statistical treatment of the data was carried out using a program package specially developed for this type of statistical analysis (Statistical Analysis System, S.A.S. Institute Inc., Cary, N.C., USA). For each of the nine compounds one-way and two-ways analyses of variance were made in order to estimate the laboratory contribution and the sampler contribution to the total variance.

RESULTS AND COMMENTS

The experiment described was conceived mainly to clarify a doubt which had been raised in a previous experiment (De Bortoli et al. 1984), i.e. whether nominally identical samplers exposed under the same conditions can give different responses. The lack of replicate analyses on the same sampler by most laboratories makes somewhat weaker the result of this test. However the analysis of variance on the data available (two laboratories) shows that for 6 out of 9 compounds the probability is 1% or smaller that the samplers are equal; for the remaining compounds (pentanal, 1.1.2. trichloroethane and 1-octene) the same probability ranges between 24 and 58%. These results point to a different response of individual OVM-3500 samplers or, in other terms, to a lack of reproducibility in the performance of the samplers. A similar conclusion was obtained by other researchers in a comparison of different brands of passive samplers, including the 3M (Stockton and Underhill, 1985). The accuracy of the samplers has been tested comparing them with the active samplers run in parallel. Table 1, reporting this comparison, shows that for two compounds (1.2-xylene and n-decane) the agreement is excellent, for six compounds (1-butanol, 1.1.2-trichloroethane, 1-octene, butylacetate, 3-heptanone and a-pinene) the active sampler values are higher than those from the passive samplers (discrepancy between 13 and 56%); for pentanal the active samplers are much higher than the passive ones.

TABLE 1   Comparison of mean concentrations and percent standard deviat - ions measured with OVM-3500 passive and active samplers

	OVM-3500 pass. samplers (6 laboratories)		act. samplers (3 laboratories)		act. samplers pass. samplers
	mean (ug/m^3)	st.dev. %	mean (ug/m^3)	st.dev. %	
1-butanol	402	21	626	37	1.56
pentanal	96	75	277	11	2.89
1.1.2-trichloroethane	875	11	984	8.5	1.13
1-octene	250	13	318	9.6	1.27
butylacetate	553	13	699	11	1.26
3-heptanone	214	13	252	10	1.18
1.2-xylene	1254	9.1	1235	4.4	0.99
a-pinene	643	9.3	801	35	1.25
n-decane	1152	11	1150	4.7	1.00

The high value of a-pinene in the active samplers is due to the results from one of the laboratories, the two others being very close to the passive samplers. The discrepancy for this compound can therefore be explained by some artifact or error at one laboratory. A t-test for the significance of the difference between two means was carried out on the other five compounds and the differences observed result highly significant ($p < 0.01$).

The desorption efficiency of pentanal for passive samplers has been found strongly dependent on the time elapsed after sampling, probably because of irreversible adsorption, and this explains the large underestimation of this compound by passive samplers. In summary it appears that the OVM-3500 samplers agree very closely with thermally desorbable active samplers for apolar compounds, whereas they tend to underestimate polar compounds.

The relative standard deviations associated with the concentrations reported in Table 1 range between 9 and 21% (if pentanal is excluded) with an average of 13%. From the statistical analysis it results that the contribution of interlaboratory deviations to the variance is 2-2.5 times that of the intralaboratory deviations.

ACKNOWLEDGEMENT

The following persons and laboratories participated: Dr. I. Johansson, Statens Miljomedicinska Laboratorium, Stockholm (Sweden); Dr. I. Kellie, Scottish Occupational and Environmental Health Service, Dundee (UK); Dr. E. Muylle, Instituut voor Hygiene en Epdidemiologie, Brussels (Belgium); Dr. K. Pannwitz; Drägerwerk Aktiengesellschaft, Lübeck (FRG); Dr. P. Wolkoff, Arbejdsmiljøinstitutet, Hellerup (Denmark).

REFERENCES

De Bortoli, M., Knöppel, H., Mølhave, L., Seifert, B., Ullrich, D. 1984. Interlaboratory comparison of passive samplers for organic vapours with respect to their applicability to indoor air pollution: a pilot study. EUR 9450 Report, CEC Luxembourg.
Stockton, S.D. and Underhill, D.W. 1985. Field evaluation of passive organic vapour samplers. Am. Ind. Hyg. Assoc. J. 46, 526-531.

FIELD COMPARISON OF CHARCOAL TUBES AND PASSIVE VAPOR MONITORS
FOR ORGANIC VAPORS

S. Hagberg, R. Nordlinder, G. Sällsten

Dept. of Occupational Medicine
Sahlgren Hospital
S:t Sigfridsgatan 85
S-412 66 Göteborg, Sweden

ABSTRACT

The conventional method used for personal sampling of organic vapors is charcoal tube with active sampling pump. In recent years a new device has been introduced: the personal passive monitor. No pump is needed as the monitor work through diffusion of molecules through a stagnant layer. We have compared charcoal tubes (SKC) and passive monitors (3M) in field measurements in three different industries: Plastic boat industry - styrene, Silk screen printing - xylene and ethylbenzene, Rotary-offset printing - ethylacetate.

Personal sampling was done with the devices placed closed together at the right hand shoulder. The sampling time was about three to four hours and the sampling flow for the tubes was 30 ml/min for styrene and 25 ml/min for the other substances. The total number of comparable measurements was 76 for styrene and 30 for the others. The amounts of different contaminants were analysed by gas-chromatography after eluation with carbon–disulfide. The results were corrected for desorptions-effiencies. The concentration was calculated by dividing the amount with corresponding sampling volume. For passive monitors the sampling rates given by the manufacturer were used.

Correlation between the different results obtained from the two methods was evaluated using linear regression analyses. X and Y corresponded to tube and monitor values, respectively. For styrene, xylene and ethylbenzene very good agreements were obtained: the slope was 0.97, 1.03 and 1.02 respectively and the correlation coefficient was 0.995, 0.978 and 0.980. The correlation coefficient was also close to unity for ethylacetate (0.992) but passive monitors gave constantly lower values (slope equal to 0.82). Possible causes to disagreement are error in the sampling rate, variations in air humidity and large variations in the concentrations. Furthermore ethylacetate is more polar than the other substances.

INTRODUCTION

Passive diffusion organic samplers were introduced in the middle of 1970 as a potential alternative to the charcoal tube method or organic vapor sampling (Tomkins and Goldsmith, 1979).

There are numerous investigations of the sampler´s capacity but most of the studies have been performed in the laboratory under controlled conditions (Rose and Perkins, 1982). There are, however, not so many field evaluations of the samplers (Stockton and Underhill, 1985). Field studies are mostly important in comparison of sampling methods since the industrial environment has great variations in air temperature and humidity, solvent concentrations and air movements.

In our study we have compared passive diffusion samplers with charcoal tubes in three different industrial environments.

MATERIALS AND METHODS

Passive diffusion organic samplers from the 3M company (3M Organic Vapor Monitors 3500) were compared with charcoal tubes from SKC, Inc.

Four different solvents were evaluated in three industries:

-   Styrene - Plastic boat industry
-   Xylene and ethylbenzene - Silk screen printing
-   Ethylacetate - Rotary-offset printing

The field measurements were done with the devices placed closed together at the right hand shoulder of the worker. The sampling time was about three to four hours with a sampling flow for the tubes of 30 ml/min for styrene and 25 ml/min for the other substances. 76 measurements were done for styrene and 30 for each of the others.

The analysis were done by gas-chromatography with FID-detector after eluation with carbon-disulfide. Desorption efficiency was determined experimentally. Sampling rates for the diffusion samplers given by the manufacturer were used.

The results were statistically treated with linear regression analysis and the correlation coefficient has been determined.

RESULTS

The desorptions efficiencies for the different methods and solvents are shown in the following table.

	Desorption efficiency %	
Solvent	Charcoal tube	Diffusive sampler
Styrene	92	92
Ethylacetate	91	97
Xylene	99	100
Ethylbenzene	100	102

The results from the measurements for the different solvents are shown in the figures below.

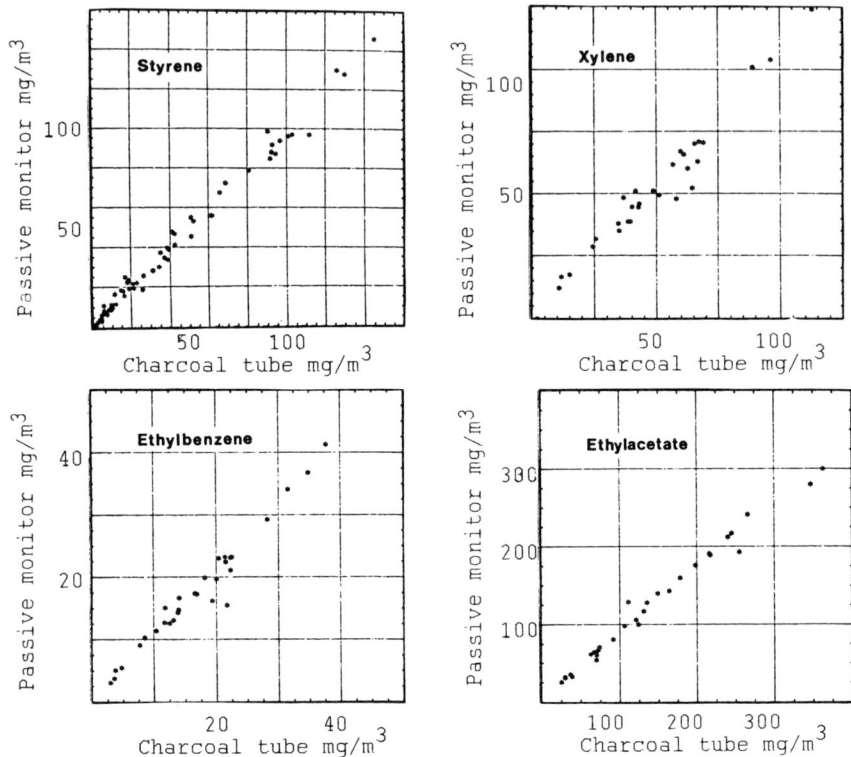

The results from the regression analyses of the two sampling methodes are presented in the table below.

Substance	Sample pairs	Range of Data(mg/m³)	Linear Regression		Correlation coefficient
			Slope& 95% CL.	Y-Intercept (mg/m³)	
Styrene	76	0.7-145	0.97±0.02	1.1	0.995
Ethylacetate	30	24.9-363	0.83±0.04	9.2	0.992
Xylene	30	8.7-113	1.03±0.09	1.3	0.978
Ethylbenzene	30	2.9- 37.7	1.02±0.08	0.5	0.980

CONCLUSIONS

There was no systematic difference found between the two methods for styrene, xylene and ethylbenzene.

The correlation was also high for ethylacetate but the values for the passive monitors gave constantly lower values.

Possible sources of errors could be

- sampling rate
- high polarity
- large variation in concentration
- air humidity

REFERENCES

Rose, V.E. and Perkins, J.L. 1982. Passive dosimetry - State of art review. Am. Ind. Hyg, Assoc. J., 43, 605-621
Stockton, S.D. and Underhill, D.W. 1985. Field evaluation of passive organic vapor samplers. Am. Ind. Hyg. Assoc. J., 46, 526-531
Tomkins, F.C. and Goldsmith, R.L. 1979. A new personal dosimeter for monitoring of industrial pollutants. Am. Ind. Hyg. Assoc. J., 38, 371-377

DYNAMIC GENERATION OF TEST GASES AND ITS APPLICATION
TO STANDARD PREPARATIONS

U. Giese, H. Stenner, A. Kettrup*
University of Paderborn, Applied Chemistry, P.O.Box 1621
D-4790 Paderborn, F.R.G.

ABSTRACT

A test generator for the dynamic generation of test gases has been developed. This device utilizes the saturation vapour pressure principle and enables the calibration of sensitive analytical methods with substance concentrations in the range of maximum acceptable limits (MAK, TRK, TLV). The test gases are producible from liquid and solid substances with finite vapour pressure. It is possible to test adsorption tubes and diffuse samplers using the test gas generator.

INTRODUCTION

In order to protect exposure to hazardous chemicals, maximum acceptable concentration (MAK-res.TRK-values) have been established (Deutsche Forschungsgemeinschaft 1985). There are several analytical methods for the determination of these chemical pollutants. In order to determine the efficiency of these methods, it is necessary to check the accuracy and recovery by means of calibration test gases (Barrat, 1982; Anderson, 1980; Nelson, 1982). Several test gas generation systems are reported in the literature. The methods are generally divided into two categories, static and dynamic methods (Nelson, 1982; Leithe, 1974; VDI-Richtlinien 3490, 1980). Construction and mode of operation of the test gas generator developed by us is described in other papers (Weber, 1986).

In this investigation, it is shown, that this test gas generator can be used for calibration of active - as well as passive samplers. The substance CH_2Cl_2 was taken as example to compare charcoal tubes (NIOSH) with diffuse samplers (Type Dupont). The concentrations were determined and results plotted against each other in a diagram. The measured values show a straight with the function y = 0.999 x + 30.5 and a correlation coefficient r = 0.9961.

COMPARISON OF ACTIVE AND PASSIVE SAMPLING IN A TEST GAS ATMOS-
PHERE OF CH_2Cl_2

1. Methods and Materials

The equipment for testing the different sampling methods
consists of a test atmosphere generator TAG 100 (Weber, Stenner, Giese, Kettrup 1986; Weber, Stenner, Kettrup 1986) and
a special glass tube (Fig. 1).

Figure 1: Calibration glasstube
1 Blower
2 Inlet for test gas from TAG 100
3 Plan for sampling

In the glass tube a test gas atmosphere is produced by
mixing a stream of air coming from the blower and high concentrated test gas coming from the test gas generator. The blower
affects a velocity of air flow, which is necessary to avoid
the improverishment at the face of the passive sampler (Blome,
H.; Hennig, U. 1985). It is necessary to have minimal velocity
of airflow to achieve a constant uptake rate (Brown, R.H.;
Harvey, R.P.; Purnell, C.J. and Saunders, K.J. 1984).

The concentration of the test gas in the glass tube can
be calculated from the flow of 17 ml/min of the saturated test
gas (260 mg/l to 1091 mg/l CH_2Cl_2) from the generator and the
air flow produced by the adjustable blower. The air flow must
be measured with an anemometer. In this way the dilution of
the concentrated test gas can be calculated. For the comparison of active and passive sampling parallel measurements were
made.

The type of diffusive sampler was the Dupont Pro tek G-
AA Organic Vapor Air Monitoring Badge. (Gregory, E.D. and
Elia, V.J. 1983; Du Pont de Nemours and Co. 1983).

For active sampling a Du Pont air sampler with a 100 mg
charcoal tube (Chrompack-NIOSH quality) was used. The sampling
parameters are shown in table 1:

TABLE 1   Sampling parameters

Parameter	passive sampling	active sampling
Temperature	23°C	
Sampling time	30 to 40 minutes	
pollutant concentration	100 to 700 mg/m³ CH_2Cl_2	
rel. humidity	50%	
uptake rate	69,2 ml/min	65 ml/min
air velocity	0,25 m/s	

2. Sample preparation and Analysis

The sample preparation consists of the desorption with 2 ml CS_2. The desorption rates, which were determined experimentally are 94 % for active sampling and 96 % for passive sampling. The analysis was made with a GC equiped with FID and Split Injection. For the quantitation a calibration curve was measured in the range of 0.1316 mg/2 ml CS_2 to 15.79 mg/2 ml CS_2 with the function y = 3503.995   x + 785.676. The correlation coefficient is r = 0.9992.

3. Reproducebility:

At a concentration of 261 mg/l CH_2Cl_2 in the saturated test gas produced by the generator, 5 measurements were made, each of which were analysed three times. The results are listed in table 2.

TABLE 2 Comparison of active and diffuse sampling with constant concentration of CH_2Cl_2

	Active sampling		passive sampling
Samples	5		5
n	15		15
x	102.11 mg/m³		27.25 mg/m³
Deviation		24,6%	
s	11%		3.7%
t_x	4.12		1.74

4. Measurements at different concentrations of CH_2Cl_2

By comparison of active and passive sampling, whereby the measured values were plotted against each other in a diagram ,

a straight with a slope of 1 and an intercept at the origin is expected in an ideal case.
A slope < 1 shows that the passive sampling has a lower sensitivity than the active sampling method. The measured values show a straight with the function $y = 0.999 x + 30.5$ and a correlation coefficient $r = 0.9961$.

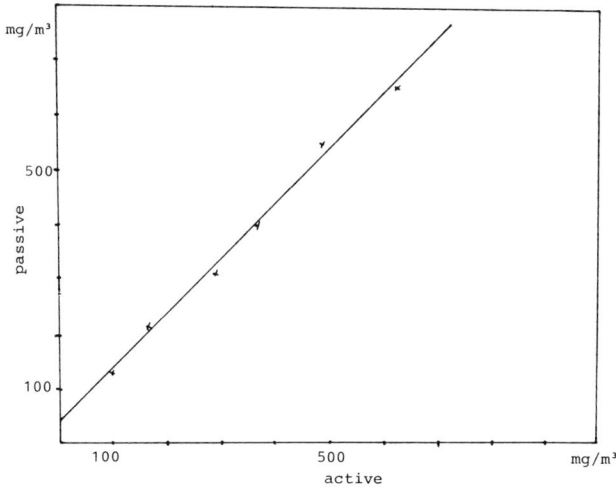

Fig.2
Comparison of active and passive sampling
Concentration range: 100 to 650 mg/m³
Function of the straight: $y = 0.999 x + 30.5$
Correlation coefficient: $r = 0.9961$

REFERENCES
Anderson, C.C. 1980. Occup.Health Chem.1
Barrat, R.S. 1981. The preparation of Standard Gas Mixtures
    Analyst: Vol.106, 817-849
Blome, H.; Hennig, U. 1985. Leistungsdaten ausgewählter Passiv-
    sammler, 1. Teil. Das Prinzip der Passivprobenahme Rein-
    haltung der Luft, Bd.45 (1985) Nr.11
Brown, R.H.; Harvey, R.P.; Purnell, C.J. and Saunders, K.J.
    1984. A Diffusive Sampler Evaluation Protocol Am.Ind.Hyg.
    Assoc.J. 45(2). 67-75
Deutsche Forschungsgemeinschaft 1985. Maximale Arbeitsplatz-
    konzentration und Biologische Arbeitsstofftoleranzwerte
    Mitteilung XXI der Senatskommission zur Prüfung gesund-

heitsschädlicher Arbeitsstoffe, Verlag Chemie, Weinheim.
Du Pont de Nemours & Co. 1983. Organic Compound Sampling Guide
    for G-AA/G.-BB Badges. Du Pont
Gregory, E.D. and Elia, V.J. 1983. Sample Retentivity Proper-
    ties of Passive Organic Vapour Samplers and Charcoal
    Tubes Under Various Conditions of Sample Loading, Rela-
    tive Humidity, Zero Exposure Level, Periods and a Compe-
    titive Solvent, Am.Ind.Hyg.Assoc.J.44 (2) 88-96
Leithe, W. 1974. Die Analyse der Luft und ihrer Verunreinigun-
    gen. Wissenschaftliche Verlagsgesellschaft mbH. Stuttgart
Nelson, G.O. 1982. Controlled Test atmospheres, Ann.Arbor.Sci.
    Publ.Inc., Ann Arbor
VDI-Richtlinie 3490, 1980. Blatt 8 und 9: Herstellung von Prüf-
    gasen
Weber, Stenner, Giese, Kettrup 1986. Apparatur zur dynamischen
    Herstellung von Prüfgasen.Chemie Technik 15. Jahrg. 1986,
    Nr.6
Weber, Stenner and Kettrup 1986. Development of an instrument
    for the dynamic generation of test gases. Fresenius Z.
    Anal.Chem. 1986, 325; 65-67

A COMPARISON OF SAMPLE GENERATION PROCEDURES FOR ESTABLISHING
A QUALITY CONTROL PROGRAMME FOR DIFFUSIVE SAMPLING/THERMAL
DESORPTION ANALYSES

T. McDonald, R. Briggs, D. Stevenson

Robens Institute of Industrial and Environmental Health and
Safety, University of Surrey, Guildford, Surrey, UK.

ABSTRACT

Two methods for preparing quality control samples for
analyses using thermal desorption were compared. Sample
production by
(a) direct syringe spiking and
(b) diffusive sampling from a dynamic atmosphere were
     compared.
     A C.V. of about ± 3% was found at the OEL for toluene
     using both methods. Diffusive sampling gave lower C.V.'s
     than direct spiking at low analyte concentrations.

INTRODUCTION

A monitoring strategy using diffusive sampling followed
by automatic thermal desorption (ATD) potentially has many
advantages over a system using pumped sampling followed by
liquid desorption. These advantages include increased sampling reliability, convenience and lower unit sample cost. For
any laboratory carrying out routine analysis it is important
to establish a quality assurance programme to monitor the
performance of particular assays. Only then can an
occupational hygienist or other professional feel confident
that the results that they are interpreting are meaningful.
A very important aspect of this is the production of batches
of quality control samples that can be included with every
batch of unknowns. The day-to-day precision of methods can
then be assessed. Such samples can be generated for internal
purposes or for inter-laboratory comparisons. The increasing
use of passive sampling/ATD creates a need for sample preparation procedures to generate quality control samples for this
type of analysis. This paper describes and compares two
different methods for producing quality control samples
suitable for ATD analysis. The two methods used were direct
spiking using a syringe and diffusive sampling from a
dynamically generated atmosphere. Toluene was chosen as a

model and the two methods of sample production were compared
at levels below and above the Occupational Exposure Limit (OEL)
for Toluene.

EXPERIMENTAL

Equipment

Thermal desorption apparatus   -   Perkin Elmer ATD 50

Gas chromatograph (for sample analysis) - Perkin Elmer 8320

Gas chromatograph (for atmospheric monitoring) - Perkin Elmer
Sigma 4 equipped with 1ml gas sampling valve.

Infra-red (IR) gas analyser   -   Miran 104

Sampling adsorbent   -   150mg of Tenax TA per tube.

Spiking syringe.   -   SGE-1B-RD3-5 1µl plunger in needle with
repeating dispenser.

Diffusive Sampling System (see Fig 1) - the atmosphere was
produced using a permeation tube generator with purified air
as a carrier.   This was then introduced into a 20 litre glass
vessel after a single stage dilution step.   The atmosphere
was exhausted from the vessel through an IR gas analyser
(for constant concentration monitoring).   The concentration
was accurately determined by gas chromatography.

Loading Methods

Analyte - toluene (AR grade)

Spiking - 1µl of methanolic solution was spiked directly onto
the back retaining gauze of the tube.

Diffusive - the concentration in the chamber was allowed to
stabilize overnight before introducing the tubes.   The
sampling time was 7 hours.   Batches of 20 tubes were exposed
simultaneously.

The system for producing diffusive samples is shown in
figure 1.

RESULTS AND DISCUSSION

1.   The % coefficient of variation (C.V.) as a function of
amount of toluene is shown for spiked samples in figure 2
and for diffusive sampling in figure 3.   In the range one
tenth the OEL to twice the OEL (equivalent to 8 to 164µg of
toluene for an 8 hour sample) both methods showed a C.V. of
less than 5%.

2. Both methods showed a similar C.V. over the range stated.

3. Direct spiking gave the lowest C.V.'s around the loading expected at the OEL (82µg for an 8 hour sample), see figure 2. There was an increase in C.V. with increasing amounts, and a marked increase at low amounts.

4. Diffusive sampling gave the lowest C.V.'s at very low loadings, increasing as the loading level increased, see figure 3.

5. For samples of below 30µg loading, using the diffusive method gave lower C.V.'s. This indicates significant differences in the way the tubes behave to different loading techniques. This may have important implications in the preparation of quality control samples for analytes with a lower OEL than toluene and consequently lower tube loadings.

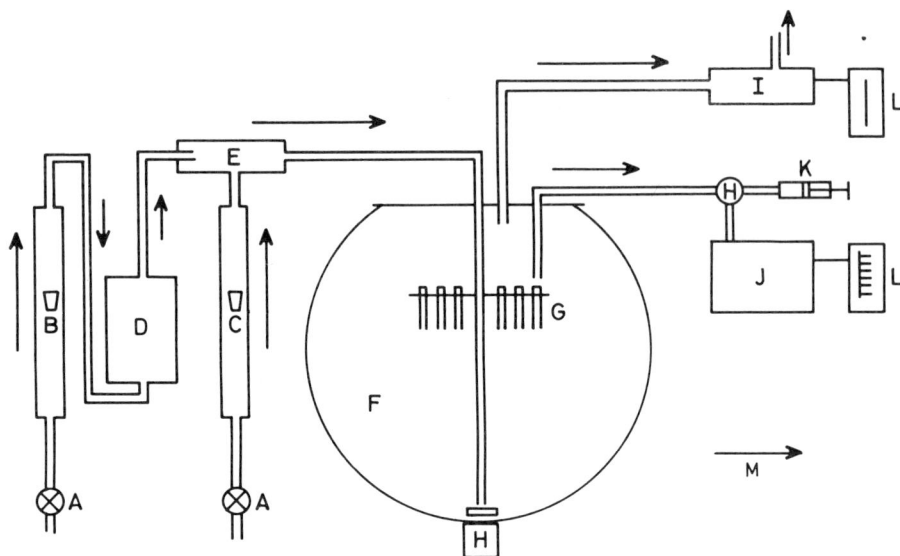

Figure 1   Diffusive sampling exposure system

Figure Key

A. Gas flow controllers
B. Primary rotameter
C. Dilution flow rotameter
D. Thermostatted permeation tube
E. Mixing vessel
F. 20l Glass exposure chamber
G. ATD diffusive sampling tubes
H. Gas sampling valve
I. IR gas analyser
J. Gas chromatograph
K. Gas sampling syringe
L. Chart recorder
M. Direction of gas flow

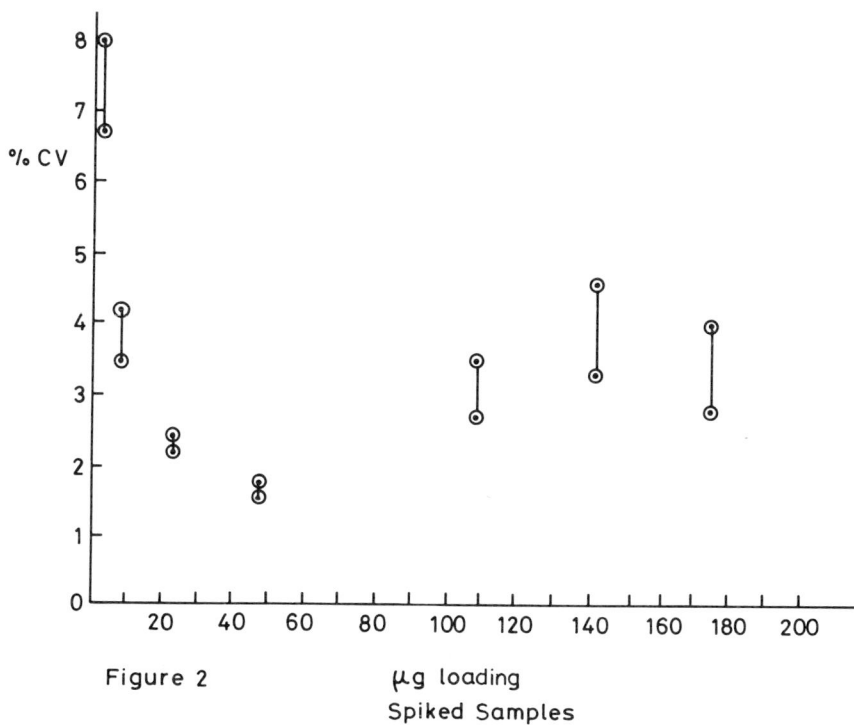

Figure 2   μg loading
                  Spiked Samples

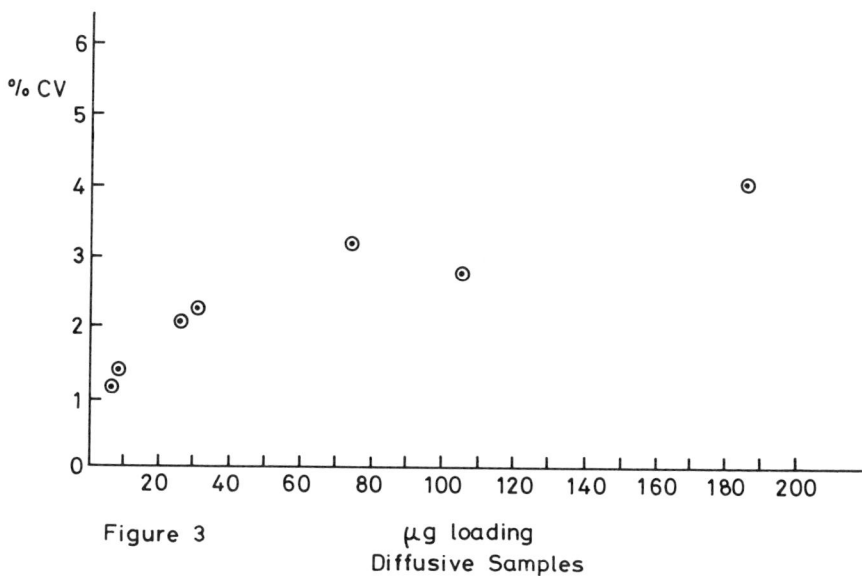

Figure 3   μg loading
                  Diffusive Samples

DIFFUSIVE SAMPLING OF HALOTHANE AND ENFLURANE

D. Norbäck, I. Michel
Department of Occupational Medicine,
Akademiska sjukhuset, S-751 85 Uppsala, Sweden.

ABSTRACT
    The halothane and enflurane massuptake of five diffusion
samplers were studied at anaesthetic work. Active sampling on
charcoal tubes and in Tedlar sample bags were used as
reference methods. Molecular sieve was found to be a less
suitable absorbent for halothane and enflurane. However, charcoal diffusion sampling showed high correlation with charcoal
tube sampling for both halothane and enflurane but the massuptake rate was influenced by the sampling time.

INTRODUCTION
    Halogenated anaesthetic gases are used together with
nitrous oxide at anaesthetic work. The aim of this study was to
evaluate various types of diffusion samplers for exposure
monitoring of halothane and enflurane.

MATERIALS AND METHODS
    Initially, three diffusion samplers of the charcoal type
(SKC-gasbadge with anasorb CA, ORSA-5 from Draeger and dosimeter no 3500 from 3-M) were selected to be tested. During the
study, one more charcoal sampler (Gasbadge GBB from Dupont)
and one molecular sieve sampler (Siemens´ dosimeter 185) were
included in the study. The dosimeters were tested with personal sampling in the range of 0.1-380 mg/m3 halothane or
enflurane at routine anaesthetic work, or at highly exposed
situations in operating-rooms without sufficient ventilation.
    Active sampling on charcoal tubes (SKC 226-01) was used
as the major reference method. Active sampling of air in
Tedlar bags was used as an additional reference method. Parallel exposure measurements were performed in the breathing zone
of the exposed personnel during either 1/4 or 2 hours periods.
    The charcoal samplers were desorbed with 2 ml carbon
disulphide and analyzed with a Hewlett-Packard 5880A gas chro-
mathograph equipped with a 10% packed squalane column (30 C
oven temp) and a flame ionization detector (FID). The detection limit was 0.5 microgram/sample. The desorption efficiency
for the various types of charcoals was determined.
    Air samples collected in Tedlar bags were analyzed within
two hours by direct injection on the gas chromatograph, or by
IR-spectroscopy for enflurane (MIRAN 104, 8.70 mikron). The
molecular sieve sampler (Siemens´ 185 dosimeter) was analyzed
by thermal desorbtion with a gas chromatograph equipped with a
packed microcolumn and an electron capture detector.

RESULTS
    Air samples collected in Tedlar bags gave lower concen-
trations of both halothane and enflurane in comparison with
the charcoal tube method, see table 1. IR-spectroscopical
determination of enflurane in the Tedlar bag air samples gave
lower concentrations than the gas chromatographical method.
The desorption efficiency for both halothane and enflurane was

higher than 95% for all types of charcoals studied. There was
a good correlation for all types of charcoal diffusion samplers between the mass uptake rate and the concentration, measured by the charcoal tube method (table 2). However, the
diffusion samplers generally showed a higher mass uptake rate
when the sampling period was 1/4 h compared to 2 h.
    In a preliminary study, it was found that the molecular
sieve dosimeter was not usable for enflurane measurements. For
halothane, the correlation between the concentration in the
air and and the mass uptake rate of this sampler was poor.
Multiple linear regression showed a significant influence of
both the sampling time ($p < 0.001$) and the relative humidity
($p < 0.001$) on the massuptake rate of halothane. A long sampling
time and a high relative humidity of the air resulted in a
decreased massuptake rate. For the charcoal dosimeters, no
significant influence of the relative humidity (7-48%) or of
the room temperature (21-26 C) on the mass uptake rate could
be detected ($p > 0.05$).
    Logaritmic regression data on the mass uptake rate are
presented in table 3. As can be seen from the table, the
correlation coefficients are of the same order of magnitude as
for the linear regressions. However, the slope of the logarit-
mic regression was generally below 1.0 for halothane.

CONCLUSION
    Diffusion sampling on activated charcoal is a reliable
method for the exposure measurements of halothane and enflurane at anaesthetic work. The results obtained are comparable to
those obtained by the conventional charcoal method, but diffusion samplers are preferred because of their simplicity. All
four types of charcoal diffusion samplers showed a high correlation with the charcoal tube method. Neither the relative
humidity nor the room temperature did influence the sampling
rate of the charcoal diffusion samplers. However, in most
cases the mass uptake was higher at the shorter sampling
period. For halothane but not for enflurane, there was a
tendency towards a decreased mass uptake rate at higher
concentrations.
    The strong influence of both the sampling time and the
relative humidity shows that molecular sieve is a less
suitable absorbent for diffusion sampling of halothane.

TABLE 1
Concentration of halothane and enflurane (mg/m3) measured in
Tedlar bag air samples as a function of the concentration
(mg/m3) measured by charcoal tube sampling.

Compound/method	N*	Slope of regression	S.D of slope	Intercept (mg/m3)	R**
HALOTHANE:					
Gas-chromatography	53	0.780	0.025	+2.4	0.974
ENFLURANE:					
Gas-chromatography	20	0.864	0.070	-2.6	0.946
IR-spectroscopy	14	0.615	0.051	+10.3	0.961

TABLE 2
The massuptake of halothane/enflurane(ng/min) as a function of
the concentration (mg/m3) measured by charcoal tube sampling.

Dosimeter type	Measuring time(hour)	N*	Slope (ml/min)	S.D of slope	Intercept (ng/min)	R**
HALOTHANE:						
no 3500:	1/4	23	28.7	1.9	-111	0.955
	2	33	23.7	1.1	+118	0.966
G-BB:	1/4	10	25.7	2.2	+279	0.970
	2	9	23.5	3.5	+361	0.930
SKC:	1/4	21	10.9	0.7	+101	0.958
	2	33	11.6	0.6	+ 25	0.961
ORSA-5:	1/4	17	6.85	0.3	+ 72	0.985
	2	31	6.16	0.3	+ 25	0.958
DOSIMETER	1/4	15	0.075	0.017	+8.5	0.777
185	2	16	0.036	0.011	+2.3	0.674
ENFLURANE:						
no 3500:	1/4	10	30.4	4.4	-617	0.926
	2	15	21.3	2.2	+181	0.935
G-BB:	1/4	11	26.6	1.4	-121	0.987
	2	9	22.1	5.4	+283	0.842
SKC:	1/4	11	11.4	1.0	- 25	0.965
	2	15	9.2	1.1	+ 76	0.917
ORSA-5:	1/4	8	7.30	0.9	- 89	0.959
	2	13	5.13	0.7	+ 47	0.917

TABLE 3
Logarithmic regression data for halothane and enflurane.

Type of dosimeter/compound		N*	Slope log(ml/min)	S.D.of slope	Intercept log(ng/min)	R**
no 3500:	Halothane	56	0.886	0.033	+1.62	0.926
	Enflurane	25	0.961	0.028	+1.48	0.990
G-BB:	Halothane	19	0.902	0.055	+1.65	0.970
	Enflurane	20	0.988	0.045	+1.43	0.982
SKC:	Halothane	54	0.884	0.039	+1.27	0.954
	Enflurane	26	0.945	0.037	+1.16	0.982
ORSA-5:	Halothane	48	0.902	0.064	+1.12	0.888
	Enflurane	21	0.765	0.044	+0.77	0.983

* N=number of measurements   ** R=Correlation coefficient
S.D. of slope = Standard deviation of the slope of regression

COMPARISON OF DIFFUSIVE SAMPLERS AND CHARCOAL TUBE METHOD
UNDER FIELD CONDITIONS

Lauri Saarinen, Maire Rothberg, Beatrice Bäck

Uusimaa Regional Institute of Occupational Health
Arinatie 3 A, SF-00370 Helsinki, Finland

ABSTRACT

Two independent adsorption methods were compared: samples were collected in pairs to examine differences in the performance of "passive" and "active" sampling with activated carbon.

MATERIAL AND METHODS

3-M diffusive samplers were used as instructed by the manufacturer, and the charcoal tube method was used as guided by NIOSH (Method 1500, 1501,1003).

The collection and desorption efficiency of toluene is high. Toluene is therefore suitable for performance testing in field conditions. 68 pairs of samples were collected at a photogravure printing shop. The exposure levels varied between 0 to 800 cm^3/m^3. The sampling time was three hours and loading levels were from 0.01 µl to overloading (more than 25 µl toluene per 150 mg charcoal).

Nine diffusive samplers were tested by parallel with a MIRAN Al infrared spectrophotometer measuring continuously and equipped with a recorder.

Effects of complex mixtures and intermittent exposure profiles were tested in car spray-painting booths. The loading levels of each analyte varied between 0.01-2 µl. The humidity conditions varied between 18-38 RH % and temperatures between 16 - 22 $^{\circ}$C.

RESULTS

The correlation between diffusive samplers and charcoal tubes is shown in figure 1. The agreement of these

two methods is good (r = 0.83). Significantly diverging
results were examined separately and found to be due to poor
calibration of the pumps, leakage in the devices or
overloading of the adsorbent.

The correlation between the results from continuously
measuring MIRAN A1 and diffusive samplers (r = 0.998) was
good (figure 2).

A paired sample test was applied on spray-painting
samples (n = 16 sample pairs, loading ranged from 0.01-2.5
l). Table 1 gives results of variation of the sample pairs
and paired t-test of two independent methods.

CONCLUSIONS
        A single adsorbate, like toluene, with good collec-
tion and desorption characteristics, does not cause serious
problems in the evaluation of exposure levels with either of
the tested methods. Complex mixtures, low or high loading
levels, and intermittent exposure profiles cause irregu-
larities and high variation coefficients between the examined
paired samples. According to paired t-tests, both methods are
comparable.

TABLE 1.   Variation of sample pairs and paired t-test of two
            independent methods

	N	CV%	Paired t-test t value
Butyl Acetate	12	23,7	-2,238
Xylene	13	7,5	-2,33
Acetone	16	19,7	-0,102
Ethanol	15	25,2	0,195
Trichloroethane	12	21	-0,609
2-Ethoxyethyl Acetate	13	19,2	-0,750
Toluene	14	15,7	0,393
i-Butanol	15	25	-1,861

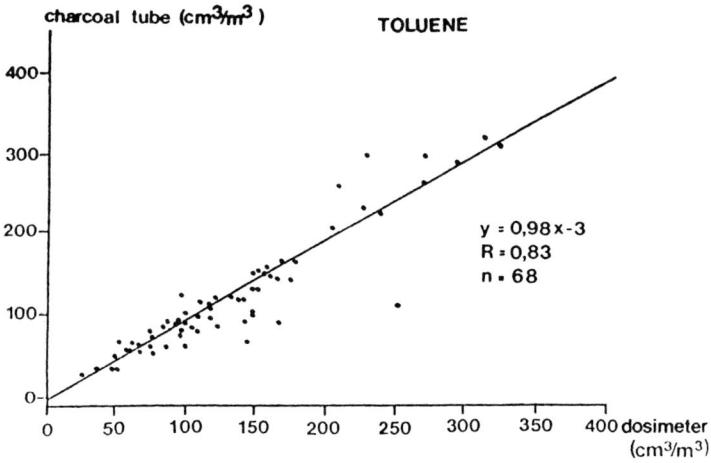

Figure 1. Correlation between charcoal tubes
and 3 M diffusive samplers

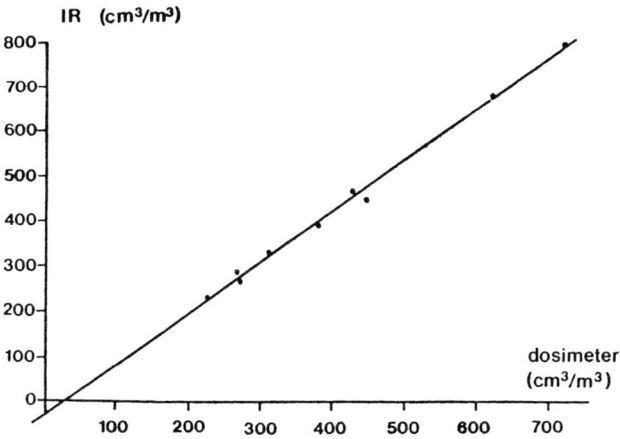

Figure 2. Correlation between MIRAN 1A and
3 M diffusive samplers

PASSIVE SAMPLING WITH LIQUID OR SOLID SUBSTRATE

C. Sala

Department of Industrial Medicine,
Lecco Hospital, Corso Promessi
Sposi 1, I-22053 Lecco

ABSTRACT

The results are given of the experiments in the laboratory and at the workplace of a new passive sampler which can be used alternately with solid and liquid media for physical and reactive sorption. The air flow can be altered by inserting cones which reduce the section of the intake in order to adapt the sampling process to the level of pollution to be monitored and to the varying duration of the tests to be carried out or of the tasks under investigation. Traditional methods of active sampling were used simultaneously with all the tests. Comparison of the results reveals a satisfactory level of agreement for tests both in the laboratory and at the workplace. Average differences were below 5% and 10% respectively.

INTRODUCTION

The growing interest in passive samplers has highlighted the need for validation tests in the laboratory and at the workplace. This is illustrated by recent papers comparing passive samplers and other sampling methods (Coyne et al., 1985; Kring et al., 1985; Stockton and Underhill, 1985; Van der Wal and Moerkerten, 1984; Hickey and Bishop, 1981) and comparing the passive samplers of various manufacturers (Cassinelli et al., 1985). These papers have generally described passive samplers which use activated carbon or a solid medium with reagent as the media for collecting pollutants. The purpose of our research was to carry out laboratory and field tests on a new conical sampler (Zambelli multi-method) which can use as collection media both solid absorbents and liquids for physical and reactive sorption. The air flow can be altered by using cones of varying intake sizes. The sampler thus has a wider range of use than others on the market. All the tests were conducted with parallel tests using active sampling methods, with standardized and proven methodology. More attention was focussed on tests using absorption liquids, as this medium has been less frequently covered in the literature.

METHODS AND EQUIPMENT

I - Laboratory tests

The compartments of the metal cylinder are fitted with one or more conical diffusive evaporators and three conical passive samplers containing activated carbon or liquid for physical or chemical sorption. The metal cylinder, which is wide enough in the upper part to allow the passage of air, is placed in a glass cylinder fitted with an intake and outflow nozzle. The air is transmitted by means of a pump or compressed-air cylinder at a controlled rate of between 2 and 5 1/min and with a linear velocity of one or a few tenths cm/s. On the external flow line are attached three parallel tubes of activated carbon or three aerators with porous baffles with the requisite battery-operated pumps, depending on the type of pollutant to be monitored. The overall volume of the transmitted air is measured by a dry meter in series with a flow meter.

Sampling times are between 100 and 150 minutes for substances with a low or medium boiling point and between 180 and 240 minutes for those with a relatively high boiling point. The atmospheric concentrations of the pollutants to be monitored can be altered over a fairly wide interval of time by varying the number of evaporators or their cone size. The flow of the pollutant into the passive samplers can also be varied by using different cones. In the case of liquid absorption, the amount of liquid used can range from 0.5 to a maximum of 2cm3 without making any appreciatle change to the diffusion chamber. The values of T, P, and H% are monitored throughout the tests.

Solvents: The solvent mass which has evaporated by diffusion is calculated by using a precision balance to weigh the cone before and after the test. Gas chromatography is used to find the solvent mass which is adsorbed in activated carbon or absorbed in liquid by active and passive sampling media.

Gas pollutants: these are produced by diffusive evaporation from solutions with a known concentration or by chemical reaction in the cone diffuser. Instrumental analysis is used to find the absorbed mass.

The test which were carried out can be shown in table form as follows:

Serie	campionamento attivo	campionamento passivo	inquinanti
A	adsorbimento su carbone attivo in Fiale	assorbimento su tessuto di carbone attivo	solventi
B	adsorbimento su carbone attivo in Fiale	assorbimento Fisico in liquido	solventi
C	assorbimento con reazione in gorgogliatori a setto poroso	assorbimento con reazione nel campionatore passivo	SO_2CH_2O, HCl ...,

II - Tests at the workplace

Parallel active and passive samples were taken, mainly in fixed positions, in accordance with series B and C above. Sampling times ranged from 120 to 300 minutes and the locations were selected on the basis of information obtained during earlier surveys using only active sampling methods.

RESULTS

After weighting for T and P, the test results were analysed for the purpose of comparing:

1) atmospheric concentrations from active and passive samples;

2) coefficients of variation between active and passive sampling series;

3) concentrations calculated by analysis and those calculated from the mass of the evaporated analyte.

In case of the first comparison, the results in the laboratory may be considered satisfactory as differences ranged between 2.1 and 8.1%; the range was greater for the tests at the workplace and covered 2.5 to 15.7%. Average values were nevertheless below 5% and 10% respectively for the two types of test. The coefficients of variation for passive sampling were much lower than those for active sampling and were again less than 5% and 10% respectively. Analysis of atmospheric concentrations in the laboratory produced results which fell between 2.2 and 14.3 % below those calculated on the basis of evaporated mass, with an average figure around 5%.

Table 1 illustrates the results of three sets of values for three commonly-used solvents jointly present at concentrations of between 120 and 232 mg/m3.

Table 1 - Comparison of results

Solvente	Differenza media attivo/passivo (%)	CV medio (%)		Sottostima media (%)	
		attivo	passivo	attivo	passivo
M.I.B.K.	3,9	10,1	3,5	9,3	4,1
Touene	2,9	7,4	2,9	8,3	3,6
n-butilacetato	6,2	11,3	6,6	11,8	6,5

It can be seen that there is close agreement between active and passive samplers both in the laboratory and at the workplace. The possibility of making specific use of various solid and liquid media, together with the possibility of varying the mass of the analyte to be sampled by using cones of different sizes, means that the passive cone sampler is a versatile tool which can be adapted to most environmental situations which have to be analysed.

BIBILIOGRAPHY

Cassinelli M.E., Hull R.D., Cuendet P.A., 1985. Performance of sulfur dioxide passive monitors. Am.Ind.Hyg.Assoc. J. 46, 599-608

Coyne L.B., Cook R.E., Mann J.R., Boyoucos S., McDonald O., Baldwin Ch. L. 1985. Formaldehyde a comparative evaluation of four monitoring Methods. Am. Ind. Hyg. Assoc. J. 46, 609-619

Kring E.V., McGibney P.D., Thornley G.D. 1985. Laboratory validation of commercially available methods for sampling ethylene oxide in air. Am. Ind. Hyg. Assoc. J. 46, 620-624.

Stockton S.D., Underhill D.W. 1985. Field of passive organic vapor samples. Am. Ind. Hyg. Assoc. J. 46, 526-531.

Van der Wal J.F., Moerken A., 1984. The performance of passive diffusion monitors for organic vapours for personal sampling of painters. Ann.Occu. Hyg. 28, 39-47.

Hickey J.L.S., Bishop C.C. (1981) Field comparison of charcoal tubes and passive monitors with mixed organic vapours. Am. Ind. Hyg. Assoc. J. 42, 264-267.

LABORATORY VALIDATION OF A DIFFUSIVELY

SAMPLING MONITOR FOR ETHYLENE OXIDE

J.M. Thompson*, R, Sithamparanadarajah**
D. John***

*Department of Chemistry, University of Birmingham
Birmingham, B15 2TT, England
**Institute of Naval Medicine, Alverstoke, Gosport
Hampshire, England
***Occupational Health and Safety Unit, University of
Aston in Birmingham, Gosta Green,
Birmingham B4 7ET, England

ABSTRACT

Humidified ethylene oxide in nitrogen standard atmospheres were prepared using a permeation oven and fed to an exposure chamber in which Simtec Adsorba Type CM monitors were exposed for standard times. The sampling rate was obtained experimentally by comparison between actively and diffusively sampled monitors. The monitors were stored for up to 15 days at 4 **degrees**, 15 degrees and 35 degrees C. Well defined decay curves and the 95% confidence envelopes about these were obtained and, from these, the correction factor curves and their 95% confidence envelopes were derived. Robust and resistant statistical techniques were used to calculate the regression equations and the confidence envelopes. The monitors were particularly tested for compliance with the U.S.O.S.H.A. limit of 1p.p.m.

INTRODUCTION

In the past, the philosophy of personal monitor validation has been based on the assumption that the adsorbed pollutant must be completely stable, in order to perform adequately (Taylor, et al; 1977). In this paper, that assumption is challenged and the alternative assumption proposed that, if the decay of trapped pollutant can be properly characterised and storage conditions defined, then such a monitor containing decaying pollutant may be quite adequate in its performance. The primary concern of the validation exercise should really be to establish that the behaviour of the monitor can be well characterised under defined conditions and that the results of monitoring exercises using such monitors can be obtained within acceptable statistical confidence limits. Brown et al (1984) in describing the U.K. Health and Safety Executive "Diffusive Sampler Evaluation Protocol" seem prepared to accept that these may decay during storage of monitors. However, they do not propose a solution to the problem other than to quote a time at which stored samples give results with a certain percentage of unstored samples.

The work described in this paper is concerned within the detailed

characterisation of the decay of ethylene oxide adsorbed from humidified
gas mixtures onto an activated charcoal and stored at various temperatures
prior to analysis by thermal desorption and gas chromatography.

EXPERIMENTAL METHODS

The Simtec Adsorba monitors used in these experiments are based on
those described by Thompson et al. (1984). The methods of exposure of
the monitors to standard atmospheres of humidified ethylene oxide and
the thermal desorption and gas chromatographic conditions have been
described elsewhere (Thompson, 1986). The monitors were exposed to
2 p.p.m. (3.6 mg m^{-3}) of ethylene oxide for 4 hours. Monitors were
stored at $4^{\circ}C$ in a refrigerator or at $15^{\circ}C$ or $35^{\circ}C$ in thermostatted
containers. A minimum of 4 monitors were tested at each storage time
for each storage temperature.

RESULTS AND DISCUSSION

Regression analysis of the decay data at each storage temperature
enabled decay functions and 95% confidence envelopes to be calculated.
A robust and resistant regression method described recently by Thompson
(1986) was used. From these functions and their confidence envelopes,
the correction curves and their confidence envelopes were calculated and
these are illustrated graphically in Figure 1.

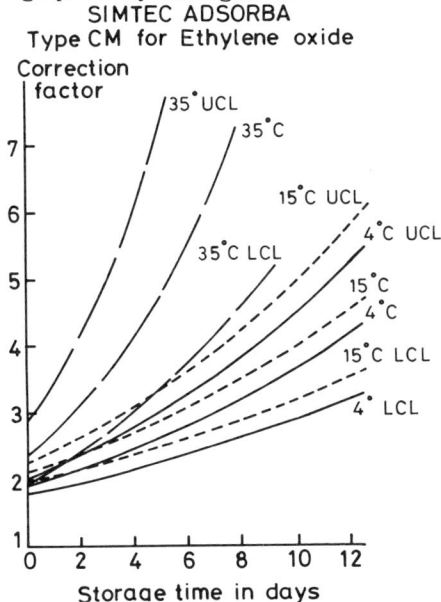

Fig. 1   Ethylene oxide decay curves & their confidence envelopes

It can be seen from these graphs that the decay is temperature dependent. Storage at $4^{\circ}C$ in a refrigerator enables one to analyse monitors up to eleven days after exposure and still comply with U.S.O.S.H.A. requirements for a 95% confidence envelope of $\pm$ 25%. Even storage at $15^{\circ}C$, enables analysis to take place up to ten days after and still be within the compliance limits. At higher temperatures, decay of ethylene oxide is much more rapid and it is preferable under such circumstances to analyse the monitors as soon as possible after exposures. However, the preferred protocol should require storage under controlled temperature conditions. If one were to store in a refrigerated container initially at $4^{\circ}C$ whose temperature drifted to 15° during storage, one could still maintain confidence in the validity of the results obtained even after many days of storage.

Because the decay process is likely to proceed by pseudo-first order kinetics, the decay function model assumes an exponential form. The regression is performed on the logarithmically transformed data. Thus the correction factor curve will not have a symmetrically disposed confidence envelope, the upper confidence envelope is wider than the lower envelope. The envelope broadens from the intercept as a result of increasing uncertainty as the decay process proceeds.

REFERENCES

Brown, R.H., Harvey, R.P., Purnell, C.J. and Saunders, K.J. 1984.
    A diffusive sampler evaluation protocol. Am. Ind. Hyg. Assoc. J.
    45, (2). 67-75.
Taylor, D.G., Kupel, R.E. and Bryant, J.M. 1977. "Documentation of the
    NIOSH Validation Tests" (U.S. D.H.E.W.). Publication No. 77-185.
Thompson, J.M. 1986. The use of a robust and resistant regression method
    for personal monitor validation with decay of trapped materials
    during storage. Anal Chemica. Acta. in Press.
Thompson, J.M., Sithamparanadarajah, R. and Stephen, W.I. 1984.
    "Adsorptive Gas Traps". U.K. Patent No. GB 2078128B.

FIELD COMPARISONS OF VARIOUS ACTIVE AND
DIFFUSIVE MONITORING METHODS FOR XYLENE
AND ISOPROPANOL

J. Thompson*, R. Sithamparanadarajah**

*Department of Chemistry,
University of Birmingham,
P.O. Box 363,
Birmingham B15 2TT
**Institute of Naval Medicine
Alverstoke, Gosport, Hampshire,
Great Britain

ABSTRACT

Field comparisons were made between NIOSH charcoal tubes and actively and diffusively sampled thermally desorbable Simtec Adsorba monitors in a histopathology laboratory. Paired comparisons of personal sampling for xylene were made using two kinds of pair : (a) actively and diffusively sampled Simtec Adsorbas, (b) NIOSH charcoal tubes and diffusively sampled Simtec Adsorbas. Comparisons for isopropranol were made at a fixed site between three kinds of monitor fixed in close proximity : NIOSH charcoal tubes and actively and diffusively sampled Simtec Adsorbas. No significant differences were found between the sampling methods either for xylene or isopropanol.

INTRODUCTION

In recent years trapping of organic vapours using tubular monitors packed with porous polymers followed by analysis using thermal desorption and gas chromatography has become increasingly popular (Gray and Thompson, 1984; Brown and Purnell, 1980; Crisp, 1980). Such monitors may be used in the active mode, coupled to a personal sampling pump, or in the diffusive (passive) mode (Gray and Thompson 1984). Recently the U.K. Health and Safety Executive have issued a protocol for validating the performance of diffusive monitors (MDHS 27). This paper describes field comparisons of NIOSH charcoal tubes and actively and diffusively sampled Simtec Adsorba Type P tubes monitoring for xylene and isopropanol. The field trials were performed in several histopathology laboratories where these solvents were the predominant ones in use.

EXPERIMENTAL

The trials involved the use of NIOSH activated carbon tubes and Simtec Adsorba Type P (Porapak-Q packaging) thermally desorbable monitors (Thorn EMI Electronics Ltd., Simtec Division, Sellers Wood Drive, Bulwell, Nottingham, U.K.). The Adsorba monitors are tubular type made from stainless steel tubing packed with adsorbent powder which is retained between two precision fitted sintered stainless steel mesh frits (Thompson, Sithamparanadarajah and Stephen, 1984). The Adsorbas were held in the

Simtec Adsorba Holder which enables a well defined diffusion path-length
to be maintained during sampling.

Field trials for xylene monitoring were of two kinds: (a) Paired
comparisons of actively and diffusively sampled Simtec Adsorba monitors
used for personal monitoring; (b) paired comparisons of diffusively
sampled Simtec Adsorba monitors and NIOSH activated charcoal tubes. The
sampling pumps used were MDA 808. Active sampling with the Adsorba
monitors was performed at 5ml min^{-1} and with NIOSH tubes at 25 ml min^{-1}.

Charcoal tubes were solvent desorbed with carbon disulphide in the
manner described by NIOSH (Manual of Analytical Methods (1977) Volume
1 and 2). Simtec Adsorba monitors were thermally desorbed at 230°C using
a Simtec Desorba thermal desorption oven.

Likewise, for isopropanol monitoring comparisons were made of NIOSH
charcoal tubes and actively and diffusively sampling Simtec Adsorba
monitors placed on a frame work in close proximity to one another at a
fixed position in the histopathology laboratory. Thermal and solvent
desorptions were performed as for xylene samples.

Chromatographic analysis for xylene was carried out using the following equipment and conditions : Pye 304 gas chromatograph with Flame
Ionization Detector (Temperature 250°C) fitted with a 50M SE 30 WCOT
Capillary column with oxygen free nitrogen carrier gas (inlet pressure
22 psi) and the injector temperature was 220°C. The temperature programme
was 2 Mins at 50°C. 7°C/Min to 200°C, 5 Mins at 200°C. For isopropanol
the variations on these conditions were : injector temperature 180°C,
temperature programme : 5 Mins at 35°C, then 5°C/Min to 50°C, hold for
0.1 Mins, then 50°C/Min to 150°C and hold for 5 Mins. Calibrations were
performed by spiking tubes in accordance with the U.K. Health and Safety
Executive protocol (MDHS 27).

RESULTS AND DISCUSSION

The paired comparisons of personal monitoring for xylene using
diffusively sampled Simtec Adsorba and actively sampled NIOSH charcoal
tubes are shown in Table One. No significant difference was found using
Wilcoxon Signed Rank Test (0.3<p<0.4) (Gibbons, 1976). The medians for
each group are very close (17.0 mg m^{-3} for the Adsorba monitors and
17.5 mg m^{-3} for the NIOSH tubes) as would be expected from the Signed
Rank Test. One can express reasonable confidence in the comparability of
results obtained using the two sampling methods.

TABLE 1 Comparison of personal monitoring for xylene using NIOSH
charcoal tubes and diffusively sampling Simtec Adsorba Tubes

Diffusive Adsorba	NIOSH Charcoal
Xylene Concentration mg m^{-3}	
14.7	17.5
13.1	23.8
15.3	17.3
21.2	18.9
18.8	18.4
49.6	25.3
12.2	12.9
34.4	41.0
18.3	15.9
12.2	17.5
3.9	9.0
23.6	18.7
10.7	9.0
9.9	8.2
90.7	108.4
23.1	20.0
17.0	13.7

TABLE 2 Comparison of personal monitoring for xylene using actively
and diffusively sampling Simtec Adsorba Monitors

Active	Diffusive
Xylene Concentration mg m^{-3}	
19.5	18.9
52.4	26.4
15.6	11.4
6.5	8.5
60.0	66.9
419	474
357	483
1.8	1.8
10.3	10.3
90.9	71.0
18.8	19.4
40.7	57.6
6.8	3.2
10.8	9.6
7.9	11.9
11.7	7.4
31.8	17.1
11.2	14.8
14.9	15.3
11.0	6.5

Paired comparisons of actively and diffusively sampled Simtec
Adsorba monitors are shown in Table Two. As with the comparison with
NIOSH tubes the diffusively sampled Adsorba shows no significantly

different behaviour from actively sampled Adsorba monitors. (Wilcoxon Signed Rank Test P = 0.5). The Medians for the groups are very close (active : 15.3 mg m^{-3} , diffusive : 15.1 mg m^{-3}).

The comparisons for isopropanol monitoring at a fixed site involving simultaneous sampling with NIOSH charcoal tubes and actively and diffusively sampled Simtec Adsorba tubes are shown in Table 3. Data analysis by Friedman's two-way analysis of variance by Ranks (Conover, 1980) show no significant differences between the sampling methods. (P > 0.25).

Thus for both xylene and isopropanol the Simtec Adsorba type P monitors were demonstrated to be equally as satisfactory whether used in the actively or diffusively sampled modes.

TABLE 3   Comparisons of NIOSH charcoal tubes and Simtec Adsorba Monitors for fixed site isopropanol monitoring

Actively Sampled Adsorba	Diffusively Sampled Adsorba	NIOSH Charcoal Tubes
6.5	8.1	3.9
11.4	18.5	17.2
0.3	13.9	15.6
17.4	9.7	19.9
60.0	44.6	85.9
28.4	27.3	25.0
37.5	22.5	49.8
38.8	20.5	39.1

REFERENCES

Brown, R.H. and Purnell, C.J. 1980. Collection and analysis of trace organic vapour pollutants in ambient atmospheres. The Performance of a Tenax-GC adsorbent tube. J. Chromatog; 178, 79-90.
Conover, W.J. 1980. "Practical non parametric Statistics" (Second Edition) (J. Wiley and Sons Inc, New York). PP 299-308.
Crisp, S. 1980. Solid Sorbent Gas Samplers. Ann. Occup. Hyg. 23, 47-76
Gibbons, J.D. 1976. "Non parametric Methods for Quantitative Analysis" (Holt, Rinehart and Winston, New York). PP 123-143
Gray, C.N. and Thompson, J.M. 1984. Passive and active atmospheric monitoring in "Recent Advances in Occupational Health". (Vol.2, Ed. J.M. Harrington)., (Churchill, Livingstone, London), Chapter 15
MDHS 27 Protocol for assessing the performance of a diffusive sampler (Health and Safety Executive, London).
NIOSH Manual of Analytical Methods. Method 127 for Xylene (Published by NISOH, USA).
NIOSH Manual of Analytical Methods. Method S65 for Isopropanol. (Published by NIOSH, USA).
Thompson, J.H., Sithamparanadarajah R. and Stephen, W.I. 1984. "Adsorptive Gas Traps" (U.K. Patent No. GB 2078128B).

MONITORING OF BENZENE, TOLUENE, XYLENE AND DIETHYLBENZENE
USING THE 3M ORGANIC VAPOUR MONITOR UNDER CONTROLLED CONDITIONS

J. Twisk[*], J. Urbanus[**]
[*]Dow Chemical (Nederland) B.V., Terneuzen, the Netherlands
[**]Dow Chemical S.P.A., Fombio (MI), Italy

ABSTRACT
     The performance of the 3M Organic Vapour Monitor (3M 3500) was tested under laboratory conditions for a mixture of benzene toluene xylene and diethyl benzene. The parameters studied were relative humidity, dosis, dose rate, sample retention, storage conditions. A charcoal tube method was used as the reference method. The results of the 3M monitor were in agreement with the results obtained with the charcoal tube method and it was concluded that the 3M monitor performed well under the conditions found in our work environment.

INTRODUCTION
     For the determination of employee exposure to airborne chemical agents the 3M organic vapour monitor is nowadays widely used within the Dow Chemical Company. Although the manufacturer (Anders, 1982) and others (Stockton, 1985) have evaluated the validity of this type of sampler, based on Dow's guidelines for the evaluation of passive dosimeters, still some essential information was missing for its use in the field. Particularly the performance of the monitor when a mixture of chemicals has to be sampled simultaneously under conditions of extreme high relative humidities as frequently found during outside operations was not sufficiently known. Therefore experiments were carried out under conditions of controlled relative humidities with a mixture of benzene, toluene, xylene and diethylbenzene. This moisture is frequently observed in our operations. Parameters studied were relative humidity, dose , dose rate, sample retention and storage conditions.

MATERIALS AND METHODS
     The experiments were carried out in a dynamic exposure chamber with a diameter of 0.09 m and a length of 0.5 m. The air velocity through the chamber was 0.15 m/s. To assure homogeneity a grid was placed at the entrance of the chamber. The monitors were placed along the axes of the chamber in a laminar position relative to the air flow. The various concentrations of benzene, toluene, xylene and diethylbenzene which were

injected in the main flow were obtained by passing relative small flows of
air either through an impinger filled with the liquid compound or through
a vessel containing small open vials with the liquid compound.

A first assessment of the concentration was made with a Miran 1A
infrared analyser. This analyser was also used to check the concentration
over the course of the experiment. The final assessment of the concentration during the experiment was carried out with a validated charcoal tube
method using pump flows of 140 ml/min in case of longterm experiments and
1000 ml/min in case of shortterm experiments.

EXPERIMENTAL DESIGN

In each experiment four monitors were exposed to the mixture at a
temperature of $20^\circ C$ and atmospheric pressure.

The concentration studied in the longterm (8 hour) experiment were
approximately 0.1, 0.5 and 1.5 times the occupational exposure limit of the
individual compounds (benzene: 10 ppm; toluene: 100 ppm; xylene: 100 ppm;
diethylbenzene: 10 ppm). The longterm experiments were carried out at 50 %
and 90 % relative humidity. The shortterm (15 min) experiments were carried
out at concentration up to 10 times the occupational exposure limit.

In the retention experiment 6 monitors were first exposed for 15
minutes at 50 % RH, then all monitors were taken out, the chamber flushed
with clean air for 15 minutes and then 3 of the 6 monitors used were
returned to be exposed to clean air for 6 hours.

In the storage experiment 3 of the 6 monitors used were analysed
immediately after exposure and 3 were analysed after 2 weeks storage at
$4^\circ C$.

RESULTS AND DISCUSSION

The results of the longterm and shortterm experiments are shown in
figure 1. As a good linear relationship between the yield of the monitor
and the dose determined by the charcoal tube method is observed under all
conditions, it can be concluded that relative humidity, dose and dose rate
do not effect the performance of the monitor. Only for benzene at the high
dose there seems to be some saturation, probably caused by competitive
co-adsorption.

The sampling rates, calculated from the slope of the lines, are in
agreement with the sampling rates reported by 3M (within 20 %) or the
sampling rate calculated by means of the Gilliland approximation (see table 1)

TABLE 1. Comparison of sampling rates found and sampling rates from 3M* or Gilliland approximation.

| | sampling rate (cc/min) | | |
	found	3M	%
Benzene	29,1	34,8	84
Toluene	26,4	29,9	88
Xylene	25,9	25,5	102
DEB	18,3	25,1**	73

* Corrected for desorption efficiency (Rodriguez 1982)
** Gilliland approximation

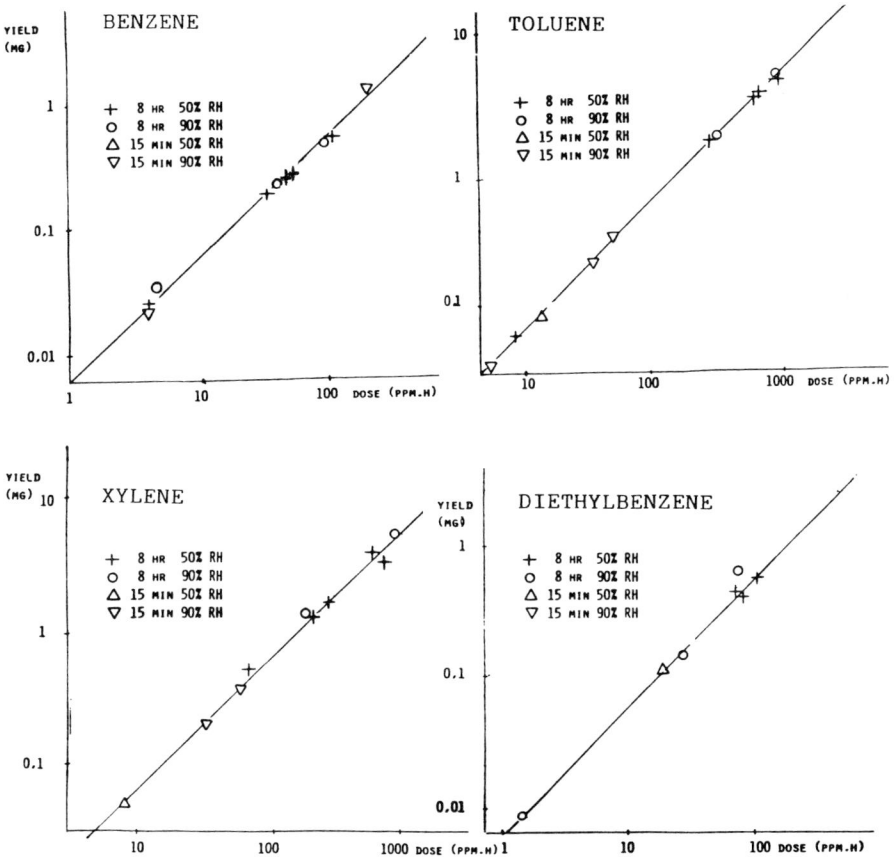

Fig. 1 Yield of the 3M monitor compared with the dose obtained by the charcoal tube method under different conditions.

It therefore can be concluded that the sampling rates determined for chemicals sampled as a mixture do not differ substantially from the sampling rates, determined when the chemicals are sampled alone. Only at high doses of all compounds the possible effect of competitive co-adsorption should be watched.

The retention experiment showed no losses for benzene and diethylbenzene but was inconclusive for xylene and toluene. For the last two compounds an increased yield was found after exposure to clean air.

The storage experiment showed that the exposed monitors can be kept in a refridgerator (4°C) easily without any loss.

CONCLUSION

Our findings indicate that the 3M monitor is a reliable device to measure occupational exposure to mixtures of benzene, toluene, xylene and diethylbenzene under the condition found in our work environment.

REFERENCES

Anders, L.W. etal. 1982. Organic Vapor Monitor with backup section # 3520 Diffusional Monitoring Semmar 3M.

Rodriguez, S.T. etal. 1982. Determination of desorption efficiencies in the 3M 3500 Organic Vapor Monitor. Am. Ind. Hyg. J. , 43 , 569-574.

Stockton, D.S. and Underhill, D.W. 1985. Field evaluation of passive organic vapor samples. Am. Ind. Hyg. Assoc. J. 46 , 526-531.

ON THE PERFORMANCE OF DIFFUSIVE SAMPLERS FOR STYRENE,
ACETONE AND METHYLENE CHLORIDE

J.F. van der Wal [1]
F.L Schulting [2]

[1] TNO Division of Technology for Society, The Netherlands
[2] TNO Medical Biological Laboratory, The Netherlands

INTRODUCTION

In the polyester industry in the Netherlands occupational health surveys are carried out. For those surveys both diffusive samplers and pumped charcoal tubes are used.

Styrene is applied as a monomer and a diluting solvent, while acetone and methylene chloride are used as solvents. For this occupational health survey, in an exposure chamber two types of diffusive samplers were investigated on the sampling performance as to styrene, acetone and methylene chloride.

MATERIALS AND METHODS

The experiments were carried out in an exposure chamber (volume 20 m^3) where a constant concentration of a vapour or a mixture of vapours could be generated and maintained. The chamber is of the dynamic type.

Two types of diffusive monitors: the ORSA-5 of Dräger and the OVM 3500 of 3M, together with one type of pumped charcoal tube, Jumbo of SKC were simultaneously exposed to known concentrations of the chemicals.

The experiments with the individual chemicals were performed according to the following scheme:

Sampling time	½ hour	2 hours	4 hours
Concentration			
0,1 TLV	x		x
1 TLV		x	
2 TLV	x		x

This scheme is a concise version of the HSE protocol [1]. Of the ORSA-5 [3] six samples were taken, and of the OVM 3500-samples [4] and Jumbo tubes three of each (sampling flow rate of Jumbo tubes 100 cm^3 min^{-1}).

With the three samplers two experiments were performed with mixtures of the three chemicals. Each chemical has the following concentrations of: 0,1 TLV, 0,5 TLV, 1 TLV and 2 TLV. For each of the four concentrations experiments were carried out for 2 and 4 hours.

The Dutch TLV's are styrene: 100 ppm; acetone: 1000 ppm and methylene chloride: 200 ppm. [2]

The concentration of vapours in the chamber was monitored with a Miran IB infrared detector (Wilks). The average concentration during the sampling period has also been determined by weighing the consumed liquid that was fed to the chamber; the average concentration is the ratio of the weight on the one hand and the product of the air flow rate and the sampling time on the other. The difference in concentration resulting from the two methods was less than ± 10%.

The samples were desorbed with CS_2 and analyzed. For this analysis a Varian gas chromotograph with a glass capillary SE-30 column (length 60 m, diameter 0,7 mm) and a flame ionization detector were used. The samples were analyzed within 3 days and storaged in a fridge (-4°C) after sampling and before analyzing.

The desorption efficiency (DE) was determined by spihing the cheer coal with the liquid chemicals. The loaded monitors were storaged at -4°C for at least 24 hours, thereafter desorbed with CS_2, and analysed.

RESULTS

The conditions during the experiments were the following: temperature: 20-25°C, R.H: 30-54%, air speed along the surface of the diffusive monitors: 15-25 cm^{-1}, chamber flow rate: 120-200 m^3 h^{-1} (ventilation rate 6-10 h^{-1}). "Starvation" at the surface is not to be expected [5,6]. The uptake rate (UR) must be calculated according to

$$U.R \quad = \quad \frac{ng \; adsorbed}{DE.c.t}$$

The desorption efficiency DE after the adsorption in the gaseous phase could be different from DE after spiking with liquid. In fact, U.R and DE are both unknown.

The results are given for the observed uptake rate (UR_{obs}), defined as

$$UR_{obs} \quad = \quad \frac{ng \; in \; eluate}{c.t.}$$

which is equal to UR.DE.

The results are summarized in Tables 1 and 2 in the Figures.
Conversion factors can be calculated from the tables, from the results derived from the uptake rate given by the manufacturers [3,4] and from the
desorption efficiency measured after spiking and will be given in Table 3.
The results indicate a low adsorption capacity for acetone and methylene
chloride particularly for the OVM 3500.
The charcoal tubes showed no break-through at the levels up to 8 TLV.h.

Table 1   Observed uptake rate of individual chemicals.

	ORSA-5	OVM 3500	Jumbo
Styrene			
UR_{obs} . ng ppm^{-1} min^{-1}	19.0	87.6	
CV, %	6.3 (n=4)	6.5 (n=3)	
DE, % (spiking)	85	94	82
UR_{obs}, ng.ppm^{-1} min^{-1}	21.5	107	
(UR manuf. DE spiking)			
Acetone			
UR_{obs} . ng . ppm^{-1} min^{-1}	13.6	56.6	
CV, %	8.5 (n=6)	14.6 (n=3)	
DE, % (spiking)	67	84	64
UR manuf. DE, ng.ppm^{-1} min^{-1}	14.4	79.9	
Methylene chloride			
UR_{obs} . ng ppm^{-1} min^{-1}	26.8	88.0	
CV.%	10.7 (n=6)	2.9 (n=2)	
DE.% (spiking)	100	100	100
UR manuf. DE, ng.ppm^{-1} min^{-1}	30.6	132	

Table 2   Observed uptake rate of mixtures.

Chemical	UR_{obs}.ng.ppm^{-1} h^{-1}		CV.%	
	ORSA-5	OVM 3500	ORSA-5	OVM 3500
Styrene	17.8	88.1	8.1	5.9
Acetone	12.6	54.9	5.8	15.5
Methylene chloride	21.2	67.9	7.4	23

Table 3   Conversion factors (95%-confidence level).

Chemical	ORSA-5	OVM 3500	Jumbo
Styrene	1.13 ± 0.14	1.22 ± 0.16	1.08 ± 0.10
Acetone	1.06 ± 0.18	1.41 ± 0.42	0.81 ± 0.13
Methylene chloride	1.14 ± 0.24	1.50 ± 0.08	1.03 ± 0.02

5.   LITERATURE REFERENCES

[1]   Health and Safety Executive: Occupational Medicine and Hygiene Laboratory.
Protocol for assessing the performance of a diffusive sampler.
MDHS 27, methods for the Deteximination of Hazardous Substances (1983).

[2]   Dutch Labour Inspectorate.
National MAC-list 1985, p. 145 (TLV..).

[3]   Manual, Dräger, ORSA 5 (1985).

[4]   Manual.
3M Organic Vapor Monitors (1981).

[5]   Gosselink, D.W. et al.,
A new personal organic vapor monitor with in situ sample eluation.
Am. Ind. Hyg. Ass. National Meeting San Francisco (1978).
Also non-dated publication by Occupational Health and Safety Products
Division, 3M Company, St. Paul, Minn.

[6]   Wal, J.F. van der, A. Moerkerken,
The performance of passive diffusion monitors for organic vapours for
personal sampling of painters.
Ann. Occup. Hyg. 28 (1984), 49-47.

FIGURES

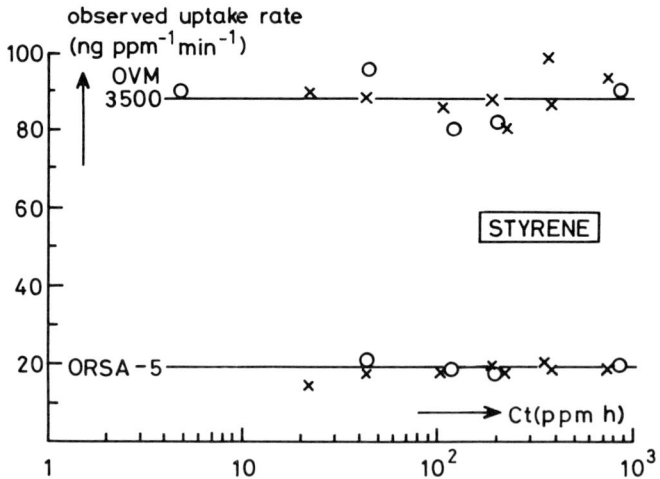

Fig. 1   Observed uptake rate of styrene.
          o individual chemical   x   in mixture

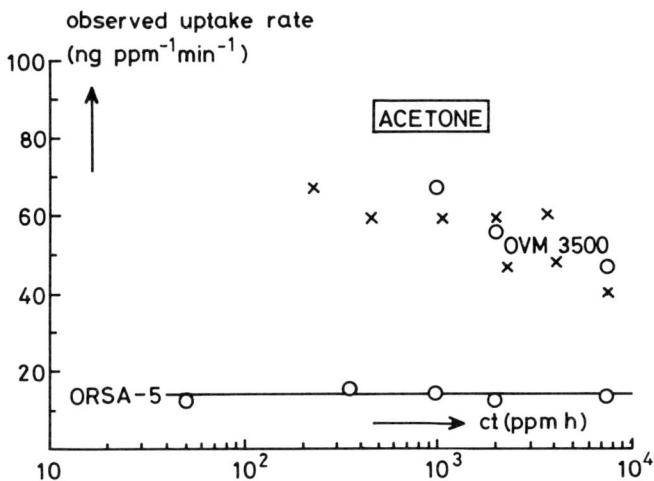

Fig. 2   Observed uptake rate of acetone.
    o   individual chemical   x   in mixture

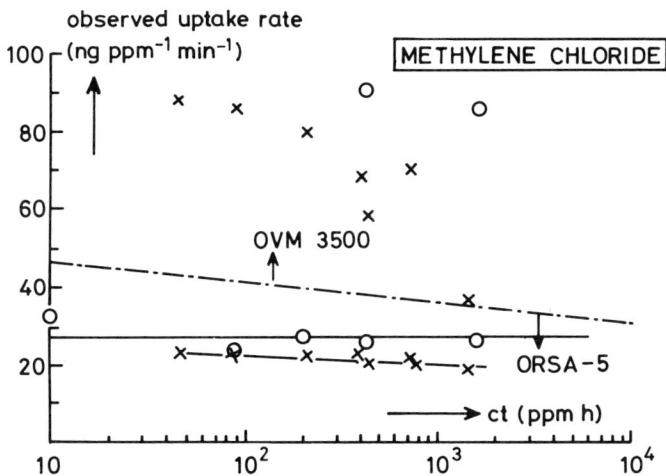

Fig. 3   Observed uptake rate of methylene chloride.
    o   individual chemical   x   in mixture

APPARATUS FOR THE PREPARATION OF ATMOSPHERES WITH KNOWN CONSTANT CONCENTRATIONS OF ORGANIC SOLVENTS.

N. Zurlo, L. Metrico, C. Sala*, F. Andreoletti**

* Industrial medicine department, Lecco Hospital, Italy
** Clinica del Lavoro "Luigi Devoto", Milan University, Italy

SUMMARY

This paper describes a simple easy-to-use apparatus for the preparation of test atmospheres of known concentrations of organic solvents. The apparatus consists of a glass cylinder with two outlets at the base and a cylindrical loading device, fitted coaxially, with seven compartments for specific evaporators equipped with diffusion regulators.

The evaporators are loaded with 2 g of solvent (one solvent per evaporator), the loading device is introduced into the glass cylinder and the sealing cap fitted. Air is passed through the cylinder at a rate of 1-10 litres min-1.

By selecting suitable regulators and flow rates, it is possible to prepare atmospheres of constant concentration at constant temperature of all the solvents tested over a range from 0 to 2 g m -3.

The delivered concentrations are calculated from diffusion coefficients, vapour pressures and the geometry of the diffusion regulators and are verified by weighing the evaporators before and after use.

ABSTRACT

A simple, easy-to-use apparatus is described for the preparation of test atmospheres of known, constant concentration (from 0 to 2 g m -3) of a given organic solvent or a mixture of up to seven components, each with the required individual concentration.

PRINCIPLE

Atmospheres of known concentrations are produced by evaporating solvents in an evaporator specially designed to ensure constant diffusion over time at constant ambient temperature and pressure. The quantity Q of solvent diffused in one hour can be calculated as follows, where Ps is the vapour pressure (mm Hg) of the solvent placed in the evaporator, M its molecular weight (g), D its diffusion coefficient (cm2 S-1) and F a coefficient, the value of which depends on the geometry and dimensions of the diffusion regulators fitted to the evaporator:

$$Q = 0.2 \ PsMDF \ (mg \ h^{-1})$$

The solvents thus released into the atmosphere are entrained by a constant flow of air, which is suitably regulated, measured and extracted for use. Dividing the quantity of solvent diffused during the flow period by the total volume of the flow yields the concentration produced in the air passing through the system.

DESCRIPTION OF APPARATUS AND OPERATION

The apparatus consists of (Fig. 1):

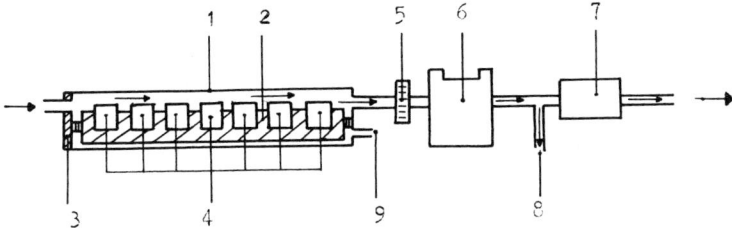

Fig. 1   Apparatus for the preparation of atmospheres of known concentration

- a cylindrical glass chamber 63 mm in diameter and 350 mm long (1);
- an aluminium cylinder with seven cylindrical cells to house the evaporators (2)
- sealing cap with air inlet (3)
- evaporators (4)
- rotameter with measuring range from 0 to 10 litres min-1 (5)
- volumetric gas-meter with temperature and pressure gauges (6)
- suction pump with throughput of 1 - 10 litres min-1 (7)
- outlet to a sampler downstream of the meter (8)
- outlet for sampling air directly from the chamber (9)

The evaporators consist of cylindrical teflon vessels, 3 cm in diameter and 2.8 cm deep, fitted with diffusion regulators in the form of truncated cones with a 60 degree aperture, or cylinders, of various cross sections, giving F values of 1 and 0.5 in the case of the conical evaporators, and 0.2, 0.08 and 0.02 in the case of the cylindrical evaporators.

In order to increase the amount of solvent evaporated, the system can be used without regulators or a suitable shim can be used to reduce the depth of the vessels, to give F values of 3 and 7 respectively.

During operation, the vessels are protected from the turbulence of the air by means of a fibre glass filter. After operation the evaporators are sealed with teflon caps.

Table 1 shows the quantities of some common solvents which evaporate in one hour at 25 degrees C from an evaporator using a truncated-cone diffusion regulator with a coefficient of F=1.

Solvent	mg	Solvent	mg
acetone	258	cellosolve acetate	3
Methylethyl ketone	126	ethyl acetate	144
amyl alcohol	3.5	methyl acetate	331
butyl alcohol	8	chloroform	412
ethyl alcohol	53	trichloroethylene	171
isopropyl alcohol	52	toluene	40
hexane	183	xylene	12

Over the temperature range which is relevant for practical purposes, the amount of solvent evaporated varies by approx. 5% per degree centigrade change in the temperature of the air.

In each case, the quantity of solvent evaporated during the air-flow period is checked by weighing the evaporator before and after, using a precision balance accurate to 0.1 mg. By selecting the number of evaporators used, the appropriate diffusion regulators and the throughput of the pump, it is possible to produce flows of air over a range of known constant concentrations.

ATMOSPHERIC HUMIDITY

During evaporation, hydrophilic solvents absorb water from the atmosphere at the rate of 0.94 F mg h -1 per mm Hg of water vapour partial pressure; for example, with a truncated-cone regulator with a coefficient of F=1, at 20 degrees C and a relative humidity of 35%, 50% and 70%, 5.8, 8.3 and 11.6 milligrams respectively of water are absorbed per hour and affect the change in weight of the evaporator.

However, this error can be prevented or offset either by drying the air upstream of the apparatus or by determining the quantity of water absorbed by a reference evaporator.

SOLVENTS IN THE GASEOUS PHASE AT AMBIENT TEMPERATURE

If the solvent(s) under consideration is in the gaseous phase at ambient temperature, a solution containing a known quantity of the solvent S dissolved in another solvent (A) is placed in the evaporator.

Given Q_s and Q_a, i.e. the quantity (mg) of the solute and solvent respectively, and MA, i.e. the molecular weight (g) of solvent A, the quantity Q of solvent S evaporated in time (t) is as follows:

$$Q = Q_s \left(1 - \exp - \left(0.2\, P_s\, \frac{M_A}{Q_A}\, DFt \right) \right)$$

where Ps and D are the vapour pressure and the diffusion coefficient of solvent S at ambient temperature and F the coefficient of the particular diffusion regulator employed.

CONCLUSIONS

The apparatus described above can be used for calibrating passive or active sampling devices, determining the collection efficiencies of solid sorbents, determining empirically the diffusion coefficients and vapour pressure of the solvents, biological testing and any other activities requiring calibrated atmospheres.

INFLUENCE OF DIFFUSIVE SAMPLING ON MONITORING STRATEGIES

J. AUFFARTH
Bundesanstalt fuer Arbeitsschutz
Republic of Germany

SUMMARY

The aim of a strategy for monitoring threshold limit values
for exposure to hazardous substances at the workplace is to
obtain reliable information on the health risks to workers at
minimum cost.   Such a strategy calls for the performance of
measurement tasks that make differing demands on the capabili-
ties of the method of measurement used. These requirements
are described and compared with the characteristic features of
diffusive samplers.   It will be shown that, in certain
circumstances, diffusive samplers can make a significant
contribution to determining exposures at the workplace.

- Monitoring Strategies, Occupational Exposure Limits, Threshold Limit Values, Diffusive Sampling

INTRODUCTION

The title contains an assumption - namely that the functioning
of a measuring instrument could provide guidelines for a
sampling strategy.

In my view, the reverse is the case. The familiar strategies
for monitoring workplaces lay down requirements against which
the suitability of the instruments selected must be assessed.
In what follows, I will discuss the consequences this has for
diffusive samplers.

THRESHOLD LIMIT VALUES

To protect workers against risks, threshold limit values are
set for exposure to hazardous substances at the workplace.
These limits provide a standard for assessing the work area.
If they are observed, it may be assumed that hardly any of the
workers in this area will suffer damage to health - even in
the long term - as a result of the substances in question.

Threshold limit values consist of numerical data on concentrations together with a set of parameters, e.g.   time scale,
rules for measurement, range of validity and group of persons
concerned.   These may differ widely depending on the country
and the safety objective pursued:

- normally, there are upper limits on average exposure over
  the course of working shift (i.e. 8 hours) - I therefore
  refer to them as <u>time-weighted averages</u>;

- maximum exposures can be limited over shorter intervals as well. Depending on the substance, these short-term limits cover periods ranging from a few seconds (ceiling limits) to several minutes or more, e.g. 5, 10, 15, 30 or 60 minutes. The minimum interval between two exposure peaks and the maximum frequency of such events per shift are also laid down (e.g. in the USA and the Federal Republic of Germany)

MONITORING STRATEGIES

Monitoring strategies have been developed to reduce the amount of measuring required (Health and Safety Executive, 1984; German Federal Ministry of Labour and Social Affairs, 1984). they demonstrate how relatively reliable data can be obtained on occupational exposures at minimum cost.

It is advisable to proceed in two stages:

- First, a work area analysis is carried out to establish which hazardous substances may be present in the ambient air and whether the current threshold limit values are being complied with.

If necessary, compliance is achieved by employing technical measures to reduce exposure.

- Subsequently, regular long-term monitoring is carried out to check for continued compliance with the threshold limit values.

The work area analysis also determines whether exposure peaks occur and checks for compliance with short-term limits.

If the concentrations of harmful substances are found to be well below the threshold limit values, at levels which are no longer significant, further measurements can be dispensed with - provided that exposure conditions in the work area do not change.

These are the essential elements of a monitoring strategy, whereby existing "a priori" knowledge should be complemented by appropriate measuring programmes to ensure reliable results.

MEASUREMENT TASKS

The procedure outlined above requires the following different tasks to be carried out:

- initial exploratory measurements at the start of the work area analysis

- precise determination of concentrations during the work area analysis

- measurement of concentration peaks.

Each task places different demands on the method of measurement used. In addition to analytical performance, consideration also needs to be given to handling and cost of operation. The criteria to be taken into account include the following:

- performance:

specificity, class interval, time resolution, measuring range, output of measuring results, operational availability;

- handling:

expertise required of measurement technician, user friendliness, display of operating states;

- cost:

investment costs, operating costs.

DIFFUSIVE SAMPLERS

What diffusive samplers can and cannot do is explained in detail at this Symposium. In terms of monitoring strategy, I would characterize them as follows:

- diffusive samplers are used chiefly for personal sampling;

- if certain conditions are met, they allow the "average" concentration of a substance in the ambient air to be determined from the amount of substance collected during the sampling period;

- they are supplied by numerous manufacturers in a wide variety of types - sampling characteristics have so far not been standardized;

- they can be used for one or more substances or groups of substances depending on type;

- they possess a sampling phase with a limited take-up capacity;

- overloading will not normally be detected;

- they can be deployed in large numbers on account of the low unit costs;

- they can be used by non-experts because of their simple construction;

- an analytical laboratory is required to evaluate the results.

PRECONDITIONS FOR SAMPLING

What role can diffusive sampling play in workplace monitoring, and how should this be taken into account in the monitoring strategy?

However it may be carried out, sampling decides whether the results of workplace measurements are valid. It is of crucial importance for monitoring threshold limit values. there is also no way of compensating for sampling errors. Responsibility for sampling should thus be in the hands of the expert who ultimately evaluates the workplace.

Sampling can only be planned by someone familiar with conditions at the workplace, which includes information on

- the substances present (chemical inventory);

- activities and procedures;

- technical safety facilities;

- time spent by workers at the workplace;

- occurrence of exposure peaks (when, where, how often?).

These data are used to delimit workplaces or group them together into work areas and to determine the nature of the measuring programme required.

Since it is now known what substances may be expected in the monitored area, information is required on the likely concentrations of these substances in the ambient air. Indications as to the levels of these concentrations are provided by:

- existing information (measurement results) on comparable workplaces,

- reliable calculations,

- exploratory exposure measurements.

Depending on whether the concentrations thus determined are

i)   - well below the threshold limit values ($\ll$ 1/10),

ii)   - in the range of the threshold limit values, or

iii) - above the threshold limit values,

i)   - further measurements can be dispensed with,

ii)   - efficient methods of measurement should be used to determine precise concentrations, or

iii) - technical measures should first be implemented to reduce exposure

EXPLORATORY MEASUREMENTS

Diffusive samplers may be useful for exploratory single estimates of exposure, since

- personal sampling is the method to be used,

- even for a large group of workers, points where high exposure levels occur can be identified by using samplers on a large scale;

- depending on the sampling system used, a large variety of substances can be recorded simultaneously, and

- no great demands need be placed on the efficiency of the method of measurement.

However, precautions must be taken to rule out incorrect negative findings (e.g. with spot checks using other methods). Close cooperation with the laboratory carrying out the analysis is essential.

Examples of such exploratory measurements are survey measurements at workplaces where substances containing solvents are hadled, such as adhesives, paints, thinners, etc. Experience has shown that activated charcoal samplers can be used to obtain an accurate picture of the level of exposure to solvent vapour mixtures, together with the variety of substances involved, at relatively low cost.

PRECISE DETERMINATION OF CONCENTRATIONS

Should the exploratory measurements indicate that concentrations may be in the range of the threshold limit values, a precise analysis of these concentrations is called for. The method of measurement used has to meet certain performance requirements as regards selectivity, accuracy, reproducibility and range of analysis. The results obtained directly affect the measures taken to protect the health of workers.

In this case, diffusive samplers are suitable for single estimation if

- the substances to be measured can be reliably recorded in the prevailing work conditions;

- it can be guaranteed that the capacity of the samplers will not be exceeded;

- the calibration function used reflects conditions correctly during sampling at the workplace, and

- comparability of the characteristics of the diffusive samplers used is ensured by standardization.

Diffusive samplers can then be employed in a way similar to collector tubes - as is already being tried in practice.

However, our experience indicates that the suitability of diffusive samplers for carrying out precise analyses of concentrations needs to be examined form case to case, e.g. by means of parallel measurements. At the same time it can also be established whether diffusive samplers are also suitalbe for follow-up monitoring, where necessary.

Here, it should be pointed out again that sampling determines the validity of the measurement results and must be carried out by an expert.

FOLLOW-UP MONITORING

If the work area analysis shows that the threshold limit values are observed, this result is checked over the long term by regular follow-up monitoring.

The details of the method to be used are laid down in advance, taking into account the prevailing work conditions. The substances present in the workplace atmosphere and their concentrations are known. The sampling interval and the number of samples to be collected are fixed so that themeasing results are representative of the exposures occurring in the monitored area. The result measured is the time-weighted average.

It is advisable to fix the measuring frequency depending on how the measured rtesults compare with the threshold limit value. The closer the measured concentrations are to the limit, the more frequently measurements need to be carried out.

If the concentrations measured are well below the limit and will remain so for the foreseeable future, regular monitoring can be dispensed with completely (apart from occasional checks).

It will thus be evident that diffusive samplers can play an important role here.

- Long-term regular monitoring must be carried out at minimum cost.

- The conditions governing the use of diffusive samplers can be precisely determined and, where appropriate, checked by means of comparative measurements.

- As the sampling procedure is defined in detail, it can also be carried out by works personnel who have received appropriate training.

- The measured result, the time-weighted average, can be compared directly with the threshold limit value and a time can then be fixed for the next monitoring programme.

Here again, however, it is essential to rule out the possibility of incorrect negative findings and to guarantee the proper functioning of the sampling system, e.g. by means of standardization.

SHORT-TERM MEASUREMENTS

It is not possible to register short-term exposure peaks with a diffusive sampler.

Opening and closing the sampler and waiting for steady operating states mean that a minimum period is required for sampling. The accuracy of the method of analysis need not necessarily be a limiting factor, since the concentration peaks of interest may be many times the threshold limit value.

It is conceivable that 15-minutes or half-hourly averages may be obtained without great difficulty. However, the points in times when exposure peaks occur will need to be determined in other ways (e.g. through knowledge of the work process or by using an instrument with a continuous display).

Measuring programmes of this kind from part of the work area analysis. It would be impractical to make short-term events the subject of regular follow-up monitoring.

CONCLUSION

Like other sampling methods, diffusive sampling has its limitations. It must be integrated within a monitoring strategy to ensure that reliable information on possible health risks to workers can be obtained through monitoring threshold limit values in the ambient air at the workplace. This strategy involves a comprehensive work area analysis to establish (and, where necessary, ensure) compliance with threshold limits and subsequent regular follow-up monitoring.

The primary goal of a monitoring strategy is to ensure that valid findings are obtained. Measurements and other sources of information all contribute to this end. The extent to which supplementary information is required depends on the effectiveness of the method of measurement used. In our view, diffusive sampling promises to provide an extensive source of information for assessing exposures at the workplace.

BIBLIOGRAPHY

Health and Safety Executive, 1984. Guidance Note EH42 Monitoring strategies for toxic substances. (HSE, London).

Bundesminister fuer Arbeit und Sozialordnung, 1984. Technische Regeln fuer gefaehrliche Arbeitsstoffe TRgA 402: Messung and Beurteilung von Konzentrationen gefaehrlicher Arbeitsstoffe in der Luft - Anwendung von Maximalen Arbeitsplatzkonzentrationen (MAK). Bundesarbeitsblatt 11/1984, 55-61.

ROLE OF DIFFUSIVE SAMPLING IN WORKPLACE MONITORING
COST-BENEFIT ANALYSIS

P.B. Meyer

TNO Division of Technology for Society
Schoemakerstraat 97 - 2628 VK Delft
The Netherlands

ABSTRACT
    Application of cost-benefit analysis to data submitted by manufacturers and occupational health services under certain stated assumptions shows that passive sampling in the personal monitoring of gases and vapour is a very attractive proposition. However, this paper recommends the simultaneous preservation of active sampling facilities.

INTRODUCTION

    The aim of cost-benefit analysis (Mishan, 1971) is to assess in terms of money all the benefits and costs of a project and to give it the go-ahead if the former exceed the latter by a sufficient margin. This is realized in practice by listing all aspects affected by the project, and later expressing them in terms of money. The cash flows resulting from this operation are used to calculate certain criteria, viz. - pay-back period, net present value, internal rate of return (see Appendix I) - in which costs and benefits are so compared that the result can be expressed in a single figure whose magnitude is an indication of the "sufficient margin" mentioned in the first sentence of this paragraph.

    What has been said so far is illustrated by the example of a health service whose management wishes to consider the implications of switching from active to passive personal monitoring of gases and vapours. Finally, it should be clear that the theory of cost-benefit analysis is only partly covered by this relatively simple practical example.

QUANTITATIVE ANALYSIS

    To provide the necessary data for the decision implied in the introductory section following aspects have to be taken into account:

A. Cost

Initial investment

In the example dealt with in this paper, we assume an initial investment of twelve instruments for active sampling. For twelve instruments the cost would be ECU 12000, comprising ten instruments in use and two as back-ups. It is assumed that, the expected lifetime of the instruments is three years, if not more than 1800 samples are taken with them annually. For 1800 - 3000 samples the expected lifetime is reduced to two years. The salvage value of the instruments is set at zero.

It should be remarked, finally, that the investment costs for passive samplers, in our present example, viz - one-time use, throw-away types of samplers, are nil.

Operating cost

- Maintenance of active samplers

This covers the costs of keeping the equipment in working order, including calibration, and is here assumed to be ECU 0.5 per sample. For passive samplers it is nil.

- Materials

This item comprises non-durable goods such as batteries, adsorption tubes and passive sampling badges.

The cost amounts to   ECU 2 per sample for active sampling
(batteries can be recharged ∿ 100 x at 0.5 ECU per sample;
adsorption tubes used in combination with
sampling pumps cost ECU 1.5 each)

                            ECU 12 per sample for passive sampling

- Salaries

These cover the salaries of the hygienists employed in carrying out and reporting the exposure measurements. If overheads are included, they are estimated to be ECU 400 per man day. It is further assumed that one hygienist can supervise at most ten simultaneous exposure determinations if the sampling is "active", and at most thirty when it is "passive".

As a first approximation the effort a hygienist spends in sampling is taken to be proportional to the number of samples taken. Strictly speaking, this is true only if the number of active samples taken ples" a multiple of approxim. 30. It should be borne in mind that the example worked out here serves as one of a technique that can readily be refined so as to take account of other restrains that practice might impose.

- Overheads
  These include, general expenses not easily attributable to any specific activities. Examples are clerical costs and costs of attendance at meetings which are of a general nature.

B. Benefits

The benefits are calculated on the basis of a fee of ECU 55 per sample. This, as well as the cost indicated in A, is based on information supplied by manufacturers and occupational health services working in the field. It should, however, be borne in mind that practice varies considerably.

C. Qualitative considerations

A passive sampler has the advantage of forming no encumbrance to the person whose exposure has to be measured and, in addition, needs no maintenance.

On the other hand, those types of passive samplers of which we have extensive experience, need more care in handling because they are more vulnerable than active samplers, and any damage is not easy to detect. Passive samplers are generally less sensitive than active samplers, and saturation of them is more difficult to establish.

Finally, passive sampling needs less supervision than active sampling as far as the functioning of the instruments is concerned. Under complicating circumstances, e.g. when correlations have to be drawn between degree of exposure and activities of exposed workers, this paucity of supervision may lead to misinterpretation of the results of measurements. These considerations are based on extensive experience and, though they are admittedly restricted to a single type of passive sampler, can be summarized as follows:

active sampling should continue to be used in those cases where low concentrations have to be monitored of, e.g., very toxic substances with a mac value of less than 5 ppm and in places exposed to very high concentrations that may cause saturation of passive samplers. All things taken together, it is assumed that an occupational health service attached to a variety of industries and encountering a variety of working conditions has to use active sampling for 20% of samples to be taken. On this assumption the cash-flow analysis set forth below is based.

Cash-flow analysis

The criteria discussed in Table 2 of this section have already been alluded to in the introduction, and are explained in appendix I. They are based on cash-flows - see appendix II - generated by the activities under consideration, and listed in Table 1.

TABLE 1   Cash Flow Analysis based on the assumption that 600, evenly distributed, samples are taken annually. Time horizon is arbitrarily taken to be three years (= 12 quarters)

A. Active sampling

costs and benefits in ECU per quarters \ year/4	Invest-ment	operating costs		fees received for sampling services	net cash flow before tax
		costs excl. of salaries	salaries		
0	12,000				- 12,000
1		$\frac{1,500}{4}$	$\frac{24,000}{4}$	$\frac{33,000}{4}$	$\frac{7,500}{4}$
2		$\frac{1,500}{4}$	$\frac{24,000}{4}$	$\frac{33,000}{4}$	$\frac{7,500}{4}$
⋮		⋮	⋮	⋮	⋮
12		$\frac{1,500}{4}$	$\frac{24,000}{4}$	$\frac{33,000}{4}$	$\frac{7,500}{4}$

B. Passive sampling

costs and benefits in ECU per quarters \ year/4	Invest-ment	operating costs		fees received for sampling services	net cash flow before tax
		costs excl. of salaries	salaries		
0	nil				nil
1		$\frac{7,200}{4}$	$\frac{8,000}{4}$	$\frac{33,000}{4}$	$\frac{17,800}{4}$
2		$\frac{7,200}{4}$	$\frac{8,000}{4}$	$\frac{33,000}{4}$	$\frac{17,800}{4}$
⋮		⋮	⋮	⋮	⋮
12		$\frac{7,200}{4}$	$\frac{8,000}{4}$	$\frac{33,000}{4}$	$\frac{17,800}{4}$

C. 20% active and 80% passive sampling

costs and benefits in ECU per quarters year/4	Invest-ment	operating costs		fees received for sampling services	net cash flow before tax
		costs excl. of salaries	salaries		
0	12,000				- 12,000
1		$\frac{6,060}{4}$	$\frac{11,200}{4}$	$\frac{33,000}{4}$	$\frac{15,740}{4}$
2		$\frac{6,060}{4}$	$\frac{11,200}{4}$	$\frac{33,000}{4}$	$\frac{15,740}{4}$
12		$\frac{6,060}{4}$	$\frac{11,200}{4}$	$\frac{33,000}{4}$	$\frac{15,740}{4}$

TABLE 2   Results of "Decision Criteria" calculations (before tax)

	pay-out time in years	net present value in ECU	internal rate of return in % per quarter
case A (active sampling)	1.6	5,800	11
case B (passive sampling)	0	42,400	∞
case C (our mixture of passive and active sampling)	0.8	25,500	31.5

Even for a non-profit organisation, whose costs are paid by clients-contributors a situation with which occupational health services are normally faced - it could be argued that at least some of their activities should show a reasonable rate of return. This is true particularly for air sampling, where it is difficult to predict the number of samples to be taken because of the possible introduction of new techniques or sudden alterations in processes. If one assumes that a quarterly 3.75%, approximately 16% on a yearly basis, before tax is a reasonable rate of return for such an organisation, Table 2 shows that from a purely economic point of view, case B (passive sampling) is by far the most attractive option, though even case A (active sampling) falls within economic bound while C (being the more realistic option as a mixture of active and passive sampling) is more attractive than A. In Fig. 1 the internal rate of return is plotted against the number of samples taken annually. It shows that under the assumption of a cut-off rate of 3.75% per quarter before tax, the minimum number of samples to be taken annually to ensure continuity of service from an economic point of view falls from some 400 to 190, increasing considerably the flexibility of the operation due to the introduction of passive samplers.

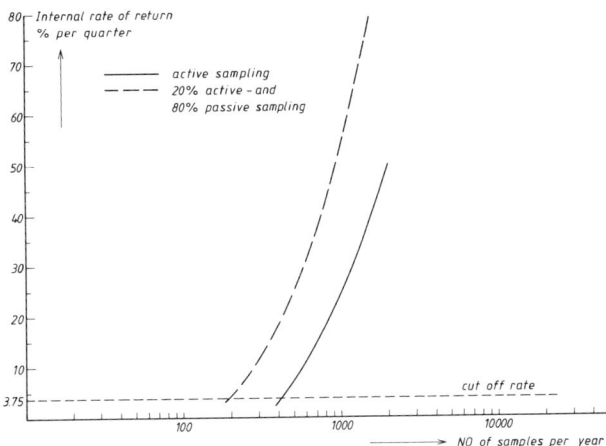

Fig. 1   Internal rate of return as a function of the number of samples taken per year.

Another possibility would be to consider reducing the sampling fee for those cases in which it can be predicted with certainty that the number of samples to be taken exceeds an annual 190. Such a price cut might constitute an incentive towards improved monitoring strategies.

COST-BENEFIT ANALYSIS

Our analysis has so far been limited to the point of view of the manager of an occupational health service. Cost-benefit analysis from a macroeconomic point of view - as alluded to in the introduction - is a somewhat different matter. Its aims are to assess the social impact of an activity, and to appraise the costs and benefits to all parties affected by that activity.

If we restrict ourselves here to hygienists and the workers they have to monitor we can make following qualitative statements:

- the number of samples to be taken is a compromise between what is economically possible and desirable from a viewpoint of occupational hygiene. Reducing the cost of monitoring, as suggested in the previous section, may prove to be an incentive towards more frequent monitoring, and thereby towards better adherence to hygienic standards.

- as far as the hygienist is concerned, the lower cost of passive sampling is due its being less time-consuming. The time saving in this way will now be spent on tasks that have so been neglected, e.g. improving the comfort of workers (temperature, humidity, etc.).

For completeness sake, it should be pointed out again that cost-benefit analysis claims to take into account all factors, e.g. the loss of sales of active sampling equipment, resulting in a loss of profit to its manufacturer, and a gain of profit to the manufacturer of passive sampling equipment is another of those factors. However, such considerations clearly fall outside the scope of this paper. We conclude with the remark that introduction of passive sampling - even under the restrictive Pareto criterion 3) - see appendix II - is an attractive proposition.

CONCLUSION

- A cost-benefit analysis based on data supplied by manufacturers and occupational health services, together with some additional assumptions, shows that the introduction of passive sampling is very attractive from an entrepreneur's point of view. Qualitative considerations show that, even in a much wider sense, passive sampling has a positive social impact under the prevailing circumstances.

- Notwithstanding the attractiveness of passive sampling, active sampling facilities should be preserved.

- The techniques of cost-benefit analysis set forth in this paper can readily be adapted to encompass other assumptions, and data, which can vary widely.

Summarizing finally, we believe we can state that, owing to its inherent simplicity, passive sampling can be regarded as a technological breakthrough in the personal sampling of people exposed to gases and vapours.

REFERENCES

Alfred, A.M. and Evans, J.B. 1971. Discounted cash flow. Chapman & Hall
   3rd ed. London.
Mishan, E.J. 1971. Cost-benefit analysis. Allen & Unwin LTD (London).
Prest, A.R. & Turvey, R. Chapter I: from Layand, R. (Ed.) 1976. Cost benefit analysis. Penguin, Harmondsworth.

APPENDIX I   METHODS FOR THE ANALYSIS OF THE RATE OF RETURN

PAY-BACK PERIOD METHOD

The simplest method is called the "payback period".
In this method, the time to return the original investment is calculated,
as the following examples show:

Investment: ECU 800,-		
Cash flows:	Project A	Project B
Year:		
1	ECU 400.-	ECU 100.-
2	ECU 400.-	ECU 200.-
3	ECU 200.-	ECU 200.-
4	ECU 100.-	ECU 200.-
5		ECU 300.-
6		ECU 400.-
Payback period:	A 2 years, B 4 1/3 years	
	A is better	

The "payback period" method has obvious drawbacks:

1) Income and costs beyond the payback period are ignored.
   However, people sometimes deliberately choose rather short periods of
   3-4 years, because they feel that forecasts beyond such a planning
   horizon will be inaccurate.

2) The interest factor is not taken into consideration. This neglects
   that certain amount of money to be received at some future date does
   not have the same "value" as the same amount received to day.

For these two reasons, the payback period favours quick returns on projects and disfavours projects beneficial only in the long run. More sophisticated methods of evaluating the future profitability of an option
have therefore been developed. These methods are based on the principle of

finding the value of future profits if an investor could realize them to-day. This principle can be approached in two ways known as "disounted cash flow" techniques.

Net Present Value (NPV)

The NPV is defined as the present values of the time sequence of project income less the present values of project costs.
This technique aims to recalculate prospective costs and revenues in terms of today's money, or, translated into a numerical example ECU 110 a year from now is worth ECU 100 today at the current interest rate of 10%. The formula for the present value

$$P = V (IF)$$

where

P = present value

V = future values

IF = interest factor

The interest factor (IF) may be calculated as follows:

$$IF = \frac{1}{(1 + i)^n}$$

where i = interest rate

n = number of year at the end of which the future amount has to be paid/received.

The computation of present values is called discounting.

In many cases an investor has to borrow money from a bank at certain interest rate. It can easily be understood that the return of such a project should not be less than the interest cost, because he has to pay interest on the banks loan at all times.

The net present value method for making investment decisions comprises the following steps:

1) Find all expected net cash flows as a function of time.

2) Present values should be calculated by recalculating these cash flows in terms of to day's money by discounting at an appropiate interest rate and subsequently summing them.

The NPV method accepts all independent projects whose NPV is greater than zero and, in ranking mutually exclusive projects, selects the project with the higher NPV.

Internal Rate of Return (IRR)

The internal rate of return is defined as the rate of discount which equates the present value of the respective time sequences of income and costs, resulting in a NPV = 0.

This is a variation of the "net present value" method, but it seeks to calculate the interest rate rather than to choose it as an assumption. Then one has to decide whether this is an adequate return.

This trial and error procedure is illustrated in Fig. 2, using the cash flows of project A (page 9).

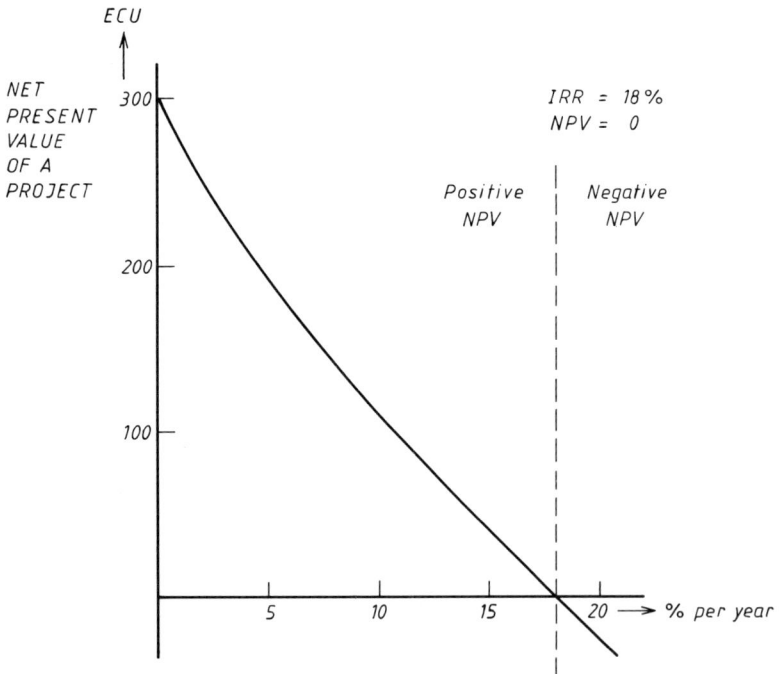

Fig. 2   Graphical determination of the internal rate of return.

APPENDIX II   GLOSSARY

Cash flow   - Refers generally to the amount of money coming
                        into and out of a business in a particular peri-
                        od. In this paper money paid in connection with
                        the investment - amortization and interest pay-
                        ments - are excluded.

Cut off rate   - E.g. figure for the internal rate of return on
                        which a go, no-go decision is based.

Discounted cash flows - The present value of future cash flows depends
                        on when - that is how soon - they are earned,
                        because the money when invested could be earning
                        interest, see appendix I.

Pareto optimum   - Named after the Italian economist Pareto. It
                        describes circumstances in which nobody can be
                        made better off without making somebody else
                        worse off. Pareto improvements ought to be pos-
                        sible, e.g. by introducing new technologies. It
                        is claimed in this paper that the introduction
                        of diffusion samplers under the discribed cir-
                        cumstances provides such an example.

SOME PRACTICAL ASPECTS OF THE USE OF DIFFUSIVE SAMPLING

IN OCCUPATIONAL HYGIENE

B.S. BORD

Technology and Air Pollution Division
Health and Safety Executive
Magdalen House Stanley Precinct
Bootle, Merseyside L20 3QY
United Kingdom

ABSTRACT

Occupational Hygienists have great needs for personal sampling techniques
of airborne contaminants which are economic, catch significant activities
and provide sufficient data. By making sampling more feasible, diffusive
methods may have an impact on the legal and moral pressures to sample.
Additionally, the quality of the data obtained may be improved by the
effect on the behaviour of people sampled. Diffusive sampling may have
some significant limitations for Occupational Hygiene but three field
examples are used to show the many benefits. There are ethical reasons
for using diffusives where possible.

SOME PRACTICAL ASPECTS OF THE USE OF DIFFUSIVE SAMPLING IN OCCUPATIONAL
HYGIENE

Introduction

Occupational Hygiene may be described as the recognition, evaluation
and control of materials which cause risks to health in the workplace. The
evaluation part of this is only effective if one may obtain high quality
data. In most cases this reduces to a need to measure concentrations of
airborne contaminants to which people are exposed. There are some
important exceptions to this where such measurements alone are not
sufficient, e.g. materials which are absorbed by or affect the skin;
carcinogens or biological agents.

Most Occupational Hygienists work in industry, for governmental
agencies or as independent professional consultants. Although their
roles vary the practical professional difficulties experienced by them
are very similar. Broadly these are:

i)   the problem of getting work done at a reasonable cost

ii)   catching the significant events

iii) obtaining enough data to make meaningful interpretation

There are of course differences in emphasis. For example,
Occupational Hygienists employed by governmental agencies about which I
speak with most direct knowledge may carry out surveys for enforcement
purposes or to gain information for national policy needs. However,

the basic field problems remain. Diffusive samplers are making a
contribution to the reduction of these problems.

Personal Sampling

As in most other countries Occupational Hygienists in the UK prefer to
use personal samples where possible. Our Occupational Exposure Limits are
usually defined on the basis of such sampling. The arguments for personal
as opposed to background sampling are well developed. Sampling in the
breathing zone of a person at work approximates most closely to his
exposure when engaged in that work. It is particularly important when
contaminants are generated by the individuals close activity in processes
such as welding, hand painting or bag opening.

Less obviously, personal sampling is necessary when a worker moves
position and his total exposure is a summation of those from different
points in space and time. An operation on a large chemical plant is a
classical example of this. The optimum strategy for such workplaces is
a combination of personal supplemented by fixed location sampling, the
latter needed to identify high concentration areas and to target those
processes requiring improvements.

Legal Frameworks

Legal frameworks differ from country to country. In British
legislation dealing with health and safety at work there is a key duty
on employers and others to ensure so far as is "reasonably practicable"
the health and safety of employees. Such an imprecise concept introduces
many difficulties but one benefit is that when technical advances are made
such as more knowledge about a hazard or improved methods of control,
there is a strong legal as well as moral pressure to use them.

Thus, since assessments are necessary prerequisites for workplace
improvements any techniques which make them easier to carry out by
reducing costs or making them more feasible are themselves tipping the
scales towards greater "reasonable practicability". In simple terms this
means that diffusive sampling may have a significance beyond its elegance
as an analytical method. It may alter not only what can be done but what
has to be done.

Sampling Hardware

Apart from the feasibility and costs aspects of diffusive sampling techniques a more subtle change in the quality of information from Occupational Hygiene surveys is achievable by unobtrusive equipment. Given the importance of personal sampling it is equally important that the operator continues to work normally. There is no point in using high precision analytical methods if the accuracy of our assessments is limited by induced changes in behaviour patterns of the operator.

Experience of sampling thousands of people in the UK by the Health and Safety Executive leads one to reflect that although very few refuse to be sampled many are affected by the act of sampling. Behaviour patterns are altered both consciously and unconsciously expecially when samplers are obtrusive. Although direct evidence of this is hard to find and without well-designed research must be anecdotal, experience leads one to suspect that the effect is important. Towards the end of a shift a re-establishment of more relaxed and normal attitudes to work are often observed and for example the abnormally increased use of controls such as respirators and ventilation equipment lapses. Workers not engaged in demanding or interesting tasks are most affected by wearing conventional pumps. It is also harder for physically slighter personnel to forget them.

Where extensive regular sampling is part of an ongoing monitoring and surveillance **requirement it is obvious that the benefits of** clip-on light weight samplers are most needed. In these cases large savings are obtainable by having less qualified assistants putting out the samplers.

Occupational Hygiene Limitations

It is important not to allow the existence of a technically attractive method such as diffusive sampling to distort the priorities of Occupational Hygiene. At present the method is not available for particulates and many of the most significant common airborne contaminants are in this category, e.g. silica, asbestos, cotton, welding fume and toxic metals. Many inorganic gases cannot yet be sampled with diffusive techniques. We may thus have high risks associated with difficult sampling.

Short term samples are also very important where some examination of brief events is necessary. At present pumped air samplers may have advantages here because the effective air volume for analysis may be critical. Thus the decision to use this or that sampling technique is unlikely to be straight forward and the Occupational Hygiene problem in hand will always have to be evaluated.

Some Field Examples

The Health and Safety Executive carries out hundreds of Occupational Hygiene investigations per year and is using diffusive samplers where appropriate. To demonstrate some of the practical aspects of diffusive sampling three real field examples have been selected. The full reports have been condensed to give something of the nature of the visits and are outlined in the Appendix.

The first example is that of a factory in which high styrene levels were suspected from the manufacture of items from glass reinforced plastic. The second is an exercise in an operating theatre which did not have an operational scavenging system for waste anaesthetic gases. The final example is that of a large road tanker depot which was chosen as part of a national survey to establish the degree of exposure to benzene in various industries.

Costs

Examination of the costs of a unit devoted to Occupational Hygiene measurements by sampling and analysis show that the capital expenditures on routinely used hardware is swamped by the costs of providing staff on site. This is due in part to the particularly long sampling periods necessitated by various national and international standards. The smallest marginal unit of staff supply is usually a man–day of scientific or Occupational Hygiene time. At a bare minimum this must be costed at £100 with double this figure a realistic possibility for some cases.

Since the arrangements for the visit, the attendance, the report writing and the subsequent discussions are necessary whatever on-site methods are used, the main cost variable is the marginal requirements to have second or third person to assist. In our first example which involved 17 personal samples, the use of long term diffusives enabled 1 person to put 10 badges on and then concentrate on the seven short term pumped samplers. In the road tanker example, two persons handled 21 samples. In both cases more personnel would have been needed with diffusives – possibly doubling the cost.

A most expensive failure of a survey is when a return visit is required because inadequate data was obtained because of pump failures. Diffusive samplers have an inherent reliability in this. In our first and third examples it is easy to see what would have been the disastrous effects of 20% pump failures.

Nor, should it be forgotten that equipment which is irritatingly buzzing and uncomforable to wear presents a temptation to any wearer to remove it, especially when working for a significant fraction of the day in isolation from the Occupational Hygienist as in the cases of the road tanker drivers. Even more mundane is the probability that samplers will be removed to facilitate basic calls of nature.

Feasibility

    There are frequent cases where personal sampling is exceedingly difficult with pumped equipment. One such example is that of the operating theatre where the problem of putting pumps onto theatre workers in a sterile workplace is almost insurmountable. Restrictions on worker movements are quite unacceptable.

    The time required to put a conventional sampler onto a worker may be significant. An experienced scientist or Occupational Hygienist would be wise to allow three minutes per sampler which could include flow adjustments as well as attention to physical considerations of personal comfort. In the road tanker depot example some 21 samplers were to be put onto the drivers and clearly the original delivery operation would have been severely disrupted.

Limitation

    In some circumstances it may be hoped that a very wide issue of clip on samplers will yield considerable data. This is probably true if personal airborne concentrations of contaminants is all that is required and the information is to be used for global statistical purposes or to inform a person when his exposure has been excessive in a similar way to radiation monitoring badges.

    However, much of Occupational Hygiene measurement has as its purpose the control of undue exposures by pinpointing those activities or areas which need improvement. Without concurrent observations of workers being sampled the results are not very meaningful. There is no more distressing result for an Occupational Hygienist than a high figure which relates to unknown events. If one of the benzene figures in the tanker driver study had been high it would have been useful data for the national survey but not easily relatable to the particular driver and his working conditions.

    In most cases the value of results are maintained only by making the Occupational Hygiene observations. The correct use of diffusive sampling is to liberate the field workers to enable them to spend relatively less time on hardware details and more on observations.

Ethics of Sampling

Finally, a brief thought should be given to the ethical aspects of sampling. Most Occupational Hygienists approach personnel on site with derived authority which is explicit or implied and induces workers at some inconvenience to themselves to co-operate with measurement exercises. As a consequence we do of course make sure that our purposes are serious and likely to be worthwhile. It should also be important to us that workers are put to a minimum of inconvenience and discomfort. This is doubly true when people are engaged in dirty or dangerous work. The professional Occupational Hygienist would consider for this reason if for no other he should use diffusive sampling if technically available.

Acknowledgements

Thanks are due to members of the Scottish Field Consultant Group of the HSE in Edinburgh who supplied field examples of the use of diffusive samplers.

APPENDIX

THREE CASES FROM THE FIELD

CASE 1   HSE SURVEY, SCOTTISH FCG,
                                ABERDEEN, SEPTEMBER 1985
                                (EXTRACT FROM FULL REPORT)

A. PROBLEM: i)   High airborne styrene levels were suspected
                        in factory when articles were made of Glass
                        Reinforced Plastic, especially buoys and
                        tanks for North Sea use.   25 people employed.

                  ii)   Processes included hand lamination, spray
                        lamination and work inside buoys.

                  iii) Factory was about 40m x 60m, 10 - 15m high.
                        General ventilation by 6 roof openings.

                  iv)   Employees worked to no fixed pattern, shifts
                        were 10½ hours to 14 hours on overtime.   RPE
                        used for some jobs included both air fed and
                        charcoal filter types.   Exhaust booths used
                        for some work.

B. SURVEY: i)   10 Workers sampled for 8 hours using diffus-
                        ion badges estimates of 10½ and 14 hours
                        exposures.

                  ii)   7 short term samples taken using battery
                        driven pumps and charcoal packed tubes.

                  iii) 4 medium term samples taken for backgrounds
                        using pumps.

                  iv)   analyses were by solvent extraction and gas
                        chromatography in HSE laboratories in
                        Edinburgh.

                  v)   One HSE Occupational Hygiene Inspector carried
                        out the sampling and investigation at the
                        workplace

C. RESULTS

   (Control Limits for styrene)   100ppm 8 hour TWA)
                                             250ppm 10 mins TWA)

Table 1   Personal Exposures to Styrene (8-hour samples)

*Samples (10 Workers' exposures)	Measured Concentration (ppm)	8-hour TWA for 10½ hours (ppm)	8-hour TWA for 14 hours (ppm)
1 Hand spray lamination	58	76	102
2 hand spray lamination	59	77	103
3 hand spray lamination	47	62	82
4 no laminating	25	33	45
5 hand laminating	16	21	28
6 hand laminating	9	12	16
7 hand laminating	27	35	47
8 hand laminating incl inside buoy	90	118	158
9 spray laminating	51	67	89
10 spray laminating	48	63	84

Table 2   Personal Exposures to Styrene (short term samples)

*Samples (7 Workers' exposures)	Period (mins)	Concentration (ppm)
11 Spray lamination	95	137
12 Lamination	105	160
13 Spray lamination	105	108
14 Spray lamination	38	40
15 Consolidation	45	52
16 Hand lamination	140	44
17 Inside buoy	116	527

Table 3   Background concentrations of styrene

Sample	Period (mins)	Concentration (ppm)
18 Mid Shop	120	7
19 Joiners' Shop	360	3
20 Spray area	420	6
21 Spray area	420	34

Table 4 Air Velocity Measurements

Location	Velocity (m/s)
1 Booth A - at face	0.30-0.40
2 Booth A - 2½m from face	0.15-0.2
3 Booth B - at face	0.10-0.15
4 Booth C - at face	0-0.30

* NOTE Names are not given in this summary but full information was
given to all persons involved.

D. RECOMMENDATIONS

i)   Reorganisation of work inside booths
needed and the establishment of booth
curtains and turntables to ensure airborne
styrene carried away from breathing zones,
checks on air flow rates needed regularly.

ii)   thorough training needed in use and
        maintenance of RPE and in confined space
        work.

E. EFFECT OF USING DIFFUSIVE SAMPLERS

i)   One HSE Inspector was able to carry out the
        survey without assistance because most
        samplers were easy to clip on and collect.
        This left him free to examine processes and
        observe the work patterns.

ii)   No samples were lost due to pump failure;

iii)   If diffusive samplers were not used a
        Scientific Officer would have been needed
        to prepare more equipment, travel to
        factory, attend for a day and return to base.

CASE 2   HSE SURVEY ; SCOTTISH FCG;
                    HOSPITAL IN EASTERN SCOTLAND, SEPTEMBER 1984
                    (EXTRACT FROM FULL REPORT)

A. PROBLEM   i)   An operating theatre was without
                           anaesthetic gas scavenging systems. Staff
                           were exposed to nitrous oxide, Halothane
                           and Ethrane. Some 12 patients were dealt
                           with during 2½ hours of urethroscopic
                           examinations.

                   ii)   Anaesthesia was induced using gases and
                           intraveneous drugs. An anaesthetic trolley
                           was used for nitrous oxide, liquid Halothane
                           and Ethrane was taken from bubblers.

                   iii)   Each urethroscopic examination took about
                           5 - 10 minutes; staff included an
                           anaesthetist, a surgeon, two nurses, a
                           theatre orderly and an auxillary. One
                           coffee break was taken.

iv) Theatre doors were opened after each test; air change designed on a basis of 16.8 per hour, i.e. supply was 1280 cfm and theatre volume 4560 cf.

v) Although no occupational exposure limits were available for anaesthetic gases recommended good practice levels have been established.

B. SURVEY

i) Personal sampling was conducted on all theatre staff and HSE personnel for the entire session. Two diffusive samplers were attached to gowns close to the breathing zone on each person except for those on the surgeon who had to scrub up and change after each test. A sampler was put on his head covering instead.

ii) Sampling was by passive tubes and buttons. Analyses were carried out by solvent and thermal desorption for the organics and nitrous oxide respectively followed by gas chromatography. This was done at the Edinburgh and London HSE Laboratories.

C. RESULTS

ppm

		Mins	Nitrous Oxide	Halothane	Ethrane
1.	Anaesthetist	165	180	9	2
2.	Surgeon	159	N/D	7	2
3.	Sister	160	38	7	2
4.	Nurse	62	32	4	1
5.	Theatre Orderly	163	77	8	2
6.	Auxillary	165	55	7	1
7.	HSE Inspector	160	6	4	1
8.	HSE Scientist	106	30	2	1

NOTES: Halothane is $CF_3CBrClH$
Ethrane is $C_3H_2ClF_5O$
* Names were not given in this summary but full information was given to to all persons involved.

<u>D.</u> <u>RECOMMENDATIONS</u>   i)   Results were not unduly high compared with
                                                similar workrooms tested elsewhere.

                                        ii)   Nitrous oxide levels below 25ppm are usually
                                                obtainable with gas scavenging and therefore
                                                it was agreed that the hospital would install
                                                a suitable system.

<u>E. EFFECT OF USING DIFFUSIVE SAMPLERS</u>

                                        i)   This visit would have been impracticable with
                                                conventional pumps, because they could not
                                                have been fixed to sterile gowns by belts.

                                        ii)   Any restrictions on the movements of staff
                                                would have been potentially hazardous in an
                                                operating theatre. Likewise adjustments to
                                                pumps and checks on air flows could have
                                                interfered with the work in progress.

                                        iii)   The diffusive samplers enabled a great deal
                                                of data to be obtained on one visit; without
                                                them two samplers per person would have been
                                                very difficult and a second visit necessary.

CASE 3   <u>HSE SURVEY, SCOTTISH FCG</u>
                                        <u>GRANGEMOUTH, FEBRUARY 1985</u>
                                        <u>(EXTRACT FROM FULL REPORT)</u>

A.   PROBLEM   i)   At a large Road Tanker Depot motor spirit
                                                tanker drivers and yard operators were
                                                required to be tested for benzene exposure
                                                for a national survey.

                                        ii)   Visit had to start at 05.30 hours to enable
                                                drivers to be reached before setting off for
                                                first deliveries.

                                        iii)   There were 5 bays for motor spirit and DERV

with additional bays for Gasoil and aviation fuel. No vapour recovery or ventilation systems installed. Fuels loaded by drivers and unloaded by them at customer. Drivers were given itineries and usually took two loads.

B. SURVEY

i) 19 Drivers and two yard employees sampled over periods 6 to 8½ hours.

ii) Passive Tenax absorption tubes used over full shifts. Analyses carried out at HSE laboratories in London using solvent extractions and gas chromatography.

C. RESULTS

(Recommended Limit for Benzene 10ppm 8 hour TWA)

	Exposure Minutes	TWA ppm
1. Driver 2 loads spirit	381	0.04
2. Driver " " "	503	0.02
3. " " " "	451	0.12
4. Driver 1 load spirit, 1 DERV	526	0.09
5. Driver 2 loads spirit	428	0.04
6. " " " "	380	0.26
7. " " " "	433	0.10
8. Driver 1 load spirit, 1 Gasoil	431	0.08
9. Driver 1 load spirit ,	389	0.04
10. Driver 2 loads spirit	397	0.12
11. Driver 1 load spirit, 1 Gasoil	442	0.04
12. Driver 2 loads spirit	497	0.21
13. Yard Operator	472	0.07
14. Driver 1 load spirit	497	0.04
15. Driver 2 loads spirit	484	0.11
16. Driver " " "	498	0.15
17. Driver 1 load spirit	457	0.04
18. Driver 2 loads spirit	490	0.20
19. " " " "	466	0.19
20. " " " "	407	0.14
21. Terminal Operator	357	0.07

* NOTE: Names are not given in this summary but full information was given to all persons involved.

D. RECOMMENDATIONS

No recommendations were made following this exercise.

E. EFFECT OF USING DIFFUSIVE SAMPLERS

i)   The important benefit was that the
men could be away without checks on
pumps being necessary. A great deal
of data was obtained.

ii)   Putting 21 samplers onto men at
05.30 hours could not have been
achieved with pumped samplers. At
3 minutes per man it would have taken
an hour and seriously delayed the
operation.

iii) People sitting in driving positions do
not like pumps attached to their waists,
the noise and the discomfort would have
been very irritating. There would have
been a risk of pumps being taken off in
the drivers' cabs.

iv)   The data was obtained with one HSE
Occupational Hygienist and a Scientific
Officer.

DISCUSSION - SESSION V

BERLIN (CEC)

Is the introduction of diffusive sampling likely to change the current practice in the workplace; in terms of monitoring stategy, better assessment, reduced costs, and so on? Or maybe more important than a reduction in cost of monitoring is an increased flexibility at the same costs and consequently better monitoring.

WEISSKOPF (FRG)

It is quite correct that flexibility is achieved, for example I don't have to look at how a pump is functioning, but I think in the end the expense as compared to passive samplers isn't very different.

COKER (UK)

I agree that the differences in a routine monitoring programme are fairly minimal in the costs involved. I can't agree with the figures in manpower involvement that Mr. Meyer presented, which were a factor of three different. In our experience the difference in manpower involved in using the different types of sampler would be quite minimal.

WOOLFENDEN (UK)

Mr. Meyer didn't include in those figures the initial validations work which have to be done.

BERLIN (CEC)

Mr. Bord mentioned the problems of distortions caused by operator behaviour during sampling. How important you think this phenomenon is in relation to somebody who is sampled very occasionally compared to somebody sampled much more frequently?

BORD (UK)

This matter has not been researched and it needs to be investigated. If people are regularly sampled I think that they will behave differently from those people who meet a hygienist once every ten years who is doing an investigation. But also it depends on the kind of work they are doing. I have a feeling that the kind of work that Mr. Coker might investigate would be quite different from the transient jobs that we have to investigate, work which is done once or twice and we don't have a second chance to get the results. People paint, they blast, they demolish, they do all kinds of things in industry and we need measurements on what they are doing and I think the quicker we can get them used to a sampler, the more accurate our evaluation is going to be of their exposure. I think some of these pumps that are available at present are extremely difficult to wear if men and women are doing arduous jobs.

BLOME (FRG)

I had the impression that Mr. Bord wanted to take as many samples as possible even if the workplaces were similar. I'm wondering whether given the high costs that are incurred through the analysis and the assessment of the results whether this is economically viable.

BORD (USA)

We are investigating particular workplaces where a potential hazard has been identified. We do as it happens take all the data back, but there is certainly not a cemetery of data. Our data base will be useful for all kinds of epidemiological purposes later.

GEIGER (Germany)

NIOSH and OCEA have come up with some very good statistical sampling strategies for reducing the number of data points for determining an exposure profile. One can typically project from 30 or 40 data points the estimated range of exposure in a certain operation for given contaminants and that will greatly reduce the burden of sampling and the associated cost.

GROSJEAN (Belgium)

Mr. Leichnitz talked about the problems which small industries face when monitoring their work. I think that the fact of providing small industries with diffusive samplers on itself is not an answer to the question. The manufacturers also should provide some kind of service such as in the field calibrations or analysis and technical advice and then the answer is: are these passive or diffusive system such really attractive from a financial point of view compared to other techniques already widespread.

BERTONI (Italy)

From my experience in field work I never failed to take active samplers with me for greater measurement accuracy. I think that there are no doubts about their greater efficiency, but what is certain is that when you carry out a survey of the environment, any kind of sampler is generally going to provide results which are going to vary greatly from one point to another; from a difference in time, from one day to the other, from morning to afternoon, and so on. But it's very easy to multiply the number of data points by using passive samplers which are very cheap if the staff are already employed. The fact that a passive sampler might to some extent under or over-estimate for reasons of accuracy a concentration in a given point is not detrimental to the workers health. If one was to use only active samplers, he would be limited by problems of time, money and apparatus. The workers in question would never have even been surveyed if we had to rely purely on active samplers.

COCHEO (Italy)

I have the impression that we are assessing costs of monitoring, we've rather under-estimated the analysis costs.

BORD (UK)

I think that the bottleneck is in the field and is not in the laboratory. It is easier to expand the laboratory than the field force.

HARPER (UK)

I would like to make the point that with direct reading colorimetry you have sampling and analysis combined together. In countries outside Europe and North America you are going to require a tremendous volume of both sampling and analysis in the future. It does seem to me that diffusive sampling with direct reading colorimetry is the way forward to take Occupational Health into the Third World.

DIFFUSIVE SAMPLING PAPER MONITORING OF

SOME HIGHLY TOXIC GASES

I. MONITORING OF PHOSGENE

Wang Xi Lin, Li Mo Rong, Yao Mei Yun
Research Institute of Chemical Defence,
P.O.Box 1044 (300), Beijing, China.

ABSTRACT

A diffusive sampling paper monitoring method for phosgene was described. A stable NBP (4'-nitro 4-benzylpyridine) reagent was used as an absorbent for phosgene , the product formed was directly determined by
spectrophotometer. Defined recovery factor was obtained through the use
of calibration curve made with standard phosgene solutions. This method
can be used for monitoring phosgene TWA concentration ranged from 0.01ppm
to 10 ppm within an inaccuracy of ± 10%. Using sampling paper of larger
area, as low as 5 ppb phosgene can be detected.

INTRODUCTION

There are many kinds of highly toxic gases in the workplace of chemical factories, such as the phosgene, cyanogen chloride, hydrogen cya-
nide,etc., they are highly toxic to human. Especially the phosgene, it is
a lung irritant and cause severe damage to the alveoli of the lungs. Because phosgene toxication has a distinct incubation period, serious symptoms may not develop until several hours exposure, a person who appears
slightly gassed immediately after exposure may become casualty several
hours later. Chinese Government always pays much attention to the safety
control of phosgene contamination of workplace environmental air. The
Ministry of Labor has recommended a Threshold Limit Value (TLV) of 0.1 ppm
for phosgene. Many kinds of monitoring method of phosgene were developed.
In this paper, we describe a new, more effective and less costly diffusive sampling paper monitoring method acceptable to determine phosgene at
TWA concentration.

SAMPLING

For monitoring of phosgene, many types of detector tube, detector
paper and other methods were developed. Detector tubes used in China's
chemical factories now are mainly short term type, it is not suitable to
determine TWA concentration. There are many methods for long term sampling
of toxic gases, e.g. NIOSH long term activated charcoal tube, ORSA 5
passive sampling method. These methods can be used for various kinds of
organic compounds, but   can not be used for phosgene, because phosgene

is highly volatile and is easily hydrolyzed on charcoal. So we have to search for new methods.

For the sampling of toxic gases, especially the volatile gases, the main point is to choose a suitable absorbent. As we know , the selection of absorbent is governed by:

a. The gas must be absorbed by absorbent physically or chemically.

b. The absorbent can react with the gas to form new compound, which is not easily volatilized.

c. The new compound is durable in the ambient conditions.

d. The new compound can be determined by simple methods.

Many reagents suitable to analyze phosgene can be used as absorbents, e.g. aniline, phenylhydrazine and other amino compounds. After experimental examinations, we find that 4'-nitro-4-benzylpyridine is the most suitable absorbent. It reacts with phosgene very quickly to form a stable compound which can be determined by spectrophotometer directly. The supporter of absorbent can be made from different materials, e.g. silica gel paper, glass fiber paper, polypropylene paper, Whatman filter paper, etc. We find that the Whatman paper is the best.

MONITORING

Analysis of phosgene by NBP method was first recommended by Witten, who used   BA (4-benzylaniline) to detect phosgene in air (Witten,1957). There were many improvements to this method, such as the use of DEP (diethyl phthalate) as solvent, accurate controlling the reagent concentration (Dixon, 1959; Linch, 1965; Madbuli, 1971). Searching for methods which can recover reaction product absorbed on paper quantitatively, we examined the extraction efficiency of different solvents. We found that the conventional DEP solvent was not satisfactory, because the product formed by phosgene-NBP reaction only dissolved in DEP slowly. But with the use of absolute ethyl alcohol, we obtained satisfactory results. We also found that the ethyl alcohol solution of phosgene-NBP reaction product had different absorption curve from its DEP solution, the former had max. absorbence at 440 mμ, and the latter at 475 mμ (see Fig.1).

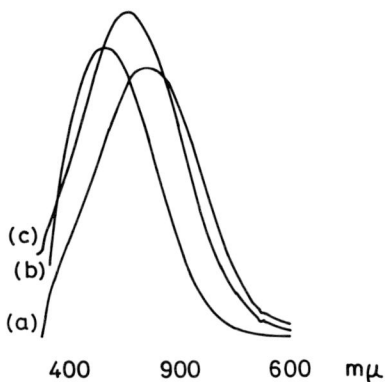

Fig. 1 Absorption curve of phosgene NBP reaction product in
different solutions
a. phosgene-NBP in DEP
b. phosgene-NBP in absolute ethyl alcohol
c. absolute ethyl alcohol solution of phosgene-NBP reaction
product on paper

In the analytical procedure, we use the absolute ethyl alcohol as
solvent and make a calibration curve with standard phosgene solutions of
different concentrationn. The recovery factor is defined as follows:

$$F_r = \frac{W}{C \times t}$$

F_r =recovery factor of sampling paper of definite area

W = dose of phosgene recovered in ethyl alcohol, µg.

C = actual concentration of phosgene in air, ppm.

t = sampling time, hr.

Table 1 gives the results we obtained from phosgene of different
concentration in air. The sampling paper (ϕ 24mm) was placed in air test
chamber filled with diluted phosgene of known concentration. After different sampling time, we extracted the colored product formed on paper
with absolute ethyl alcohol and measured its absorbence at 460 mµ with
spectrophotometer. The recovery factor was calculated from phosgene concentration, sampling time, and the dose of phosgene recovered in ethyl
alcohol.

Table 1   Recovery factor obtained from different concentration
           of phosgene

phosgene concentration	diameter of sampling paper	sampling time	dose of phosgene recovered in ethyl alcohol	recovery factor
ppm	mm	hr.	µg	µg/ppm hr
1.77±0.15	24	0.5	16.4	18.5
1.79±0.11	24	1	35.8	20.0
1.67±0.16	24	2	68.0	20.4
1.68±0.11	24	4	136	20.2
1.64±0.16	24	8	232	17.7
0.51±0.23	24	1	9.2	18.0
0.79±0.24	24	1	15.1	19.1
0.92±0.44	24	1	18.4	20.0
1.46±0.19	24	1	28.6	19.6
2.20±0.62	24	1	41.0	18.6
7.41±1.26	24	0.5	80.0	21.6
10.05±3.04	24	0.5	106	21.1

The results show that this method has good advantages in simultaneous sampling and determination of phosgene in air over a long period. We have determined the recovery factors for different time periods ranging from 1 to 8 hours. Their recovery factors show no difference within an inaccuracy of ±10%. It is satisfactory to the safety control requirements. We supposed that the random errors should come from various sources, such as temperature, moisture, air convection, and fluctuation of phogene concentration. This method also has the advantage that it is not affected by chlorine, which isa raw material in the production of phosgene,and is always coexistent with the phosgene.

As a matter of fact,the recovery factor determined here is actually a measurement of the rate of diffusion of phosgene molecules. Phosgene has a molecular mass of 99, its diffusion coefficient is approximately 0.08 cm^2/s. Then we can calculate its diffusion rate from Fick's law theoretically. Pannwitz recommended a method of calculating diffusion rate of organic compounds in diffusion collector ORSA 5, we suppose it also fit for our sampling paper (Pannwitz, 1984). But we have difficulties to define the exact value of diffusion section 1, so we can not check our recovery factor with theoretical value.

APPLICATION

This sampling paper has a good response to low concentration of phosgene ranged from 0.01 ppm to 10 ppm. At different concentrations we can use paper of different area. As we tested, sampling paper of larger area has higher recovery factor and better sensitivity to low concentration. Sampling paper of 36 mm diameter can be used to determine as low as 5 ppb phosgene within 8 hours, but under this condition the inaccuracy is rather higher. With sampling paper of different areas, we can get a series of calibration curve, each one has different recovery factor. We usually use the sampling paper of 24 mm diameter, it is satisfactory for safety control with high accuracy (see Fig. 2).

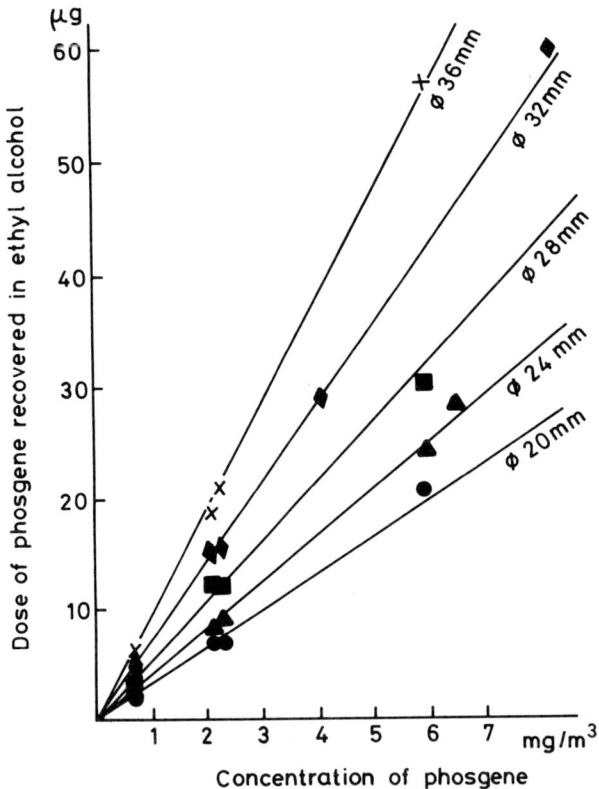

Fig. 2 Calibration curves of sampling paper of different area

This sampling paper is mainly used for personal monitoring, but it can also be used to make environmental evaluation of workplace and its circumstances. For the convenience of use, we made a sampling paper holder, which can be attached to a clothes button at a site as near to the nose as possible. The wearer can estimate the dose of phosgene inhaled during the working period by observing the appearance of the color on the paper. There is a warning color on the paper holder to show the threshold value of 6 ppm-hour. The exact dose inhaled must be determined by an industrial hygiene chemist.

REFERENCES

AIHA Analytical Guides 1969. Phosgene. Amer. Ind. Hyg. Assoc. J. 30, 192.
Dixon,B.E. 1959. A field method for the determination of phosgene. Analyst 84, 463.
Linch,A.L. 1965. Phosgene in air Development of improved detector procedure. Amer. Ind. Hyg. Assoc. J. 26, 465
Madbuli,H.N. 1971. An improved method for determination of phosgene in air Amer. Ind. Hyg. Assoc. J. 32, 163
Pannwitz,K.H. 1984. Diffusion coefficients. Drager review. 52, 1.
Witten,B. 1957. Sensitive detector crayons for phosgene. Anal. Chem. 29, 885.

FIELD MEASUREMENTS OF FORMALDEHYDE
FOR WORKPLACE MONITORING,
USING VARIOUS ACTIVE AND PASSIVE METHODS
FOR PERSONAL AND AREA SAMPLING

G.J. Wentrup
TUV Hannover e.V., D-3000 Hannover 81,
F.R. Brenk
M. Wenzel
B. Striefler
Niedersachsisches Landesamt fur Immissionsschutz,

At various workplaces in the furniture, textile and chipboard-manufacturing industries measurements were taken of formaldehyde using the following active and passive sampling methods:

ACTIVE SAMPLERS

- gas washing bottles filled with an aqueous, hydrochloric 2.4-dinitrophenylhydrazine solution (area and personal DNPH)

- glass-fibre filters impregnated with 2.4-dinitrophenylhydrazine and phosphoric acid (personal DNPH)

- sampling tubes filled with silica gel (DFG method and silica gel type B)

PASSIVE SAMPLERS

- diffusive samplers with coated stainless-steel screens (triethanolamine)
- diffusive samplers filled with silica gel (Orsa 5S)
- permeative/diffusive samplers with impregnated glass-fibre filters (2.4-dinitrophenylhydrazine and phosphoric acid)

In the presence of 2.4-dinitrophenylhydrazine, aldehydes and ketones react to form corresponding hydrazones. These are extracted using carbon tetrachloride, washed with 2N hydrochloric acid (to remove excess 2.4-dinitrophenylhydrazine), evaporated until dry and then taken up in methanol. These hydrazone solutions undergo HPLC (High-Performance Liquid Chromatography) analysis to determine the individual quantities of aldehydes and ketones (UV detection at 365 nm).

With the silica gel method, the formaldehyde adsorbed on the silica gel is desorbed by water free of formaldehyde and then mixed with solium tetrachloromercurate, sodium sulphite and pararosaniline. The formaldehyde content is subsequently determined photometrically, on the basis of the red-violet dye formed, at a wavelength of 570 nm. The triethanolamine diffusive samplers are analysed in the same fashion.

SUMMARY

Tables 1 to 3 summarize the results of the field measurements. A comparison of the results obtained with different methods at the same measuring location shows that although they correlate very well in some cases, they exhibit significant differences in others. This can doubtless be explained by the distribution of the measuring points within a measuring location and the different forms of sampling employed (area or personal), not to mention systematic errors in the methods used (e.g. non-representative samples).

For the above reasons, active sampling using DNPH-impregnated filters consistently yielded excessively low values and was therefore not included in the tables. Passive sampling with triethanolamine was also excluded on account of widely varying results with parallel samples.

Table 1 : measurement results - furniture industry

Measuring Location N°	Work area	Sampling method						
		Area DNPH active	DNPH passive	Personal DNPH active	DNPH passive	Area silicagel Type B active	Orsa 5S passive	Personal Silicage Type B active
		Formaldehyde concentration ————————mg/m3————————————————————						
1	Board store	0.05(4) 0.15(4) ——— 0.10	–	–	–	–	0,02(8)	0,06(8)
2	Board saw	0.12(4) 0.14(4) ——— 0.13	–	–	0,15(8)	–	0,02(8)	0,05(8)
3	Veneering station	0.24(1) 0.24(1) 0.27(1) 0.08(1) 0.47(1) 0.28(1) ——— 0.26	–	0,26(7)	0,24(8)	0,05(8) 0,06(8) 0,06		0,09(8)
4	Veneering press	0.18(3) 0.21(4) ——— 0.20	–	–	–	0.09(7)	0.04(7)-	
5	Veneering press output	0.20(1) 0.21(1) 0.27(1) 0.27(1) 0.20(1) 0.14(1) ——— 0.22	–	–	–	–	0.06(8) 0.02(7)	–
6	Thermoplastic edge glueing equipment (dual) output	0.07(4) 0.11(4) ——— 0.09 0.13(4)	0.18(8)	0.11(7)	0.10(7)	–	0.03(7) 0.03 0.03(7)	
7	Thermoplastic edge glueing equipment (single) output	0.18(4) ——— 0.15 0.12(3)	–	0.09(8)	0.16(8)	–	0.02(8)	–
8	Interim store for parts requiring assembly	0.18(1) 0.14(2) 0.14(2) ——— 0.15	–	–	0.27(8)	–	0.01(8)	–

1. Figures in brackets = sampling period in hours
2. "-" means no measurement taken, or value below detection limit
3. Figures below the line in a column represent arithmetic means.

Table 2 : measurement results - textile industry

Measuring location N°	Work area	Area DNPH active	Area DNPH passive	Personal DNPH active	Personal DNPH passive	Area DFG method Type B active	Area Orsa 5S passive	Personal silica gel Type B active	Personal Orsa5S passive
					Formaldehyde concentration) mg/m³				
1	tenter frame 8	0.05(4)					0.02(4)		
	input	0.06(4)					0.02(4)		
		0.06	0.07(8)	0.10(8)	-	0.03(½)	0.02	-	0.03(8)
2	tenter frame 8	0.05(4)					0.02(4)		
	output	0.06(4)					0.03(4)		
		0.06	-	-	-	0.03(½)	0.03	0.03(8)	0.03(8)
3	tenter frame10	0.06(4)					0.01(4)		
	input	0.05(4)					0.01(4)		
		0.06	-	-	-	0.04(½)	0.01	0.02(8)	0.04(8)
4	tenter frame10	0.04(4)							
	output	0.05(4)							
		0.05	-	-	-	0.03(½)	-	-	-
5	tenter frame12	0.14(4)							
	input	0.32(4)							
		0.23	-	0.09(8)	-	0.07(½)	0./04(8)	-	0.03(8)
6	tenter frame12	0.35(4)					0.02(4)		
	output	0.28					0.02(4)		
		0.32	-	-	-	0.03(½)	0.02	0.03(8)	0.01(8)
7	inspecting	0.08(4)							
	machine	0.08(4)							
		0.08	-	-	-	0.06(½)	0.03(8)	-	0.01(8)
8	inspecting	0.09(4)					0.02(4)		
	and rolling	0.09(4)					0.03(4)		
	machine	0.09	-	-	0.09(8)	0.06(½)	0.03	0.05(8)	-
9	underground	0.10(4)							
	store centre	0.13(4)							
	gangway	0.12	-	0.07(8)	-	0.18(½)	0.07(8)	0.08(8)	0.01(8)
10	underground	0.20(4)							
	store side	0.13(4)							
	passage	0.16	-	-	-	-	-	-	-
	above-ground	0.21(4)							
	store centre	0.18(4)							
	gangway	0.20	-	-	-	0.02(½)	-	-	-

1. Figures in brackets = sampling period in hours
2. "-" means no measurement taken, or value below detection limit
3. Figures below the line in a column represent arithmetic means.

Table 3 : measurement results - chipboard manufacturing

Measuring Location N°	Work area (overall C content)	Area DNPH active	DNPH passive	Personal DNPH active	DNPH passive	Area silicagel Orsa 5S Type B active	passive	Personal Silicagel Orsa Type B active	passi
				Formaldehyde concentration) mg/m^3					
1	Continuous press (10 mg/m^3)	0.88(1) 1.03(1) 1.13(1) 0.77(1) 0.92(1) 0.77(1) 1.62(1) 2.61(1) $\overline{1.22}$	1.10(8)	0.68(8)	-	1.33(8)	0.36(4) 0.41(4) $\overline{0.39}$	-	0.32(8) -
2	Discontinuous press (17 mg/m^3)	0.84(1/2) 1.10(1/2) 3.07(1/3) 3.61(1/3) 4.49(1/3) 6.00(1/3) $\overline{3.20}$	-	-	-	-	-	1.04(8)	0.30(8)
3	Sanding belt	0.06(4) 0.13(4) $\overline{0.10}$	-	0.08(8)	0.19(8)	-	-	-	0.05(8)
4 + 5	Panel saw	-	-	-	-	-	-	0.05(8)	0.04(8)
6	Cutting-to-size	0.10(2) 0.07(2) 0.04(2) $\overline{0.07}$	-	0.04(8)	0.05(8)	-	0.02(8)	-	-
7	Board inspection	0.10(4) 0.08(4) $\overline{0.09}$	-	-	0.19(8)	-	0.04(8) -	0.04(8)	0.03(8) 0.03(8)

1. Figures in brackets = sampling period in hours
2. "-" means no measurement taken, or value below detection limit
3. Figures below the line in a column represent arithmetic means.

TRENDS IN ANALYTICAL REQUIREMENTS OF

DIFFUSIVE SAMPLING SYSTEMS

J. Kristensson

Institute of Analytical Chemistry
University of Stockholm
S-106 91 Stockholm, Sweden

ABSTRACT

Diffusive sampling offers a simple and effective method for taking samples of volatile compounds. Thermal desorption with cryofocusing and/or cyogenic trapping gives high sensitivity because the sample is injected as a narrow band and without dilution. Fused silica capillary columns give effective separation of complex mixtures. Detection by a mass selectiv detector gives both universal and selective detection of the compounds in the sample. The mass selective detector can be used both for quantitativ and qualitative analysis by positive identification. The analytical requirements on the system and its different parts are discussed.

INTRODUCTION

Gas chromatography is the analytical technique mainly applied to analysis of volatile compounds that has been concentrated by sampling on solid adsorbents.

The analytical requirements to be discussed is thermal desorption of adsorbents followed by a cryofocusing step.

Split injection onto the separation column.

Choice of different separation columns and detection by universal and specific detectors including mass selective detectors.

THERMAL DESORPTION

In thermal desorption the sample is heated off the adsorbent and an inert gas stream sweeps the sample in to the analytical system.

Before heating, air in the adsorbent should be replaced by inert gas to avoid oxidation, both of the compounds to be analysed and the adsorbent.

The temperature is increased to a level where the sample is released from the adsorbent. The temperature is not increased to a level where thermal breakdown of sample and adsorbent occurs.

The maximum temperature depends mainly on the used adsorbent. The recommended maximum temepratures are given by the manufacturers. Thermal breakdown of adsorbents can occur at temperatures well below the recommended maximum temperatures. For instance Porapak Q has a recommended maximum temperature of 250°C but disturbing peaks in the chromatogram from thermal breakdown has been reported at a temperature of 170°C (Krumperman, 1982).

Thermal breakdown of adsorbents and breakthrough volumes are influenced by interfering compounds (Pellizarri et.al., 1975, 1976. Russell, 1975. Barens et.al., 1981. Karasek et.al., 1981).

If high concentrations of nitrogen oxides was present at the sampling, breakedown of Tenax GC occured at thermal desorption (Hansson et.al., 1981).

The use of microwave heating for thermal desorption has been reported (Rektorik, 1982. Neu et.al., 1982). Activated carbon was used as adsorbent and the microwave heating was extremely rapid, only a few seconds. In many applications the use of a second cryofocusing concentration step was unnecessary.

New adsorbents which are thermally stable at very high temperatures have been developed. Thermosorb and Tenax TA have maximum temperatures of 500°C respectively 400°C. These adsorbents is not yet fully evaluated, but lower breakthrough volumes has been reported (Zlatkis et.al., 1985). With new adsorbents it might in the future be possible to use higher desorption temperatures.

When changing the parameters for thermal desorption, e.g. desorption temperatures and times, it is obvious that any undesirable effects must be controlled by blank samples.

CRYOFOCUSING

Cryofocusing uses low temperatures to focus the sample into a plug at or near the head of the separation column. The purpose is to improve the peak shape. Cryofocusing is particularly advantageous when a sample is to be thermally desorbed from a solid adsorbent onto a capillary column. Usually high flow rates, 30 - 50 ml/min, are required at the thermal desorption of samples from porous polymers. Thus a combination of cryogenic trapping and cryofocusing is used.

To avoid overload of the cryogenic trap and to decrease the breakthrough volume of volatile compounds a split system can be installed before the cryogenic trap. In that way only a fraction of the gas flow through the adsorbent will pass through the cold trap. (Broadway et.al., 1986).

Cryofocusing can be used in two ways, "on-column cryofocusing" or "external cryofocusing". On-column cryofocusing means that volatile compounds are trapped as narrow bands on the column or a pre-column in the oven. External cryofocusing means that the volatile compounds are trapped as narrow bands outside the oven compartment, either in the injector or completely outside the gas chromatograph. Both systems are used in instruments for thermal desorption of adsorbents.

The major advantage of cryofocusing is that the compounds of interest are placed as narrow bands onto the separation column. Thus the best separation efficiency of the chromatographic column can be achieved (Brettell et.al., 1986).

The function of cryofocusing is to slow down the migration along the column by increasing the capacity factor. Thus the chromatographic performance is changed, the sample is not "freezed out" in the cold trap. Increase of the capacity factor can be achieved by lowering the temperature or by changing the phase ratio, eg increasing the film thickness of the liquid phase. (Kolb et.al., 1986).

At cryofocusing the trap can be either a fused silica or a packed cold trap. The cooling can be achieved by different cooling mixtures, Table 1, or electronic cooling by peltier cooler.

Simple fused silica traps, eg uncoated fused silica 20-25 cm x 0.32 mm, at liquid nitrogen temperatures cannot quantitatively trap low boiling compounds (bp. $< 70°C$). (Graydon et.al. 1983). By increasing the length of the trap or by increasing the film thickness, compounds with lower boiling points can be trapped. (Kolb et.al. 1986). By applying a carefully arranged negative cryo gradient over the trap, low boiling compounds can be trapped (Jacobsson, 1984).

TABLE 1.   Cooling mixtures.

Mixture	Temperature (°C)
Ice/NaCl	-21
Ice/CaCl$_2$	-40 to -55
Solid CO$_2$	-78
Liquid N$_2$/Acetone	-95
Liquid N$_2$/Ethanol	-120
Liquid N$_2$	-196

The use of packed cold traps have the advantages of high capacity, high breakthrough volume, cryo gradient is almost unimportant and the trap will not be plugged. Packed cold traps are preferred at cryofocusing and cryogenic trapping of low boiling compounds.

INJECTION

The sample is injected from the cold trap onto the separation column by a rapid increase of the temperature. If capillary columns are used for the separation the temperature raise must be rapid to get the sample transferred as a narrow band to the column. Usually a split system is installed between the trap and the column head. The advantage of the split system is not only to increase the transfer speed

and give the optimum gas flow through the column, but one
can also recollect the split effluent on a new adsorbent
sampler. Thus a second injection can be made of the same sample and consequently the main disadvantage, the so called
one shot analysis, with thermal desorption is avoided.
(Kristensson, 1984).

SEPARATION COLUMN

Diffusive sampling offers a simple and effective method
to take samples in occupational hygiene measurements.
The samples are usually complex mixtures of several compounds with different volatility and with very wide concentration ranges.
To be able to separate these complex mixtures, high efficiency columns must be used. Thus capillary columns are
usually applied.

1956, 30 years ago, open tubular columns or capillary
columns were introduced by Golay.
During these 30 years the development of capillary columns
has given
  - fused silica columns with different internal
    diameters,
  - effective deactivation, persilylation,
  - bonded phases (crosslinked or nonextractable, here no
               differences between these columns are
               made),
  - different film thickness of liquid phase.
The fused silica columns are flexible and easy to handle. The use of capillaries is possible to any chromatogra-
pher.

High separation efficiency, inertness and reproducibility of the columns make the use of several different
selective liquid phases unnecessary. A few liquid phases
with different polarity can be used to solve most of the
separation problems.

Bonded phases give low bleeding columns which can be
used both at low temperatures, subambient, and at high
temperatures, up to 400°C.

The columns are rinsable with solvents, if nonvolatile compounds have been deposited in the column.

Different film thicknesses of the liquid phases give possibilities to change sample capacity and analysis time. The phase ratio, β, can be expressed by the formula

$$\beta \approx \frac{V_G}{V_L} = \frac{r_c}{2d_f}$$

V_G= volume of gas phase in the column
V_L= volume of liquid phase in the column
r_c= inside radius of the column tube
d_f= thickness of the liquid film

The phase ratio can be lowered by increasing the film thickness and hence the sample capacity of the column is increased. In table 2 phase ratios at different column diameters are calculated.

TABLE 2.   Phase ratio calculations at different diameters and film thickness.

Column id.	Film thickness (μm)			
(mm)	0.2	1.0	3.0	5.0
0.25	312.5	62.5	20.8	12.5
0.32	400.0	80.0	26.7	16.0
0.53	662.5	132.5	44.2	26.5
0.75	937.5	187.5	62.5	37.5

There is no definition of sample capacity, but one can approximately use the amount of a compound decreasing the plate number by 20% as the capacity of the column. The capacity will of course depend on the compound. In table 3. approximate capacities are given for some different columns.

Diffusive sampling of toluene on ATD-50 diffusive sampler (Perkin-Elmer Ltd.) on Tenax GC has an uptake rate of 1.7 ng/ppm/min. Sampling at 80 ppm for 8 hours gives 65 μg toluene. Only one system in table 3 could be used without overloading the column. To be able to use the system in

table 3 with the lowest sample capacity, a split ratio of
1:200 would be necessary.

TABLE 3.   Approximate capacities in micrograms for
            different columns.

Column id. (mm)	Film thickness (μm)	Sample capacity (μg)
0.25	0.25	0.1 - 0.4
0.32	0.25	0.5 - 1
0.53	1.0	1 - 2
0.53	5.0	100
0.75	1.0	10 - 30

(Seferovic et.al., 1986, Supelco Reporter, 1986).

In general when analyzing volatile organic compounds
with capillary columns one should use columns with thick
liquid films, 1 - 5 μm, and with large internal diameters,
0.32 - 0.53 mm.

The effluent from the cold trap can of course not only
be splitted but also divided onto two or more columns with
different liquid phases and a selective separation can be
achieved.

Of course one should choose the column best fitted to
solve the separation problem. If only a few compounds which
are easy to separate are present in the sample, one should
not hesitate to use packed columns. The advantages of packed
columns are simplicity to use, no problems with sample
capacity and the columns are cheap to buy.

Packed columns can be exchanged to wide bore columns,
0.54 - 0.75 mm id, without changing injectors or detectors
in the instrument. Usually analysis speed is lowered by approximately 30%.

For special purposes porous layer open tubular columns,
PLOT columns, have been developed (Chrompack News, 1984,
Chrompack News, 1985).
Analysis of nitrous oxide with the use of micro packed
columns, eg. packed capillaries, has been reported.
(Kristensson, 1984).

DETECTION

Diffusive sampling offers a simple and effective method to take samples of volatile compounds.
Thermal desorption with cryofocusing and/or cyogenic trapping gives high sensitivity because the sample is injected as a narrow band and without dilution.
Fused silica capillary columns give effective separation of complex mixtures.

To be able to fully use these benefits the use of a sophisticated detection system is necessary.
In gas chromatography several different detectors are used. The detectors are either universal or selective for certain compounds. The response, sensitivity and selectivity vary between different detectors and different compounds.

A universal detector like the flame ionization detector, FID, has a very good response for organic compounds. In general analysis of organic compounds FID is mostly used.

In the photo ionization detector, PID, a UV-lamp is used to ionize the copmounds. UV-lamps with different energies can be used to achive certain limited selectivity.
PID with 10.2 eV UV-lamp acts almost like a FID. The PID has high response to conjugated systems, especially aromatic compounds.
If UV-lamps with lower energies, eg. 9.2 or 9.7 eV, are used the detector becomes selective to compounds with lower ionization potentials, for instance amines.

The PID can be considered as a nondestructive detector. The detector can easily be connected in series with other detectors. In this way both universal and selective detectors can be used on the same sample.

A list of some other detectors and their selectivity is given in table 4.
TCD and to a certain extent the ECD can be considered as nondestructive and used in series with other detectors.
An other method to be able to use different detectors is to connect two or more columns in parallel and split the sample between the different columns.

TABLE 4.   Selective detectors.

Detector		Applicability
Electron capture,	ECD	halogens, oxygen
Flame photometric,	PID	sulphur, phosphor
Thermoionic,	TID	nitrogen, phosphor
Thermal conductivity,	TCD	permanent gases

If a mass selective detector or a mass spectrometer
is used both universal and specific detection can be achieved at the same time. The total ion current gives universal
detection similar to FID. Registration of certain m/e-values
during a chromatogram, so called multiple ion detection
(MID) or single ion monitoring (SIM), gives selective detection. An example of the use of a mass selective detector is
given in figure 1.
A mass selective detector can give positive identification at
qualitative analysis of complex mixtures. Mass spectrum of
all peaks during a chromatogram can be collected and compared to referece mass spectrum of pure compounds stored in a
computer library.

If more structural information from the detector is needed, for instance at separation of isomers, GC-FTIR can be
used. To get enough amount of sample out of the column the
use of wide bore columns with thick films is recommended.
The sample from the column is transferred into a heated
light-pipe in which the IR-beam passes through the compound
in gas phase. GC-FTIR is a complement to the mass selective
detector.

A good computer system is almost a necessity in handling data from a multi-component analysis of a complex mixture by high resolution gas chromatography.

CONCLUSTION

Diffusive sampling offers a simple and effective method
to take samples of volatile compounds.
Thermal desorption with cryofocusing and/or cyogenic trapping gives high sensitivity because the sample is injected
as a narrow band and without dilution.

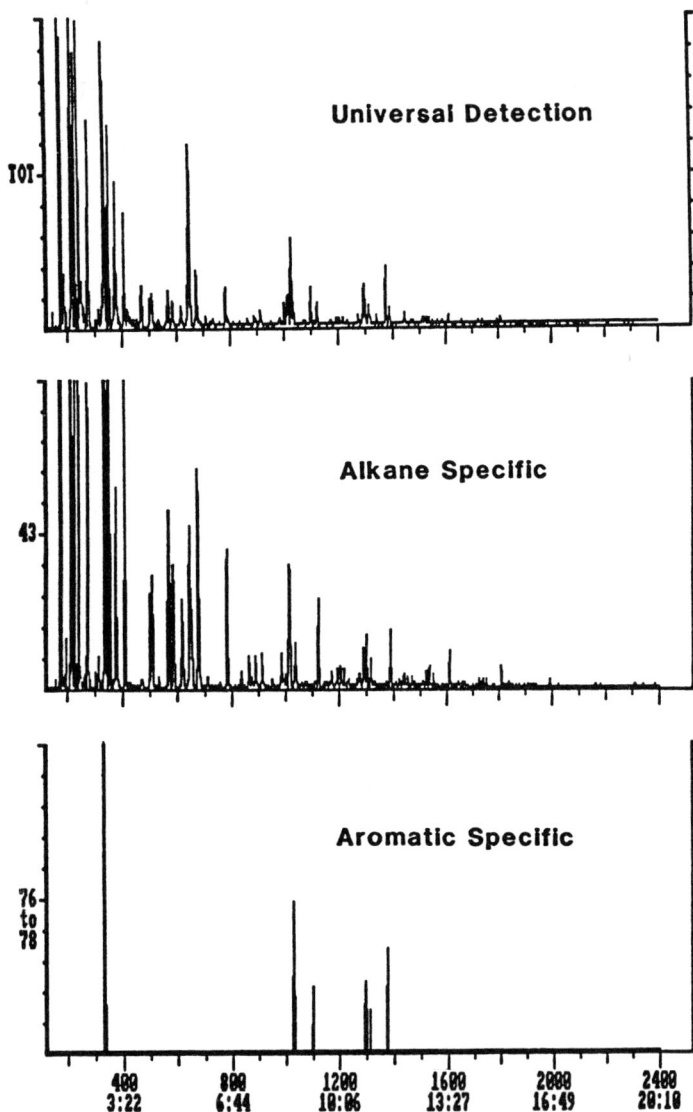

Figure 1. The use of Ion Trap mass selective detector, ITυS, as universal and specific detector. Sample is gasoline.

Fused silica capillary columns give effective separation of complex mixtures.
Detection by a mass selective detector gives both universal and selective detection of the compounds in the sample. The mass selective detector can be used both for quantitative and qualitative analysis with positive identification.
A system that fulfils these parts, and especially if the system is automatic and connected to a computer, is today the most optimal system to be used for sampling and analysis in occupational hygiene measurements.

REFERENCES

Barnes, R.D., Law, L.M., MacLeod, A.J., 1981.
    Analyst, 106, 412.
Bettell, T.A., Grob, R.L.,1986
    Int. Lab., April, 30.
Broadway, G., Tipler, A., Rendle, M., 1986
    "Application of multiple splitting techniques in two stage thermal desorption gas chromatography".
    Presentation at the 37th Pittsburgh Conference.
Chrompack News, 1984. Vol.II. No. 3E, 6.
Chrompack News, 1985. Vol.12. No. 1, 1.
Ettre, L.S., 1986. "Recent developments in the field of capillary column gas chromatography".
    Presentation at the 37th Pittsburg conference.
Hansson, R.L., Clark, C.R., Carpenter, R.L., Hobbs, C.H., 1981. Environ. Sci. Technol., 15, 701.
Karasek, F.W., Clement, R.E., Sweetman, J.A., 1981.
    Anal. Chem., 53, 1050 A.
Kolb. B., Ettre, L.S., Liebhardt, B., 1986.
    "Headspace cryofocusing in gas chromatography".
    Presentation at the 37th Pittsburg conference.
Kristensson, J., 1984. Analysis of volatiles. Methods and applications, 109, Editor P. Schreier,
    Walter de Gruyter & Co.
Krumperman, P.H., 1982. J. Agr. Food Chem., 20, 909.
Neu, H.J., Merz, W., Panzel, H., 1982.
    HRC&CC, 7, 382.
Pellizarri, E.D., Bunch, J.E., Berkerley, R.E., 1976.
    Anal. Lett., 9, 45.
Pellizarri, E.D., Carpenter, B.H., Bunch, J.E., 1975.
    Environ. Sci. Technol., 9, 556.
Pellizarri, E.D., Carpenter, B.H., Sawicki, E. 1975.
    Environ. Sci. Technol., 9, 552.
Rektorik, J., 1982., Chimie Magazine, 5, 42.
Russell, J.W., 1975. Environ. Sci. Technol., 9, 1175.

Seferovic, W., Hinshaw Jr., J.BV., Ettre, L.S., 1986.
"Comparative data on GC capillary columns having
various diameter and film thickness".
Presentation at the 37th Pittsburg conference.

Supelco Reporter, 1986. Vol.V, No.1, 1.

Zlatkis, A., Ghaoui, L., Shanfield, H., 1985.
Chromatographia, 20, 343.

A DIFFUSIVE SAMPLER FOR SUB-PARTS-PER-MILLION LEVELS OF FORMALDEHYDE IN AIR
USING CHEMOSORPTION ON 2,4-DINITROPHENYLHYDRAZINE-COATED GLASS FIBER FILTERS

J.-O. Levin, R. Lindahl, K. Andersson
National Board of Occupational Safety and Health, Research Department in
Umeå, Box 6104, S-900 06   Umeå, Sweden

ABSTRACT
       Formaldehyde is sampled from air by diffusion into a 37-mm glass fiber
filter impregnated with 2,4-dinitrophenylhydrazine and phosphoric acid
mounted in a modified standard aerosol air sampling cassette. The formaldehyde hydrazone is desorbed from the filter with acetonitrile and determined
by high-performance liquid chromatography using UV detection at 365 nm. The
sampling rate of the diffusive sampler was determined to 61 mL/min with a
standard deviation of 5 mL/min using standard atmospheres of formaldehyde
generated from paraformaldehyde. Sampling rate was independent of formaldehyde concentrations between 0.2 and 5 mg/m^3. The effect of air velocity was
studied, and the sampling rate was found to be constant between 0.1 and
0.5 m/s. The sampling rate was also found to be independent of the sampling
time from 15 min to 8 hours. The sensitivity of the diffusive method is
approximately 0.2 mg/m^3, using a sampling time of 15 min. In an 8-hour sample
approximately 5 µg/m^3 (5 ppb) can be detected.

INTRODUCTION
       A number of analytical methods for the determination of airborne form-
aldehyde have been published, including spectrophotometric and chromatographic methods. One of the most rapid and sensitive methods is high-performance liquid chromatography on the 2,4-dinitrophenylhydrazone of formaldehyde, a method used by several investigators (Kuwata et al., 1979;
Andersson et al. 1979; Fung and Grosjean, 1981). For sampling of ambient air,
absorber solutions have been most frequently used; these are not suited for
field work, however, especially not in personal monitoring. We have previously reported the use of XAD-2 coated with 2,4-dinitrophenylhydrazine
(DNPH) for the sampling of several aldehydes, including formaldehyde, acrolein and glutaraldehyde (Andersson et al. 1979; Andersson et al. 1981).
Other workers have subsequently used other DNPH-coated solid sorbents
(Beasley et al. 1980, Grosjean and Fung 1982; Kuwata et al. 1983; Lipari
and Swarin 1985). We recently reported an improved method, utilizing DNPH-
coated glass fiber filters. In this sampling system, air can be sampled at
a rate of up to 1 L/min, affording a sensitivity of about 0.001 mg/m^3,
based on a 50-L air sample (Levin et al. 1985). Coated filters of the same
type were tested for diffusive sampling. We now wish to report an extensive

evaluation of this diffusive sampler. We have determined the sampling rate and the effect of formaldehyde concentration, sampling time, air velocity and air humidity on the sampling rate.

MATERIALS AND METHODS
Coated filters for active sampling.
   The coated filters were prepared according to a previously described procedure (Levin et al. 1985). This procedure was modified to include glycerin in the coating solution. In this way, water is better retained in the coated filters when sampling is performed in extremely dry atmospheres, i. e. below 10% relative humidity. The filters were coated in the following manner: To 300 mg of DNPH, recrystallized twice from 4 M HCl and dried for only a few minutes, was added 0.5 mL 85% phosphoric acid (Merck, p.a.), 1.5 mL 20% glycerin (May & Baker, p.a.) in ethanol and 9.0 mL of acetonitrile. It is important to use formaldehyde-free acetonitrile; Rathburn HPLC Grade S acetonitrile is essentially formaldehyde-free. 13-mm-diameter glass fiber filters, organic- and binder-free (Type AE, 0.3 μm pore size, SKC, Inc.), were immersed in the cold solution for a few seconds. The filters were then allowed to dry for 2 hours at room temperature, after which they were stored in a desiccator over saturated sodium chloride solution to assure constant relative humidity. It is possible to store the filters in this way for several months. Filters coated in this manner contain approximately 2 mg DNPH. The formaldehyde blank was always less than 0.05 μg/filter. The filters were used in two-section polypropylene filter holders (No. 225-32, SKC, Inc.), and the collection efficiency of the filters was determined as previously described (Levin et al. 1985).
Coated filters for diffusive sampling.
   To 300 mg of DNPH HCl was added 5 mL of 10% phosphoric acid and 25 mL of acetonitrile. No glycerin was added since there is no risk of the filters drying out when used for passive sampling. A 0.5-mL portion of the solution (5 mg of DNPH) was added with a 0.5 mL Voll pipette to each 37-mm-diameter glass fiber filter, organic- and binder-free (Type AE, 0.3 μm pore size, SKC, Inc.). The filters were dried for 30 min at room temperature and conditioned in the same way as the filters for active sampling. The formaldehyde blank was always less than 0.15 μg/filter.

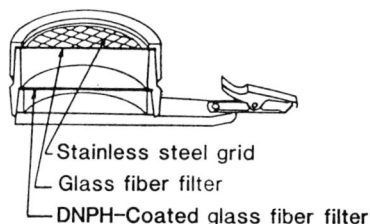

Stainless steel grid
Glass fiber filter
DNPH–Coated glass fiber filter

Figure 1. Filter cassette for 37-mm filters modified for diffusive sampling.

Dosimeter for diffusive sampling.

The coated filters for diffusive sampling were used in a modified 37-mm-filter holder, shown in Figure 1. One section of the standard two-section cassette (No M0037AD, Millipore Corp., MA) was cut into two parts, and one part was used as the front section of the sampler to create a controlled diffusion barrier consisting of an uncoated glass fiber filter between two stainless steel grids (No 7908/B30, Casella London Ltd., G.B.), according to Figure 1. The other part of the section was pasted on a piece of poly-acrylic ester plastic equipped with a clip. The front cross-sectional area of the sampler was 7.1 cm^2 and the length of the controlled diffusion path was 10 mm.

Generation of standard atmospheres of formaldehyde.

Formaldehyde was generated in a dynamic system by decomposition of para-formaldehyde at 25-30°. The generation system is described in the original article (in press).

Liquid chromatography.

The formaldehyde 2,4-dinitrophenylhydrazone was eluted from both 13- and 37-mm filters by shaking for 1 min with 3 mL of acetonitrile in a 4-mL glass vial. The solution was filtered through a Millex-SR filter and 10 µL was injected into the liquid chromatograph. Details of the liquid chromatographic determination were described previously (Levin, et al. 1985).

RESULTS AND DISCUSSION

The active sampling method was used as reference method in all experiments with the diffusive sampler. The DNPH-coated 13-mm glass fiber filters used in the active mode permitted determination of formaldehyde down to approximately 1 µg/m^3 in a 50-L air sample. The formaldehyde blank of 0.05 µg/filter corresponds to 1 µg/m^3 in a 50-L air sample.

Sampling rate of diffusive sampler.

The sampling rate was experimentally determined by exposing six samplers at a time in the dynamic generation system while six active samples were taken simultaneously with a sampling rate of 100 mL/min. 79 experiments with the diffusive sampler gave a mean sampling rate of 61 mL/min, with a standard deviation of 3.7 mL/min.

Effect of sampling time on sampling rate.

Figure 2 shows the effect of sampling time on the sampling rate. It can be seen from the graph that the sampling rate is constant when using sampling times between 15 and 420 min.

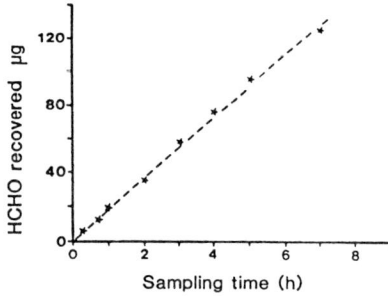

Figure 2. Effect of sampling time on sampling rate of diffusive sampler. Formaldehyde concentration 5 mg/m^3, relative humidity 50%, and face velocity 0.3 m/s. The dotted line represents a sampling rate of 61 mL/min. Each point represents at least three determinations.

Effect of formaldehyde concentration on sampling rate.

The diffusive sampler was exposed to formaldehyde levels from 0.1 to 5.0 mg/m^3. Table I shows the sampling rate of the sampler at different formaldehyde concentrations with sampling times of 15 and 240 minutes, and with a relative humidity of 20%. As the table shows, the sampling rate is not influenced by the formaldehyde concentration within the range studied.

TABLE I. Effect of formaldehyde concentration on sampling rate of diffusive sampler. Relative humidity 20% and face velocity 0.3 m/s. Rsd = relative standard deviation. N = number of experiments.

HCHO conc. (mg/m^3)	Sampling time (min)	Sampling rate (mL/min)	Rsd (%)	N
0.1	240	60	5	6
0.2	15	59	4	6
1.0	15	60	4	3
5.0	15	60	3	6
5.0	240	62	2	3

TABLE II. Effect of relative humidity on sampling rate of diffusive sampler. Formaldehyde concentration 5 mg/m^3 and face velocity 0.3 m/s.

Rh (%)	Sampling time (min)	Sampling rate (mL/min)	Rsd (%)	N
8	15	61	11	6
20	15	60	3	6
20	240	62	2	3
50	15	62	8	3
50	240	63	3	3
85	15	60	6	12
85	240	62	4	6

Effect of relative humidity on sampling rate.

The sampler was evaluated at relative humidities of 8, 20, 50, and 85%. Sampling times of 15 and 240 min were used. The results are shown in Table II. It can be seen that the sampling rate is not affected by the relative humidity of the sampled air.

Effect of face velocity on sampling rate.

The effect of air velocity on the sampling rate was studied by varying the face velocity between 0.05 and 1.0 m/s. The results of these studies are shown in Figure 3. As shown in the figure, the sampling rate decreases at 0.05 m/s, and increases at 0.7 m/s. However, between 0.1 and 0.5 m/s the sampling rate is 61 mL/min with a standard deviation of 5 mL/min. Most diffusive samplers require an air velocity of approximately 0.1 m/s which is maintained under most personal sampling conditions.

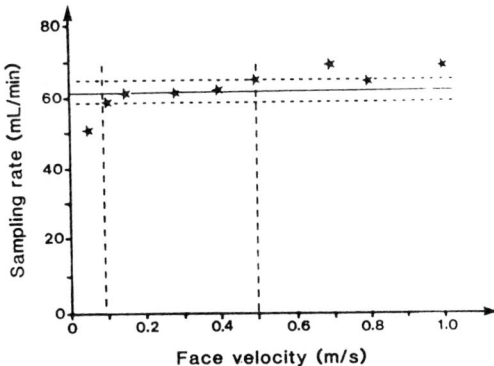

Figure 3. Effect of face velocity on sampling rate of diffusive sampler. Formaldehyde concentration 1 mg/m^3, relative humidity 50% and sampling time 20 min. The area within the dotted lines represents a sampling rate of 61 mL/min with a standard deviation of 5 mL/min. Each point represents at least three determinations.

Storage stability of diffusive sampler.

The unexposed samplers were stored individually in sealed bags made of laminated aluminium. The formaldehyde blank of the samplers did not increase after two months of storage in these bags. The sampling rate of the diffusive sampler was not affected by two months storing of the unexposed sampler. Once formed the hydrazone is stable on the filter since no decrease in recovery of formaldehyde hydrazone could be noted after storing the exposed samplers for one month.

Sensitivity of diffusive sampler.

Since the diffusive sampler has a sampling rate as high as 61 mL/min, the sampler can be used for short-time sampling. With a formaldehyde blank of 0.15 µg/filter, the sensitivity of a 15 min sample is approximately 0.2 mg/m^3. In an 8-hour sample the sensitivity will become 0.005 mg/m^3 (5 ppb).

REFERENCES

Andersson, K., Andersson, G., Nilsson, C.-A., Levin, J.-O. 1979. Chemosorption of formaldehyde on Amberlite XAD-2 coated with 2,4dinitrophenylhydrazine. Chemosphere 8, 823-827.
Andersson, K., Hallgren, C., Levin, J.-O., Nilsson, C.-A. 1981. Chemosorption sampling and analysis of formaldehyde in air. Influence on recovery during simultaneous sampling of formaldehyde, phenol, furfural and furfuryl alcohol. Scand. J. Work Environ. Health 7, 282-289.
Andersson, K., Hallgren, C., Levin, J.-O., Nilsson, C.-A. 1981. Solid chemosorbent for sub-ppm levels of acrolein and glutaraldehydein air. Chemosphere 10, 275-280.
Beasley, R.K., Hoffman, C.-E., Rueppel, M.L., Worley, J.W. 1980. Sampling of formaldehyde in air with coated solid sorbent and determination by high performance liquid chromatography. Anal. Chem. 52, 1110-1114.
Grosjean, D., Fung, K. 1982. Collection efficiencies of cartridges and microimpingers for sampling of aldehydes in air as 2,4-dinitrophenylhydrazones. Anal. Chem. 54, 1221-1224.
Kuwata, K., Uebori, M., Yamasaki, H., Kuge, Y. 1983. Determination of aliphatic aldehydes in air by liquid chromatography. Anal. Chem. 55, 2013-2016.
Levin, J.-O., Andersson, K., Lindahl, R., Nilsson, C.-A. 1985. Determination of sub-parts-per-million levels of formaldehyde in air using active or passive sampling on 2,4-dinitrophenylhydrazine-coated glass fiber filters and high-performance liquid chromatography. Anal. Chem. 57, 1032-1035.
Lipari, F., Swarin, S.J. 1985. 2,4-Dinitrophenylhydrazine-Coated florisil sampling cartridges for the determination of formaldehyde in air. Environ. Sci. Technol. 19, 70-74.

DIFFUSION TUBE TO DATA RECORDING: A COMPLETE SYSTEM FOR PERSONAL MONITORING

G. M. Broadway

Perkin-Elmer Limited
Beaconsfield, Bucks., England

ABSTRACT

The diffusive sampler provides a cost-effective means for routine workplace air monitoring and is especially effective in personal monitoring studies. However, sample collection by a diffusive monitor is just the first step in the analytical process. Further cost reductions may be achieved with reusable samplers and automation of the analytical process.

An Automated Thermal Desorption system capable of processing up to 50 sample batches is connected directly to a gas chromatograph for analysis. Several choices for data processing are available. This may be carried out either at the gas chromatograph or by a personal computer or by a mainframe system depending on the number of samples to be processed and the archival facilities required.

INTRODUCTION

In the late 1970s an ad hoc committee of the Health and Safety Executive in London approached Perkin-Elmer to develop an automated system for the analysis of exposed diffusive monitors. In response to this, a total analytical system was developed using a thermal desorption procedure to remove the components from the sampler, to provide a cost-effective means of analysis. This paper describes the instrumental requirements for a completely automated system for personal monitoring from the diffusion tube to the data recording.

THE DIFFUSIVE SAMPLER

The first step in workplace air monitoring is the collection of the sample. This has been carried out for many years by drawing air through a bed of charcoal using a portable volumetric pump. It is highly desirable that each worker is monitored for his individual exposure to toxic vapors rather than measuring the general level of toxic vapors in a workplace. The drawback with pumped sampling, however, is the high cost involved in mass personal screening projects. Good quality volumetric pumps are extremely expensive and one pump is required for each person involved. It is also necessary to have spare pumps in case of breakdown and the pumps require servicing at regular intervals. An alternative means of monitoring is therefore desirable for cost reduction purposes and recently, diffusive samplers have been used (Palmes and Gunnison, 1973; Ferber, Sharp and Freedman, 1976; Bailey and Hollingdale-Smith, 1977).

The Perkin-Elmer diffusive sampler was developed after consultations with members of a Health and Safety Committee on Personal Monitoring which later became Working Group 5 of the Health and Safety Committee on Analytical Requirements.

A tube-type sampler (Fig.1) was adopted primarily because it is an ideal shape for automated analysis (Brown, Charlton and Saunders, 1981). The sampler is filled with a suitable adsorbent which is retained by two stainless steel 100 mesh gauzes. Diffusion of organic vapors from the environment to the adsorbent occurs according to Ficks 1st Law along an air gap, which is incorporated at one end of the tube. The dimensions of the air gap are strictly controlled in the manufacturing procedures and are subject to a UK patent.

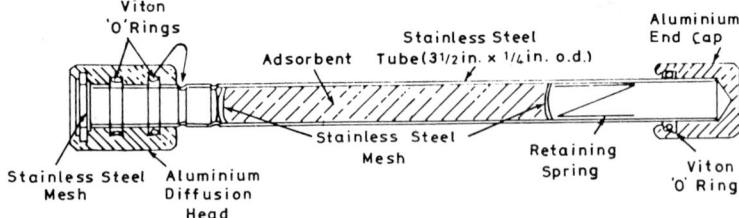

Fig.1   Tube-form diffusive sampler for organic vapors.

Variations from sampler to sampler are less than 1%. During sampling, the storage cap is removed from the air gap end of the sampler and replaced by a diffusion cap. Two types of cap are available. The first contains a 60 mesh stainless steel gauze which defines the start of the diffusion (air) gap. The second type also includes a silicone membrane under the gauze to inhibit ingress of water vapor. The purpose of the end cap is threefold. Firstly, it defines the edge of the air gap, secondly it minimizes turbulence within the air gap and thirdly it prevents, to a certain extent, particulate matter entering the tube.

TABLE 1   Typical uptake rates

		Uptake Rate	
	Adsorbent	ng/ppm/min	ml/min
Acrylonitrile	Porapak N	1.2	0.55
Benzene	Tenax	1.4	0.43
Butadiene	Molecular Sieve 13x	1.77	0.88
Halothane	Spherocarb	4.3	0.53
Nitrous Oxide	Molecular Sieve 5A	1.07	0.59
Styrene	Tenax	2.0	0.47
Toluene	Tenax	1.7	0.45

A fairly high aspect ratio for the air gap is used; that is to say, the length is greater than the cross-sectional area. The advantage of this construction over badge-type samplers which have a low aspect ratio is that the effect of air flow across the face of the sampler is minimized (Brown, 1981) making the samplers ideal for static point sampling as well as personal monitoring. The dimensions of the air gap are such that diffusive sampling takes place at a rate roughly equivalent to pumped sampling at 0.5 ml/min. The sampler is approximately the size of a pen and may be conveniently worn in the pocket. A pen clip is provided for this purpose and is large enough to contain information for the identity of the wearer. Uptake rates for some common compounds are given in Table 1 above.

ANALYSIS OF THE DIFFUSIVE SAMPLER

Once the sample has been collected by the diffusive sampler, it is returned to the laboratory for analysis. The most common technique used for this purpose is gas chromatography and traditionally this has been performed by solvent extraction of the volatile organics from the adsorbent using 1-5 ml of solvent and injecting an aliquot (1-2 ul) directly into the gas chromatograph. This technique suffers from several drawbacks including poor extraction efficiencies for certain compounds, dilution of the sample leading to loss of sensitivity, use of solvents which are often toxic and flammable and destruction of the sampler. Its main disadvantage, however, is that it is an extremely time-consuming and labor-intensive technique. Diffusive sampling enables mass screening of the workforce to be realised at relatively low cost but analytical techniques based on solvent extraction cannot cope with the increased workload that mass screening would involve.

Thermal desorption, where the diffusive sampler is heated in a stream of inert gas to remove the volatiles, is an attractive alternative technique. The extraction efficiencies are often greater than 95% (Barnes, Law and Macleod 1981; William, Bombaugh, Lewis and Ogle 1980). The whole sample is available for analysis, no toxic solvents are required and the sampler may be reused many times. Thermal desorption is also readily automated thus enabling large numbers of diffusive samples to be analyzed day and night.

The Perkin-Elmer ATD 50 (Automated Thermal Desorption system) was designed (after consultations with members of Working Group 5) to automate completely the analysis of diffusive sampling tubes. The analytical sequence is microprocessor-controlled and repeatable from sample to sample. A schematic of the system is shown in Fig.2.

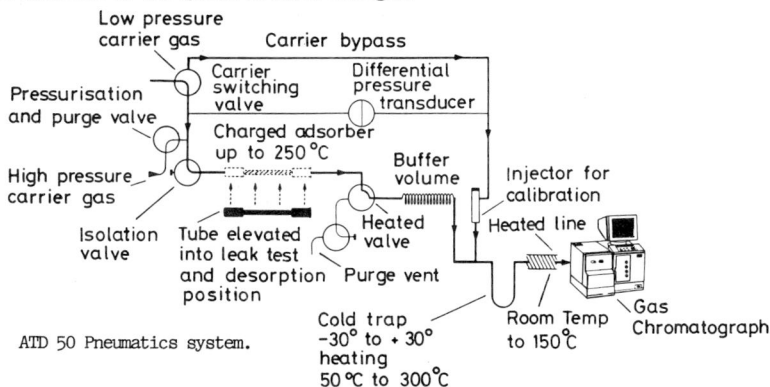

Fig.2   ATD 50 Pneumatics system.

Up to 50 tube-type diffusive samplers may be loaded into the instrument. Each sampler is taken in turn and connected to an inert gas supply. The tube is purged of air and then heated to a preset temperature for a controlled time with inert gas flowing through the adsorbent. The sampler has a fairly high thermal mass with the result that complete desorption of the tube may take several minutes. A low mass cold trap is therefore included in the system to preconcentrate the components prior to analysis by gas chromatography. The cold trap, which is electrically cooled to -30°C, is capable of heating to 300°C at a rate in excess of 1200 degC/min. All that the operator is required to do is to remove the storage caps from the diffusive samplers, replace them with caps designed to open only whilst the sampler is under analysis, and note the position in which the samplers are placed in the carousel.

A major advantage in thermal desorption is the high sensitivity of the technique because all of the sampler may be analyzed, compared with solvent extraction where approximately only one thousandth of the extracted sample may be injected into the gas chromatograph. However, this means that each sample can only be analyzed once by thermal desorption. For this reason, the ATD 50 automatically carries out a number of safety checks which ensure the presence of carrier gas and that the sampler is sealed into the system without leaks before heat is applied to the sampler. If any sampler fails the pre-analysis check routine for any reason, it is returned to the turntable and the reason for its failure logged. The integrity of the sample is therefore maintained at all times.

Thermal desorption has been used for a large number of pollutants ranging from nitrous oxide with a boiling point of -88°C (O'Sullivan, Holdsworth and Musgrave 1982) to nicotine with a boiling point in excess of 250°C (Bell 1985). Table 2 shows the precision attained from various compounds using diffusive sampling in conjunction with thermal desorption. In most cases, precision better than ±10% can be achieved.

TABLE 2   Precision of the diffusive sampling/thermal desorption system.

Component	Adsorbent	Sample Range	RSD
Acetone	Spherocarb	100–1000 ppm	±10.0%
Benzene*	Tenax	68ug	±3.97%
Dichloromethane	Spherocarb	20–200 ppm	±6.0%
Ethanol	Spherocarb	100–1000 ppm	±10.0%
Ethylene oxide	Spherocarb	5–25 ppm	±12.0%
Isopropyl alcohol	Spherocarb	40–400 ppm	±10.0%
Nitrous oxide	Molecular sieve 5A	32–1130 ppm	±17.3%
Styrene	Tenax	18 ppm	±3.47%
Vinyl chloride monomer	Spherocarb	9–16 ppm	±6.2%

*Sample placed in diffusive monitor by direct injection.

Thermal desorption is most commonly used in conjunction with gas chromatography and the two techniques are well suited. In thermal desorption, the volatiles are released into a stream of inert gas which is also used as the carrier gas for the gas chromatograph. Gas chromatography is a powerful separation and quantitative tool and it is possible to detect routinely sub-nanogram quantities of compounds. This may increase to sub-picogram levels with certain selective detectors. However, gas chromatography is not generally regarded as a qualitative technique and so absolute identification of unknown components is not feasible with conventional detector systems. Gas chromatography can be used in conjunction with Fourier transform infrared (Hurrell 1986) and also with mass spectrometers. The combination of these techniques provides an extremely powerful system for the separation and identification of unknown volatile components.

DATA PROCESSING OPTIONS

Diffusive samplers, when used in conjunction with automated thermal desorption, give a technique for large mass screening projects of personnel exposed to hazardous vapors at relatively low costs. But mass screening projects will also produce an enormous increase in the amount of data produced. Therefore, an efficient data processing system, capable of interpreting the raw data generated by the gas chromatograph into a format which is informative to the industrial hygienist is required. The first level of data processing may occur at the gas chromatograph .

Perkin-Elmer 8000 Series gas chromatographs are compatible with the ATD 50, Ion Trap (mass spectrometric) and FTIR detectors and in addition allow storage, in battery backed-up memory, of up to ten GC and data handling methods. The addition of a single PCB allows complete automation of the ATD 50/GC combination. At a prescribed sample position, a method containing all the GC and data handling parameters for a batch of samples is recalled from memory. When the next batch of samples is ready for analysis, a new method is set up and so on. A facility is included for automatic calibration from standards at set positions in the sequence. Corrections to the time weighted average exposure level may also be made automatically for samplers that have not been exposed for the standard period. This type of system is ideal for a small laboratory involved in environmental monitoring. It is totally automatic, relatively low cost and provides the report directly in ppm for time weighted exposures. Each sampler is identified in the report by its position on the turntable and by the name or code of the worker. However, such a system does not have the facility for bulk storage of reports.

If storage and archival facilities are required, two options exist. For smaller laboratories, where the number of samples processed is not high, the gas chromatograph may be linked to a personal computer. Processing of the raw data still takes place at the gas chromatograph but the reports are transmitted automatically to the computer. This system allows the operator bulk storage and some statistical processing of the results. Day to day trends may be monitored and alarm thresholds set to highlight deviations from expected levels.

For the larger laboratory, processing large numbers of samplers, the data handling facilities in the gas chromatograph may be enhanced by an option which also allows transmission of the raw signal to a LIMS 2000 computer.

Not only is a bulk storage facility available but results may be reprocessed at a later date, trends in exposure levels monitored and data can be searched, sorted and selected by any criteria. The data from personal monitors may be combined with health records of the personnel to obtain a complete picture of the overall welfare of the workforce.

CONCLUSIONS

The mass screening of workers exposed to hazardous vapors is necessary to obtain an overall picture of the effectiveness of safety and welfare procedures in present-day working environments. Personal diffusive samplers have distinct cost advantages over pumped samplers because the necessity of expensive volumetric pumps is eliminated. Analysis by thermal desorption instead of solvent extraction techniques also gives added cost benefits by using reusable monitors and by automation, thus reducing the running costs of the analysis. Diffusive sampling and thermal desorption techniques may be used for a wide range of organic vapors and the precision obtained is usually better than $\pm 10\%$, some $2\frac{1}{2}$ times better than alternative methods.

The options available to process the data from the large number of samples produced from diffusive sampling and thermal desorption depend on the requirements of each laboratory. They range from a simple chromatographic data processing to large LIMS 2000 computers which can perform many functions simultaneously.

Diffusive sampling and automated thermal desorption provide the only method for mass screening of workers exposed to organic vapors, at a reasonable cost.

REFERENCES

Bailey, A; Hollingdale-Smith, P.A., 1977: A personal diffusion sampler for evaluating time weighted exposure to organic gases and vapors, Am.Occup.Hyg. 20 345-356.
Barnes, R.D.; Law, L.M.; Macleod, A.J., 1981: Analyst, 106 412-418.
Bell, R, 1985: Perkin-Elmer ATD 50 Users Group Meeting, Beaconsfield, Bucks, England.
Brown, R.H., 1981: Present state of the art on passive sampling, Occupational Hygiene Residential Conference, Gt.Malvern, 5-7 October 1981.
Brown, R.H.; Charlton, J.; Saunders, K.J., 1981: The development of an improved diffusive sampler, Am.Ind.Hyg.Assoc. J 42 865-869.
Ferber, B.I.; Sharp, F.A.; Freedman, R.W., 1976: Dosimeter for oxides of nitrogen, Am.Ind.Hyg. Assoc. J 37 32-36.
Hurrell, R.A., 1986: Chromatographic considerations in the optimization of GC/IR performance; Proceedings of the 7th International Symposium on Capillary Chromatography, p.404-414, Pub. University of Nagoya Press.
O'Sullivan, J.; Holdsworth, B.; Musgrove, M., 1982: Passive monitors for the determination of personal nitrous oxide exposure levels, Anaesthesia 37 4 467-468.
Palmes, E.D. and Gunnison, A.F., 1973: Personal monitoring device for gaseous contaminants, Am.Ind.Hyg.Assoc. J 34 78-81.
Williams, C.H. Jr.; Bombaugh, K.J.; Lewis, D.S.; Ogle, L.D., 1980: Comparison of thermal desorption and solvent extraction for the analysis of organic air contaminants adsorbed on Tenax. Am.Ind.Hyg.Conf., Houston, May 1980, Abstract 235.

A DIFFUSIVE SAMPLER FOR MONITORING MERCURY VAPOUR
IN WORKPLACE AIR

H.L. Horton[f], C.J. Jackson*, B. Miller[f], C.J. Purnell[ø] and N.G. West*
*Occupational Medicine and Hygiene Laboratory, London, UK
[f]ICI Mond Division, Runcorn, UK
[ø]London School of Hygiene and Tropical Medicine, London, UK

ABSTRACT

    A new monitoring method for mercury vapour in workplace air has been
developed, based on a badge-type diffusive sampler. The absorbent employed
in the sampler is a paper impregnated with cuprous iodide, a mercury specific colorimetric reagent. This enables semi-quantitative colorimetric
analysis to be performed at the factory site with the option to return the
exposed papers to the laboratory for further analysis using instrumental
techniques. The performance of the sampler has been assessed both in
laboratory experiments and factory trials. The average uptake rate of the
sampler has been found to be 98 cm^3 min^{-1} (800 ng ppm^{-1} min^{-1}) when exposed
for periods of between two and eight hours in an atmosphere containing a
mercury vapour concentration in the range 0.005 to 0.15 mg m^{-3}. The random
error in the uptake rate was found to be 11% expressed as a coefficient of
variation. In the factory trials, a good correlation was found between
results obtained using the reference pumped sampling method and the new
diffusive sampler. There was no systematic difference between the two
methods.

INTRODUCTION

    Mercury is the only pure metal which is a liquid at ambient tempera-

ture and, because of its special properties, is used widely in industry and

dentistry. Since it has an appreciable vapour pressure at temperatures

as low as 20°C, any process involving the use of metallic mercury can

result in worker exposure to the vapour. The toxic effects of exposure are

severe and the recommended limits for occupational exposure have accord-

ingly been set at very low levels. In the UK the eight hour time-weighted

average exposure limit is presently set at 0.05 mgm^{-3} (0.006 ppm) (Health

and Safety Executive, 1985).

    The established methods for monitoring mercury vapour in the workplace

require an air sample to be drawn by a pump through a solid or liquid

absorbent as, for example, in the work of Rathje and Marcero, 1976. After

sampling, the mercury is released from the absorbent and determined by

cold vapour atomic absorption spectrometry. The objective of the present

work was to develop an alternative approach based on a diffusive sampler

and using an absorbent which would react colorimetrically with the mercury

vapour. This would enable semi-quantitative measurements to be made on-site followed, if required, by accurate quantitative analysis using instrumental techniques in the laboratory.

EXPERIMENTAL

Choice of diffusive sampler

The method was developed using the Porton Down diffusive sampler, Figure 1, (Bailey and Hollingdale-Smith, 1977) with a cuprous iodide impregnated paper as the absorbent.

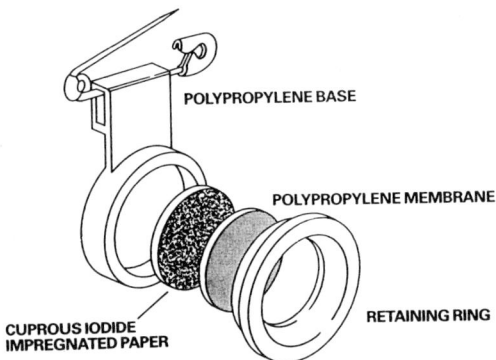

POLYPROPYLENE BASE

POLYPROPYLENE MEMBRANE

RETAINING RING

CUPROUS IODIDE
IMPREGNATED PAPER

Fig. 1   Porton Down diffusive sampler

Mercury vapour reacts with cuprous iodide to form a pink-orange coloured complex which may be used for semi-quantitative determination (Crisp et al, 1981). Other designs of diffusive sampler may be employed providing that their uptake rates are compatible with the sensitivity of the analytical technique employed.

Preparation of cuprous iodide impregnated papers

Reagents. Cupric sulphate solution: dissolve 125 g $CuSO_4.5H_2O$ in 500 cm^3 water. Potassium iodide solution: dissolve 165 g KI in 500 cm^3 water. Sodium sulphite solution: dissolve 40 g Na_2SO_3 in 500 cm^3 water. Whatman SG81 silica-gel loaded filter paper.

Procedure. Pour the cupric sulphate, potassium iodide and sodium sulphite solutions into three separate shallow trays and fill a fourth with water. Cut the silica-gel paper into strips measuring 10 x 20 cm and place each strip in sequence in the three solutions and then in the water. The strip should remain in each solution for two minutes. Finally hang the strip up to dry in a dark place and then cut into pieces to fit the diffusive sampler. Store in a dark, airtight container.

Exposure chamber

For the laboratory experimentation, an exposure chamber was set up in which known concentrations of mercury vapour could be generated. The chamber was cylindrical, measuring 25 cm in diameter and 25 cm in height. Mercury vapour was generated by continuous reduction of mercuric nitrate by stannous chloride in solution and, after dilution with air, was fed into the base of the chamber at a rate of 10 litres min^{-1}. Two reference pumped sampling methods were used to check the mercury vapour concentration. The methods used acidic potassium permanganate solution and the solid absorbent Hopcalite respectively to trap the mercury vapour. In addition, the chamber was continuously monitored using a mercury vapour meter.

Colorimetric analysis

Reagents. Metanil yellow solution: dissolve 0.083 g Metanil yellow (BDH Chemicals pH indicator) in 100 cm^3 water. Biebrich scarlet solution: dissolve 0.044 g Biebrich scarlet (BDH Chemicals stain) in 100 cm^3 water.

Colour standards. Colour standards were prepared by suspending strips of silica-gel loaded filter paper in an aqueous mixture of dyes then allowing the papers to dry in air. The compositions of the dyeing solutions were determined by trial and error using cuprous iodide impregnated papers, exposed in a standard mercury atmosphere, for reference. The compositions listed in Table 1 are for guidance only and should be checked in the users own laboratory.

TABLE 1   Colour Standards

Exposure mgm^{-3} x hours	Metanil yellow solution (cm^3)	Biebrich scarlet solution (cm^3)	Distilled water (cm^3)
0.1	3	1	100
0.2	3	1	76
0.4	2	1	30

Analysis. After sampling, the exposure of the paper to mercury vapour is estimated by simple visual comparison with the colour standards.

Quantitative analysis

Three instrumental techniques have been used successfully for quantitative analysis of the absorbent papers; cold vapour atomic absorption
spectrometry, X-ray fluorescence spectrometry and inductively coupled
plasma emission spectrometry. The cold vapour atomic absorption spectrometry method is as described for the determination of mercury in Hopcalite
(Health and Safety Executive, 1983a) except that the reduction of ionic
mercury is performed in an alkaline rather than an acidic medium. A 10 cm^3
aliquot of a solution containing 200 g sodium hydroxide per litre, is added
to the 5 cm^3 sample or standard solution prior to the addition of the stannous chloride solution.

The X-ray fluorescence spectrometric analysis is performed directly
on the exposed papers using the $HgL\alpha$ emission line (Purnell, West and Brown,
1981). Inductively coupled plasma emission spectrometry may be used with
either direct nebulization or continuous reduction as the means of sample
introduction. The mercury emission line at 184.95 nm is the preferred
wavelength.

ASSESSMENT OF SAMPLE PERFORMANCE

The performance of the new sampler was assessed by laboratory and
field experiment, using for guidance, a published standard experimental
protocol (Health and Safety Executive, 1983b).

Laboratory experiments

It was known from previous work (Purnell, Wright and Brown, 1981) that
the uptake rate of the Porton Down sampler for styrene is reduced when the
air velocity across the face of the sampler is less than 0.1 ms^{-1}.
Preliminary experimentation confirmed that the same minimum air velocity
applied in the case of mercury vapour and all laboratory experimentation
was therefore performed with an air velocity past the sampler of 0.25 ms^{-1}.

The effect of exposure time and concentration on the sampler uptake
rate was determined by a series of experiments conducted to the design
shown in Table 2.

TABLE 2   Observed uptake rates (ng ppm^{-1} min^{-1})

Concentration	Time 120 min	240 min	480 min
Low (~0.1 OEL)	420	866	774
	557	774	819
	435	860	774
	680	679	821
	689	700	936
	562	664	833
Medium (~1.0 OEL)	667	611	869
	670	995	772
	689	873	883
	632	750	715
	693	962	711
	657	669	730
High (~2.0 OEL)	587	883	772
	682	794	885
	628	825	723
	596	860	793
	687	895	740
	722	823	714

OEL - Occupational Exposure Limit 0.05 mgm^{-3}

For each combination of time and concentration, a set of six diffusive samplers was exposed in the test chamber and the mass of mercury absorbed on the impregnated papers was subsequently determined by cold vapour atomic absorption spectrometry. An analysis of variance on the data in Table 2 showed that there was a statistically significant effect of time on the uptake rate but that there was no effect of concentration and no concentration x time interaction. The random sampling analytical error was 11% expressed as a coefficient of variation. Further experimentation at 180, 300 and 360 minutes indicated that the uptake rate was independent of exposure time for periods greater than 120 minutes, with an average uptake rate of 800 ng ppm^{-1} min^{-1} (98 cm^{-3} min^{-1}).

The effect of changes in humidity on the uptake rate was also studied experimentally and no significant effect was found over the range 20 to 80% relative humidity at ambient temperature. In addition, storage of exposed papers for a period of four weeks prior to analysis had no effect on the measured uptake rate.

Field experiments

Field comparisons of the diffusive sampling method with the pumped Hopcalite tube reference method were undertaken to assess the performance of the new sampler in the workplace environment. Paired measurements were made of the personal exposures of fifty five workers in four different industries using metallic mercury. The relationship between the two sampling methods was examined by linear regression analysis of the logarithmic transformed data (Figure 2) with the following results:

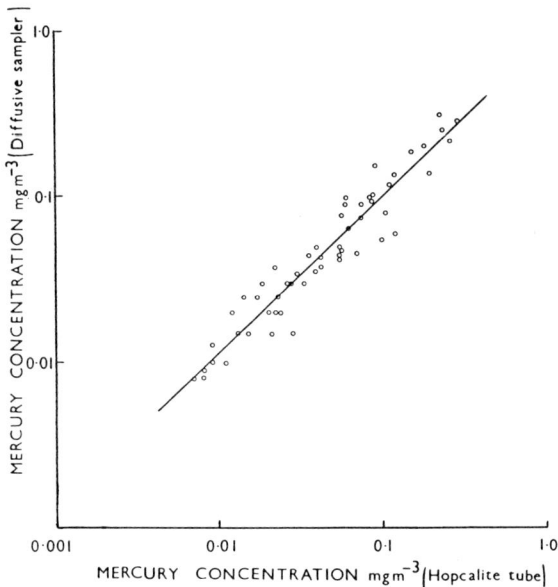

Fig. 2   Relationship between diffusive and reference sampling methods.

Slope = 0.939   Standard of error of slope = 0.040

Intercept = -0.058   Standard error of intercept = 0.057

Correlation coefficient = 0.955

These results show that there was a good correlation between the two sampling methods for mercury vapour and no evidence of any systematic difference between them. All the analytical data for this field comparison were obtained using either cold vapour atomic absorption spectrometry or

X-ray fluorescence spectrometry. Other comparison exercises using on-site colorimetric analysis also showed good agreement with the reference method.

COMPARISON OF ALTERNATIVE ANALYTICAL TECHNIQUES

Colorimetric analysis

Using simple visual colour comparison, only a semi-quantitative estimate of the mercury concentration can be made. For a four hour exposure it was possible to categorise the exposure in terms of the occupational exposure limit as follows; < 0.5, 0.5 to 1.0, 1.0 to 2.0 and >2.0. Although the reaction between mercury and cuprous iodide is highly specific with no known interferences, the presence of chlorine in the atmosphere can mask the development of the pink-orange colour due to the displacement of iodine from the cuprous iodide. However, if the sampler is removed to a chlorine free atmosphere then the brown iodine stain disappears leaving the original pink-orange coloration.

Quantitative analysis

Over most of the range of potential usage of the new sampler, all three instrumental analysis techniques are capable of accurate analysis. However, at the lower end of the exposure range, the techniques are operating at or near their limits of sensitivity. Over a four hour sampling period, the detection limits* for cold vapour atomic absorption spectrometry, plasma emission spectrometry and X-ray fluorescence spectrometry correspond to mercury vapour concentrations of 0.005, 0.003 and 0.009 mgm^{-3} respectively. The quantitation limits, at which the analytical precision is ten percent (coefficient of variation) are approximately three times the above detection limits. Although X-ray fluorescence spectrometry is the least sensitive of the three analytical techniques, it has the great advantage that no sample preparation is required, making it very cost effective. All three instrumental techniques are essentially free from interferences.

DISCUSSION

The new diffusive sampling system has been shown to be applicable over the normal range of mercury vapour concentrations found in the workplace. The random error in the determination is comparable with that found using

*Defined as 3.29σ where σ is the standard deviation of results on blank samples (Currie, 1977).

pumped systems and field trials have indicated good agreement with a reference pumped sampling method. The minimum recommended sampling period for
the diffusive sampler is two hours; over two hours, the uptake rate is constant but below two hours there is an apparent reduction. The reason for
this variation in uptake rate is unclear at present, but possibly a small
mass of mercury is adsorbed onto the polypropylene diffusion barrier.

One potential problem with the system was the very low mass of mercury
absorbed on the impregnated paper, typically 100 to 3000 ng. This makes it
essential that great care is taken to avoid contamination during all stages
of sampling and analysis.

CONCLUSION

It has been demonstrated that a diffusive sampling system, based on
the Porton Down sampler, can be used for monitoring vapour in workplace air.

The diffusive system is simpler, cheaper and analytically more
flexible, than existing pumped sampling systems. A detailed methodology
for the new system is in the course of publication (Health and Safety
Executive, 198-).

REFERENCES

Bailey, A. and Hollingdale-Smith, P.A. 1977. A personal diffusion sampler
    for evaluating time weighted exposure to organic gases and vapours.
    Ann. Occup. Hyg., 20 345-356.
Crisp, S., Meddle, D.W., Nunan, J.M. and Smith, A.F. 1981. A Copper(I)
    iodide paper for the detection and determination of the concentration
    of mercury in the workplace atmosphere. Analyst, 106, 1318-1325.
Currie, L.A. 1977. Detection and quantitation in X-ray fluorescence
    spectrometry. In "X-ray fluorescence analysis of environmental
    samples" (Ed. T.G. Dzubay). (Ann Arbor Science) pp 289-306.
Health and Safety Executive 1983a. Methods for the determination of
    hazardous substances. Mercury vapour in air. MDHS 16. HSE, London
Health and Safety Executive 1983b. Methods for the determination of
    hazardous substances. Protocol for assessing the performance of a
    diffusive sampler. MDHS 27. HSE, London.
Health and Safety Executive 1985. Occupational exposure limits 1985.
    Guidance note EH 40. HMSO, London.
Health and Safety Executive 198-. Methods for the determination of
    hazardous substances. Mercury vapour in air using a diffusive sampler.
    MDHS-. HSE, London.
Rathje, A.O. and Marcero, D.H. 1976. Improved hopcalite procedure for the
    determination of mercury vapour in air by flameless atomic absorption.
    Am. Ind. Hyg. Assoc. J., 37, 311-314.
Purnell, C.J., West, N.G. and Brown, R.H. 1981. Colorimetric and X-ray
    Analysis of gases collected on diffusive samplers. Chemistry and
    Industry, September 1981, 594-599.
Purnell, C.J., Wright, M.D. and Brown, R.H. 1981. Performance of the
    Porton Down charcoal cloth diffusive sampler. Analyst, 106, 590-598.

AUTOMATED ANALYSIS OF MONITORS

Maire Rothberg, Lauri Saarinen
Uusimaa Regional Institute of Occupational Health
Arinatie 3 A, SF-00370 Helsinki, Finland

ABSTRACT

An automated procedure has been developed for the identification and quantitative analysis of organic vapors collected by monitors. With the help of our program the data of actual chromatograms are compared with the corresponding data of reference compounds, measured under the same analytical conditions. One internal standard, methoxyflurane, is used for the calculation of relative response and relative retention indices. Information on nearly 30 common solvents and complex mixtures of hydrocarbons used in Finnish industry is available in our data library.

INTRODUCTION

The use of passive diffusion badges for sampling of organic vapors has recently gained great popularity. They are easy to use compared to other field sampling techniques. At the same time the demand for the measurement of workers' exposure to airborne organic vapors has increased significantly. Fast and flexible methods for the processing of chromatograms and the preparation of customer reports must therefore be available.

Chromatograms of organic vapors include peaks produced by single components and complex mixtures of hydrocarbons. As sufficient separations on the conventional packed columns are difficult to achieve, capillary columns are being used increasingly in routine analysis. However, the separating power of capillary column has made identification even more difficult, because determinations require the processing of large amounts of data produced by GC analysis of complex samples. Most data systems interfaced to commercial GC systems are powerful enough to handle this task. To implement rapid and correct applications, special software is necessary. With the help of our program the data on unknown compounds are compared with the corresponding information of reference data, measured under the same analytical conditions. One internal standard, methoxyflurane, is used for the calculation of relative response and relative retention indices. The internal standard of known concentration added to each sample which is analyzed, corrects also the possible changes in the detector response and sample injection volume between analyses.

METHOD

The desorption solvent depends on the characteristics of the sample, the most common being CS_2. The analytical instrumentation consists of a Gas chromatograph (Hewlett Packard 5890A) with an automatic sampler (HP 7672A), integrator (HP 3388A) and capillary columns (25 m x 0.31 mm SE 54 or 50 m x 0.22 mm FFAP). The signals from the flame ionization detector are integrated and stored in the tape unit. All data are then sent by RS 232 interface to a microcomputer (HP 200 series model 9920). The microcomputer has a 917 kbyte RAM, a flexible disc drive 270 kbyte, one 14 mbyte hard disc unit, a printer/plotter and a visual display unit. The software is implemented in Basic language. The program offers three menus with the following functions:

- manipulation of the sample report data
- printed results
- data library manipulation

For indentification and calculation the analyst gives the following information by manual input:

- file name of the data library
- file names of analysis reports
- name of workplace, date of exposure, sampling time (etc.)
- the mode of results, calculated in cm^3/m^3 or mgs

RESULTS

The program (Figure 1, flow chart) is used to aid in the identification and quantitation analysis of monitor samples as follows. The first step is to select the data library file which corresponds to the column and temperature conditions used for analysis. There is a reference library for each column and each temperature condition. The data library contains also the names of compounds, molecular weights, densities, relative response and retention indices, diffusion coefficients, desorption efficiencies and TLV values. The library can be easily modified by the user.

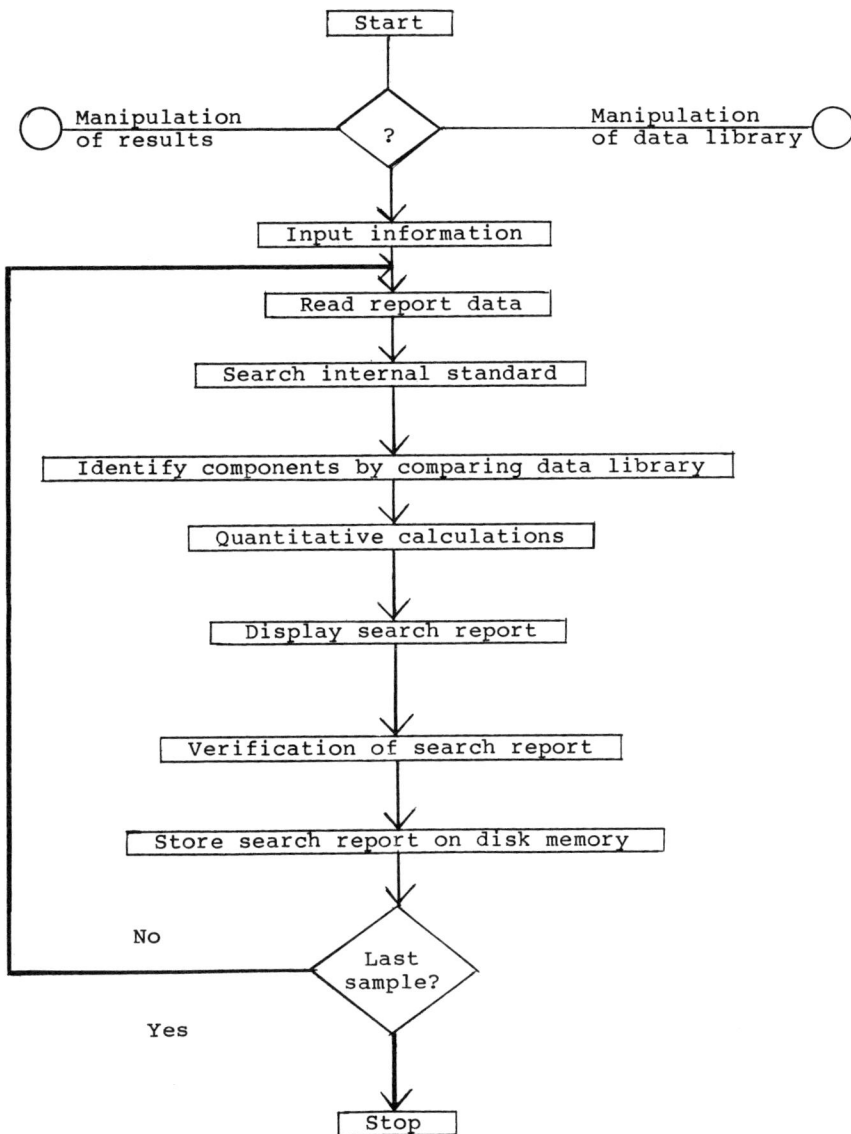

```
                    ┌─────────┐
                    │  Start  │
                    └─────────┘
                         │
                         ▼
  ○ Manipulation    ┌─────────┐    Manipulation    ○
    of results      │    ?    │    of data library
                    └─────────┘
                         │
                         ▼
              ┌──────────────────────┐
              │   Input information   │
              └──────────────────────┘
                         │
                         ▼
              ┌──────────────────────┐
              │   Read report data    │
              └──────────────────────┘
                         │
                         ▼
          ┌──────────────────────────────┐
          │   Search internal standard    │
          └──────────────────────────────┘
                         │
                         ▼
  ┌──────────────────────────────────────────────────┐
  │  Identify components by comparing data library    │
  └──────────────────────────────────────────────────┘
                         │
                         ▼
          ┌──────────────────────────────┐
          │   Quantitative calculations   │
          └──────────────────────────────┘
                         │
                         ▼
          ┌──────────────────────────────┐
          │    Display search report      │
          └──────────────────────────────┘
                         │
                         ▼
      ┌──────────────────────────────────────┐
      │    Verification of search report      │
      └──────────────────────────────────────┘
                         │
                         ▼
  ┌──────────────────────────────────────────────────┐
  │   Store search report on disk memory              │
  └──────────────────────────────────────────────────┘
                         │
        No               ▼
                    ◇ Last ◇
                    sample?
        Yes              │
                         ▼
                    ┌─────────┐
                    │  Stop   │
                    └─────────┘
```

Figure 1

The next step is to give report data files of actual samples, name of workplace, date of exposure, sampling time and mode of the final results. An internal standard is sought. If it is not found, the search is stopped immediately. After the identification process is completed for all compounds, quantitation calculations are performed. A search report is displayed on the screen for further verification. Human interaction is necessary, because several compounds can be matched from the library into the tolerance window of 0.5 % and a new tolerance value may be entered, or the false identification may be manually deleted. The approved results are stored on the flexible disk unit. In the analysis of complex mixtures of hydrocarbons (solvent naphta mixtures) the cross-correlation technique is utilized. After identification of the finger prints of known references, quantitative results are calculated peak by peak. The program is capable of recognizing several mixtures contained in a sample.

The automated procedure has been developed to simplify the interpretation of data produced by capillary column GC analysis. Both the hardware and the columns have been developed so that the data obtained on a one-time basis can be used to calibrate data obtained several weeks later. The results of investigations on response indices are presented in Table 1.
The response indices have been obtained by analysing two mixtures of equal composition during a six-week period at intervals of one week.

TABLE 1 Relative standard deviation of response indices

Compound	RSD, % n = 12
Acetone	3.7
Ethyl alcohol	1.8
Isopropyl alcohol	1.7
n-Hexane	2.2
Ethylacetate	1.1
Isobutyl alcohol	1.3
Toluene	2.4
Zylene	3.8

The index measurements are reproducible and save a considerabe amount of time during analysis, as daily standardization is eliminated. The automated procedures described here provide not only a time-effective approach to organic vapor determinations from passive diffusion badges, but also accurate results.

DISCUSSION - SESSION VI

CRABLE (USA)

How many laboratories are involved in the BCR programme, and how frequently are the samples sent out?

GRIEPINK (CEC)

In this last exercise 12 laboratories were involved mainly from industry and a few from government.

CRABLE (USA)

I would like to offer you some encouragement. The proficiency analytical testing programmes which was set up by NIOSH in 1972-1973 now has some 300 labs in it. Originally we found the same scatter that you have found. People began to look more closely at the quality assurance and the payoff came a little later.

WEST (UK)

First a comment on those results you showed. I notice that there are only 10 or so laboratories. I suspect that they are some of the best laboratories in Europe and the scatter that you would find if you were to include a large selection of laboratories would obviously be a lot greater. Another point. In the UK, we do have a scheme called AQUA which is a quality assurance scheme in occupational hygiene measurement. We started it originally for the Health and Safety Laboratories but we've now expanded it to involve some 30 or 40 other laboratories. At the moment it is restricted to the UK but we hope that it might expand perhaps to other laboratories in Europe in due course.

TINDLE (UK)

Mr. Broadway has pointed out all of the advantages of automated thermal desorption and none of the disadvantages. I feel that instead of automated analyses we can have automated loss. To avoid this, you need very sophisticated computer handling.

BROADWAY (UK)

Yes, there are limitations and I think that we recognize some of the limitations of the system. I know for example it would be no good for hydrogen sulphide or for formaldehyde.

AUFFARTH (FRG)

I would like to ask what provisions have been made for a user of a diffusive sampler to see whether or not the sampler is overloaded. There are some diffusive samplers that we've seen here which have two or three layers of active carbon, one behind the other and these individual layers have to be analysed separately to see whether or not the sampler has been overloaded. In the case of many other samplers, such a monitoring control system does not seem to exist at all.

FIRTH (UK)

Properly designed diffusive samplers are aimed at measuring concentrations around the occupational exposure limit over the required exposure period and they should be able to deal with levels that are two or three times the value without overloading. Therefore I would suggest that if users are getting gross overloading of their diffusion samplers then they shouldn't be concerned about getting an accurate result so much as being concerned about what the regulatory authorities are going to be saying about it.

NORBACK (Sweden)

I would argue that the overall performance of badge-type carbon systems with an automatic sampler and carbon disulphide desorption is maybe the best system. The tube-type sampler with an ATD 50 system or similar has many advantages, but it can be more tricky to handle in practice. It also has a larger capital cost and it needs centralisation of the analysis to be really effective.

BERTONI (Italy)

With thermal desorption, everything depends on the technical knowhow of the analyst. But there is no additional difficulty in the use of thermal desorption techniques providing you have the knowhow regarding the apparatus. There are no problems of blanks. As regards costs, thermal desorption tubes can be reused so the costs for these are lower than for active charcoal.

COCHEO (Italy)

I'm in favour of thermal desorption, I've been using this for 7 years. The main limitation I feel is the non-repeatability of the analysis. Se can't make use of the double column analysis. On the one hand the thermal desorption technique is very advantageous as it's possible to automate it, and it gives a better recovery yield in relation to solvent desorption. On the other hand, we need to have available an apparatus which would enable us somehow or other to repeat the analysis on the same sample.

BERTONI (Italy)

I'd like to reply to my friend Cocheo. I think that we run the risks of getting off the track here because problems connected to thermal desorption are common both to active and to passive sampling. The apparatus which Mr. Cocheo referred to would be very useful if downstream you use capillary columns where it's necessary to inject a small amount of the sample but otherwise in some cases it could lead to a loss of the advantage of thermal desorption in which is that you can have a large number of samples and a very high sensitivity.

COKER (UK)

So far, we've only heard of one type of thermal desorption. There is also an indirect type which desorbs into a reservoir. This has some additional advantages that you've retained the sample and you can repeat analyses if you wish.

SAUNDERS (UK)

We haven't really considered the automation of solvent desorption. With the advent of robotics this has high potential for cost saving.

BERLIN (CEC)

How far will the authorities go in accepting data that hasn't been confirmed by mass spectrometry?

FIRTH (UK)

In effect the regulatory bodies will accept data which they regard as being of a suitable quality. If you look at the majority of the exposure limits, they are relating to single compounds. There is, at the present time, no clear cut regulatory procedure for assessing the risk of mixtures of compounds. So, from a regulatory point of view, we're dealing with relatively simple systems which can be coped with by most straight-forward laboratory analytical systems such as chromatography.

FIELDS (UK)

On the subject of using an FID rather than a mass spectometer, if you do this, you will invariably err on the side of safety in that if the compound you are looking for is there, then you will see it and if there is another compound there that you are not looking for, you will simply get a higher result.

BROWN (UK)

Obviously it's nice to have a mass spectrometer to give you qualitative information on a gas chromatographic analysis, but this is just another extra expense for most people. One system which we are currently using at the Health and Safety Laboratories for thermal desorption analysis has two capillaries in the gas chromatograph: one polar and one non-polar and both leading to an FID. This enables us to have some sort of cross-check with a relatively simple system as to what the components are in a complex analysis. We still see the major role of diffusive samplers in large surveys where we in fact know what we are looking for before we start. In these cases, the qualitative information is not nearly so important, and even where we don't have this information, it's relatively simple to take some additional samples, maybe with a hand pump, to check out the gas chromatographic analysis before you put the real samples through.

MILLER (UK)

I have a thermal desorber connected to a gas chromatograph and split the gas at the column exit and send the bulk of it through to an FID or a nitrogen-selective detector for quantification and the rest of the split goes through to a mass spectrometer. This enables me to give my customers more confidence in the results they get and also enables me to pick up unknowns that appear on the samplers.

KENNEDY (UK)

We've talked about the possibilities of identifying unknown peaks on gas chromatographes. We think it is very important to have information about the composition of the product that is being used. If we don't have that information, then we probably won't make the correct choice of passive sampler or active sampler and we may well not make the correct choice of analytical method.

BERLIN (CEC)

This allows you to put greater pressure on the supplier to provide you with certificates of analysis. Is that the approach?

KENNEDY (UK)

We ask our suppliers to give us this information. Clearly in the UK we have the Section 6 legislation which does not require full disclosure, but obviously we have commercial pressures and if a supplier is selling us a hazardous product and won't tell us what is in it for whatever reason we always have the option of refusing to buy his product.

BROWN (UK)

When looking at costs, we've only been looking at the high tech end of diffusive sampling where there are very high capital costs in buying the thermal desorbers, the GCs, data systems, etc. However, there are very adequate techniques using colorimetric dosimeters to provide occupational health data and these are going to have very low capital cost outlay and only a small unit cost and I'd just like to make that point so that some people working for small companies aren't put off using diffusive samplers because of what would appear to be very high capital costs required to use some of the techniques.

LABORATORY TESTING OF A THERMALLY DESORBABLE PASSIVE SAMPLER

FOR VOLATILE ORGANIC COMPOUNDS IN INDOOR AIR

M. De Bortoli, H. Knöppel, E. Pecchio, H. Vissers

CEC, Joint Research Centre, Ispra (Italy)

ABSTRACT

The features of a small (160x6 mm) inexpensive diffusion sampler employing Tenax as adsorbent to be used in indoor air quality investigations have been studied in a glass test cell with vapours of benzene, toluene, hexanal, 1.3-xylene, n-decane, limonene and naphtalene. Mean sampling flows over one week of 5.2-7.6 cm^3/h have been observed, independently of air humidity (40-80% R.H.) and air velocity (down to 0.001 m/sec.). The tightness of different cap types was determined employing perfluorocarbons as tracers and in few cases found as low as 0.1 cm^3/week. The sampler is amenable to completely automatic analysis.

INTRODUCTION

During the last few years several types of passive samplers for volatile organic compounds (VOC) have been developed and made commercially available. They were the answer to the need for light, inexpensive samplers felt by those responsible for the industrial hygiene monitoring of workers, where sampling times of few hours are required to measure relatively high concentrations. A different requirement is that of monitoring the exposure to VOC in non industrial indoor spaces, like homes, offices, schools, etc., where concentrations are much lower and may vary to a much larger degree. Hence sampling times of few days are necessary, in order to provide a representative measure, taking into account also personal behaviour patterns. The experience with charcoal passive samplers of different brands (De Bortoli et al., 1984) showed that they are not suited to this purpose for insufficient sensitivity, due to the dilution in the solvent and for incomplete extraction by the solvent of polar compounds. In principle a thermally desorbable passive sampler could overcome these drawbacks and would present the additional advantage of being amenable to complete automation of the analysis, which may become an important feature when a large number of samplers has to be processed. The positive experience in the analysis of a variety of VOC in ambient air by active sampling on Tenax (Versino et al., 1974) led to the adoption of this polymer as adsorbent.

EXPERIMENTAL SECTION

The geometry of the sampler was determined by the thermal desorption cold trap injector (Chrompack, Middelburg, The Netherlands) already described elsewhere (De Bortoli and Pecchio, 1985). The glass tubes (6 mm o.d., 3 mm i.d., 160 mm length) were packed with about 100 mg of Tenax TA (60-80 mesh); the air-adsorbent interface was defined by a stainless steel gauze diaphragm,which was located at 25 mm from the tube opening (diffusion length). The operating conditions of the thermal desorption cold trap injector were the following: desorption temperature $250^{\circ}C$ and time 15 min.; intermediate trapping at $-120^{\circ}C$; splitless injection.

The analyses were carried out using a Hewlett-Packard 5880 gas chromatograph equipped with a glass capillary column (0.32 mm i.d. x 25 m, coated with SE-52 0.25 μm by MEGA, S. Giorgio su Legnano, Italy) and a flame ionisation detector.

The exposure of the samplers to the VOC was carried out in an all glass test cell (450 1 capacity) which was supplied with a controlled atmosphere at a rate of about 1 air change per hour. The organic vapours, released by permeation tubes, were diluted with air supplied by a clean air generator. Humidity was also controlled. A small fan homogenized the air inside the cell. During the passive sampling experiments the vapour concentrations in the cell were periodically checked by active sampling.

The determination of the infiltration rate (or tightness) with caps of different types was carried out exposing the capped tubes to known concentrations of perfluorocarbons in order to have a better sensitivity and to discriminate against vapours emitted by the caps themselves.

RESULTS AND COMMENTS

The features required for a diffusion passive sampler to be employed in indoor air pollution investigations may be summarized as follows: the sampler should collect the VOC present in a representative and reproducible way, both qualitatively and quantitatively, over a few days, possibly one week. The sampling rate should be independent of the air velocity and of relative humidity over the range occuring in living spaces. The sensitivity of the whole system (sampling and analysis) should be adequate to measure few ng/l of single compounds. When the sampler is capped the caps should effectively seal it, so that the volume sampled over one month through the caps be negligeable compared with the volume collected during operation of the sampler. Experiments aimed at characterizing the above mentioned sampler with respect to these features gave the following results:

The mean sampling flows measured over one week (in 2 cases even 11 days) in seven experiments are reported in Table 1 along with standard deviations; sampling flows (S) calculated using the formula S=D.A/L (where D is the diffusion coefficient, A the sampling section and L the diffusion length) are also reported for the compounds of which diffusion coefficients are known.

TABLE 1   Sampling flows of diffusion passive sampler (cm^3/h)

	mean observed	std. dev.	calculated	obs./calc.
benzene	5.20	0.59	9.49	0.55
toluene	6.79	0.60	8.64	0.79
hexanal	6.01	0.55	–	–
1.3-xylene	7.20	1.16	7.00	1.03
n-decane	5.97	0.80	6.11	0.98
limonene	5.76	0.90	–	–
naphtalene	7.61	1.18	6.11	1.25

From the last column it appears that the most volatile compounds present mean sampling flows smaller than the calculated ones. To investigate this finding several samplers were exposed to the same atmosphere for increasing times: the results for benzene and n-decane are shown in Fig. 1.

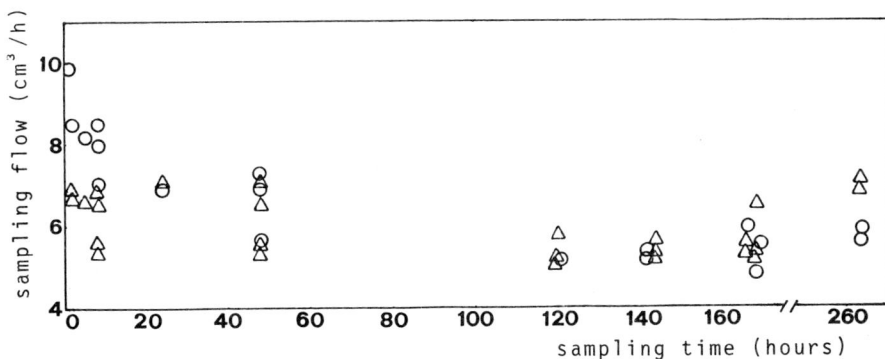

Fig.1 Sampling flow of passive sampler versus time for benzene (circles) and n-decane (triangles).

Whereas for n-decane the sampling flow is substantially independent of sampling time, for benzene a marked decrease, especially in the first hours is evident. This effect already observed over 48 hours (Brown and Walkin, 1981) can be explained by the presence of a not negligable compound concentration on the air adsorbent interface, which decreases the concentration gradient and hence the driving force for adsorption (Coutant et al., 1985). Since the mean value is reproducible the variation of sampling flow with time observed for benzene is not a limitation in the practical use of the sampler, provided a standard sampling time is adopted.

Concerning the influence of humidity on the sampling flow, no significant difference was observed between 40% and 80% R.H.

Also no influence of air velocity was found, no significant difference being observed in the sampling flows between high velocity (fan running, about 0.5 m/sec.) and low velocity (no fan, less than 0.001 m/sec.), as could be expected in view of the very low sampling flow of this sampler type.

The sensitivity of the analytical equipment is sufficient to measure few (1-10) ng/l of the compounds of interest in indoor air pollution collecting the vapours present in few hundreds of cubic centimeters of air.

The last aspect investigated was the effective tightness of different cap types used to seal the sampler before and after sampling. The perfluorocarbon technique used for this test can evidence the infiltration of $0.1 \text{ cm}^3/$ week and some of the caps tested effectively presented such a tightness. However, only those made out of metal (both swagelok type and with O-ring seal) were acceptable due to VOC emission by the others (plastic or rubber).

REFERENCES

Brown, R.H. and Walkin, K.T. 1981. Performance of a tube-type diffusive sampler for organic vapours in air. In proceedings of the 5th International S.A.C. Conference, May 1981, Cambridge (G.B.), 205-208.
Coutant, W.R., Lewis, G.R. and Mulik, J. 1985. Passive sampling device with reversible adsorption. Anal.Chem. 57, 219-223.
De Bortoli, M., Knöppel, H. Mølhave, L., Seifert, B., Ullrich, D. 1984. Interlaboratory comparison of passive samplers for organic vapours with respect to their applicability to indoor air pollution: a pilot study. EUR 9450 Report, CEC Luxembourg.
De Bortoli, M. and Pecchio, E. 1985. Sensitive determination of perfluorocarbon tracers in air by intermediate trapping and GC-ECD analysis. J. High Resol. Chromat. and Chromat. Comm. 8, 422-425.
Versino, B., de Groot, M. and Geiss, F. 1974. Air pollution: sampling by adsorption columns. Chromatographia 7, 302-304.

A DIFFUSIVE SAMPLER - STAIN TUBE EXPOSURE MONITOR

S J Gentry and P T Walsh

Health and Safety Executive
Broad Lane
Sheffield S3 7HQ
UK

ABSTRACT

A simple, versatile and low-cost method of personal monitoring has been developed. The technique involves the collection of the analyte on a diffusive sampler followed by thermal desorption onto a short term stain-tube. Three chlorinated and three aromatic hydrocarbons were used to demonstrate the validity of the method. Using currently available stain-tubes, the technique has the potential to monitor exposure to some 40 analytes at levels around their 8-hour limits.

INTRODUCTION

The increasing requirement for the measurement of time weighted average (TWA) exposure of an individual to long term toxic hazards demands the development of suitable methods of personal monitoring. Ideally the method should be non-intrusive, easy to operate, versatile and inexpensive, in addition to having the usual analytical requirements of accuracy, reliability and selectivity. Current methods of personal monitoring based on diffusive sampling followed by analysis using chromatography, spectrometry etc. satisfy most of the above criteria. However the analysis stage can be complex and requires considerable expenditure on equipment and labour. This paper discusses a novel, low cost monitoring system based on the combination of diffusive sampler and an existing field-analysis technique.

METHOD AND PROCEDURE

Analyte is collected by a diffusive sampler so that the uptake is proportional to exposure. The analyte is then thermally desorbed into an air stream passing directly through a short-term stain tube. In this way the total mass of analyte collected by the sampler is presented to the stain tube as a varying concentration-time profile dependent on the initial uptake, the desorption rate and air flow rate. For such a technique to behave as an exposure monitor the stain length on analysis should be proportional to the mass of analyte collected and thus, via the diffusive sampler mechanism, the total exposure (concentration x time).

Three chlorinated hydrocarbons, 1,1,1-trichloroethane, trichloroethylene, tetrachloroethylene, and three aromatic hydrocarbons, benzene, toluene, m-xylene, were used as test analytes to validate the technique experimentally. Perkin Elmer diffusive sampler tubes were loaded with Chromosorb 102 as adsorbent for the chlorinated hydrocarbons and Tenax for the aromatic hydrocarbons. The tubes were then exposed to atmospheres containing the above vapours at a range of concentrations and for a range of times such that the exposures varied from about 10% to 200% of the 8-hour exposure limits. The analytes were thermally desorbed at 200°C (except 1,1,1-trichloroethane which was desorbed at 150°C) onto the appropriate Draeger or Kitagawa stain tubes. The stain lengths were measured using the existing scales.

RESULTS

For the chlorinated hydrocarbons it was found that, for exposures up to about 150% of the recommended limits, the scale reading was a linear function of exposure. For the aromatic hydrocarbons the scale reading was linear with exposure over the whole of the measured range. Consequently the data were fitted to the function y = m.x where y is the mean stain length taken from a number of samples, x is the exposure and the slope m represents the mean uptake rate in arbitrary units for each analyte. Values obtained for m together with the relative standard error in m are shown in the table.

From the manufacturers recommended sampling volume for conventional use of the stain tube, the scale reading may be converted into mass units and hence the mean uptake rate may be expressed in conventional units of ng/ppm/min. The figures thus obtained are included in the table together with values obtained from other work using an automatic thermal desorber (Perkin Elmer ATD50) and analysis by gas chromatography. It can be seen that there is reasonable agreement between the two analytical techniques.

TABLE 1   Analytical Data for chlorinated and aromatic hydrocarbons

Analyte	m* ($\times 10^3$)	Relative standard error in m (%)	Uptake rate (ng/ppm/ min)	Uptake rate from other work (ng/ppm/ min)
1,1,1-trichloroethane	1.53	1.6	1.7	
trichloroethylene	1.11	2.7	1.8	2.2
tetrachloroethylene	0.44	8.5	3.0	2.5
benzene	2.10	5.9	1.4	1.4
toluene	4.40	1.7	1.7	1.7
m-xylene	1.90	2.6	1.7	

* Calibration equation: mean stain-tube scale length = m. exposure (ppm min)

DISCUSSION

The results presented here clearly demonstrate the validity of the technique provided that certain criteria are met. These are: (i) that the diffusive sampler can take up a mass of analyte proportional to exposure over the range necessary for personal monitoring and (ii) that the analyte can be efficiently desorbed onto an appropriate stain tube to produce a stain length proportional to its mass. The Perkin Elmer diffusive sampler tube has been shown (Cox et aL, 1984; Perkin Elmer) to satisfy the first criterion for many organic vapours. Indeed, from the published data it can be seen that to a first approximation uptake rate is independent of both adsorber and analyte and may be taken as about 2 ng/ppm/min. Using this figure it is then possible to calculate the mass adsorbed on 8-hour exposure of the sampler to the recommended exposure limit of any gas or vapour. This figure can then be compared with the equivalent mass range of the appropriate stain tube to evaluate the potential of the technique for a range of gases and vapours. Using the acceptance condition that the stain tube should have a detection limit of better than one quarter of the 8-hour TWA exposure limit, it can be shown that this technique is potentially suitable, subject to the above criteria, for about 40 gases and vapours. These analytes are listed in Table 2.

TABLE 2   Gases and vapours potentially suitable for analysis by the technique using currently available stain tubes

Acetone	1,2-dichloroethylene	Isopropyl acetate
Ammonia	Dichloromethane	Methyl ethyl ketone
Benzene	Ethyl acetate	Nitrogen dioxide
Bromochloromethane	Ethanol	Pentane
Butadiene	2-ethoxyethanol	Phenol
Butan-2-ol	2-ethoxyethyl acetate	Phosgene
Butyl acetate	Ethyl benzene	Phosphine
Carbon dioxide	Formaldehyde	Styrene
Carbon monoxide	n-heptane	Sulphur dioxide
Carbon tetrachloride	n-hexane	Tetrachloroethylene
Chlorobenzene	Hydrogen chloride	Toluene
Chloroethane	Hydrogen cyanide	1,1,1-trichloroethane
Chloroprene	Hydrogen sulphide	Trichloroethylene
Cresol	Isoamyl acetate	Vinyl chloride
Cyclohexane	Isobutyl acetate	Xylene
1,1-dichloroethane		

Clearly, by changing the geometry of the diffusive sampler, and thus varying the mass collected, the useful range of the technique can be extended to take advantage of other available stain tubes.

REFERENCES

Cox P C, Walkin K T and Wright M D. 1984. A diffusive sampler
    handbook. Health and Safety Executive. Internal Report
    IR/L/AO/84/16.
Perkin-Elmer Ltd. Model ATD50 Thermal desorption applications, No 11.

A PASSIVE SAMPLER FOR THE COLORIMETRIC

MEASUREMENT OF TDI

J A Groves, P A Ellwood and J S Szikora

Health and Safety Executive

London, UK

ABSTRACT

A colorimetric passive sampler for the determination of toluene diisocyanate is described. The sampler is constructed from commercially available components, responds to both the 2,4- and 2,6-isomers of toluene diisocyanate and can be used in the range 5 - 40 ug/m^3 of NCO group for an 8 hour sample.

INTRODUCTION

The UK threshold limit values (TLVs) for isocyanates were replaced in 1983 with a control limit (Health and Safety Executive, 1984). The control limit (20 ug/m^3 for an 8 hour time-weighted average and 70 ug/m^3 for a 10 minute time-weighted average) applies to all chemicals having an isocyanate, -NCO, group. Thus, unlike the TLVs for isocyanates (Health and Safety Executive, 1980) which applied to specific monomers, the control limits apply to monomers, prepolymers, oligomers and partially reacted species.

At present there is only one analytical method available, using HPLC with electrochemical and UV detection, (Health and Safety Executive, 1983) which gives the same response for all isocyanate species likely to be found in an air sample. While this method is perfectly satisfactory for the determination of toluene diisocyanate (TDI), it is not a field method and, since it employs sample collection into toluene, the problem of solvent evaporation makes 8 hour sampling inconvenient.

One conseqûence of the introduction of the control limit based on total -NCO is that the popular modified Marcali field method (HMSO, 1973) is no longer valid for the determination of TDI monomer since the 2,4- and 2,6- isomers of TDI give different responses in the analysis. Thus the need exists for a field method for TDI monomers.

Recent tests on a number of paper tape monitors for isocyanates (Groves et al, 1985) showed that two commercially available colorimetric tapes have almost identical responses to the two TDI isomers and thus are able to provide a measure of total -NCO concentration in air derived from mixtures of the isomers. Reaction with TDI produces a coloured stain almost immediately on the tape and so forms the basis of an on-site monitoring technique for TDI requring no chemical post sampling treatment but merely assessment of colour density on the tape.

These tapes are produced for use with static background monitors, but it should be possible to evaluate personal exposure to TDI, both over 10 minutes, by pumping air through the tape, and over 8 hours by either low flow rate pumping or passive sampling techniques. This paper describes the development of a passive TDI sampler based on one of these tapes.

METHOD

The isocyanate tape manufactured by GMD Systems Inc. (known as SKC in the UK) was selected for this work for the following reasons:

1. The tape response to the two TDI isomers is known to be almost identical (Groves et al, 1985).

2. Colour formation is rapid.

3. The coloured stain is essentially permanent.

Separate standard atmospheres of both 2,4- and 2,6-TDI were prepared using permeation tubes (Dharmarajan and Rando, 1979) and the concentrations were checked by the modified Marcali method and the HPLC procedure. Since each calibration took 8 hours, the modified Marcali method was used to monitor the atmosphere concentrations in preference to the HPLC method because it provided a rapid estimate of the concentration.

The sampling chamber was equipped with a revolving carousel to which up to 6 samplers could be attached. The carousel ensured all sampling devices were exposed on average to the same atmosphere with an average face velocity across the samplers of about 0.10 m/s.

After initial experiments with various commercially available diffusive samplers, the 3M Type 3550 badge was selected for further investigation. The charcoal cloth insert was removed and the inside of the badge was packed with card and a 2 cm strip of the paper tape placed on the card such that when the polypropylene cover was positioned in the normal way, it was in contact with the tape.

The samplers were attached to the carousel and exposed to the atmospheres, in batches of six, for periods up to 8 hours.

The density of the stain was evaluated with a reflectance paper tape reader originally marketed by MDA (Poole, Dorset) for use with a tape for the detection of vinyl chloride.

RESULTS

The relationships between paper tape stain intensity and isocyanate atmosphere concentration for an 8 hour exposure for each TDI isomer are shown in Figure 1. There is a fair degree of scatter, reflecting the difficulty associated with assessing the standard atmosphere concentration accurately at such low levels, but generally the data for the 2,4- and 2,6-TDI isomers are reasonably close: in the range 0-2 times the 8 hour control limit (0 - 40 ug NCO/m^3) the sampling device appears to perform acceptably, especially in view of its cheapness and simplicity.

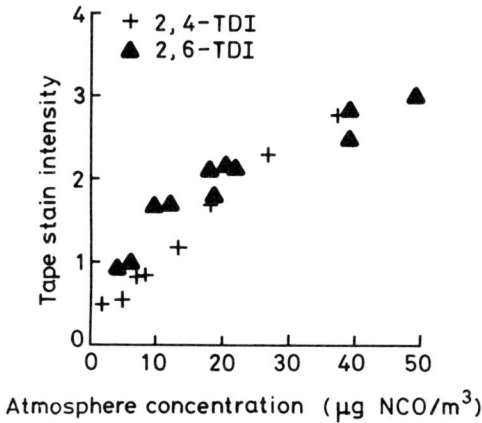

Figure 1 Tape stain intensity vs concentration for 8 hour samples

The relationships between paper tape stain intensity and exposure time for a given isocyanate concentration for each isomer are given in Figure 2. In this case the agreement between the data for the two isomers is very good, and this probably reflects the greater accuracy involved in measuring time rather than concentration. In Figure 3, the relevant portion of Figure 1 is plotted on a scale such that direct comparison with Figure 2 is possible on a 'ug hour' basis. The close agreement observed between Figures 2 and 3 indicate that the paper tape stain intensity obtained in the passive samplers can be used to evaluate total dose of TDI for any period up to 8 hours.

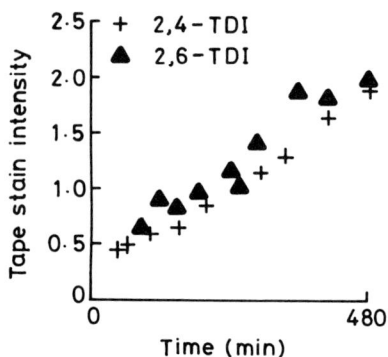

Figure 2   Tape stain intensity
vs Time at fixed concn. of
20.4 ug/m³ NCO.

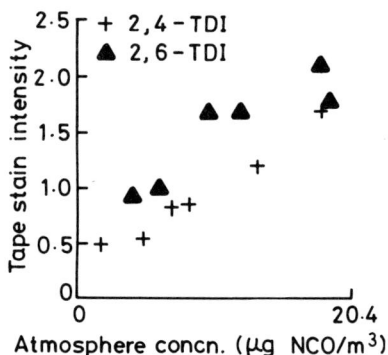

Figure 3   Tape stain intensity vs
concn. for 8 hour samples.

CONCLUSIONS

The passive sampler described should enable compliance with the UK 8
hour control limit to be assessed on-site where TDI is in use in the range
5 - 40 ug NCO/m^3 without the need for chemical work up. Concentrations
of higher levels of TDI can be determined over shorter time periods.
Future work includes the testing of standard atmospheres containing
mixtures of the 2,4- and 2,6-TDI isomers and field testing of the samplers.

REFERENCES

Dharmarajan, V. and Rando, R.J. 1979. A new method for the generation
    of standard atmospheres of organo isocyanates
    Am. Ind. Hyg. Assoc. J., 40, 870-876.
Groves, J.A., Brown, R.H., Szikora, J.S., Bagon, D.A., Glover, R. and
    Bugler, J. 1985. A Laboratory study of impinger efficiencies and
    a comparison of the Marcali method, MDHS 25 and paper tape monitors
    for the measurement of 2,4- and 2,6-toluene diisocyanate.
    Health and Safety Executive Internal Report, IR/L/AO/85/24.
Health and Safety Executive. 1980. Threshold Limit Values 1980.
    Guidance note EH15/80, HMSO, ISBN 011 883379 0.
Health and Safety Executive. 1983. Methods for the Determination of
    Hazardous Substances, Organic Isocyanates in Air,
    MDHS 25, HMSO, ISBN 0 7176 0143 9.
Health and Safety Executive, 1984. Occupational Exposure Limits 1984.
    Guidance note EH40, HMSO, ISBN 011 883560 2.
HMSO. 1973. Methods for the Detection of Toxic Substances in air,
    Booklet No 20, Aromatic Isocyanates.

PASSIVE SAMPLING OF NITROGEN DIOXIDE AND SULFUR DIOXIDE

IN AMBIENT AIR

Markus Hangartner and Peter Burri

Federal Institute of Technology
Department of Hygiene and Applied Ergonomics
8092 Zurich, Switzerland

ABSTRACT

Sulfur dioxide and nitrogen dioxide have a great importance as ambient air pollutants because of their effects on
the respiratory system of both men and plants. In the Swiss
Clean Air Act the air quality standards for these pollutants
are promulgated and therefore the surveillance is necessary.
With the help of continuous monitoring systems a precise monitoring can be done though this is restricted to only a few
points. On the other hand, passive samplers are devices, which
can easily be transported and run without electricity. The ap-
plicabilty of NO_2 and SO_2 passive samplers for ambient monitoring was evaluated with respect to accuracy, precision, influence of weather and storability. Nitrogen dioxide samplers
were found applicable over a wide range of ambient measurements; sulfur dioxide samplers described could be used only
with certain reservations.

INTRODUCTION

The main source of nitrogen dioxide in Switzerland is
road traffic and that of sulfur dioxide is space heating. Thus
these pollutants have importance for the whole region and
hence the Swiss Clean Air Act has selected them as criteria
pollutants.

Statistical Definition	Nitrogen Dioxide	Sulfur Dioxide
Annual Average	30 $\mu g/m^3$	30 $\mu g/m^3$
95% Value (of all 1/2-hr values)	100 $\mu g/m^3$	100 $\mu g/m^3$
24-hr Average (exceeding not more than once a year)	80 $\mu g/m^3$	100 $\mu g/m^3$

Table 1: Air quality standards for NO_2 and SO_2 in Switzerland

In order to control these ambient air quality standards simple instruments are needed which can be used for the whole area. Passive samplers fulfil this requirement. They are based on the principle of passive diffusion of a pollutant to an adsorbing medium. Thereby a total dose integrated over the sampling time is obtained. The use of such a passive sampler depends upon the detection limit of the analytical method as compared to the relevant standard. Further it depends upon its agreement with other usual methods of ambient air quality measurements, the influence of weather conditions on the sampling efficiency and also the question of storing the exposed samplers. This paper describes the study on the use of passive samplers for the ambient air quality measurements.

MATERIALS AND METHODS

The passive samplers used in this study were developed by PALMES (1) and have already been used by several authors (2,3) until now. They consist of small acrylic tubing 7.4 cm long, 1 cm in diameter having 3 stainless steel mesh as support for adsorbing material at its one end. Triethanolamine was used as adsorbent for nitrogen dioxide and a triethanolamine/glycole mixture (3:1) was used for sulfur dioxide.

In order to check the conformity with the continuously measuring instruments a Chemiluminescent Nitrogen Oxide Analyzer (Monitor Labs Inc Model 8840) and a Fluorescent SO_2 Analyzer (Monitor Labs Inc Model 8850) were used. For calibrating the monitors and preparing the test atmosphere for the exposure of passive samplers the calibrator of Monitor Labs (Model 8550) was used. The conformity under field conditions and the interfering effects of weather were studied by sampling simultaneously outdoors and drawing the same air indoors and sampling it in the environmental chamber under controlled conditions.

RESULTS

The agreement of passive sampler values for NO_2 and for SO_2 with those obtained by the monitors in the laboratory and under field conditions are shown in Figure 1:

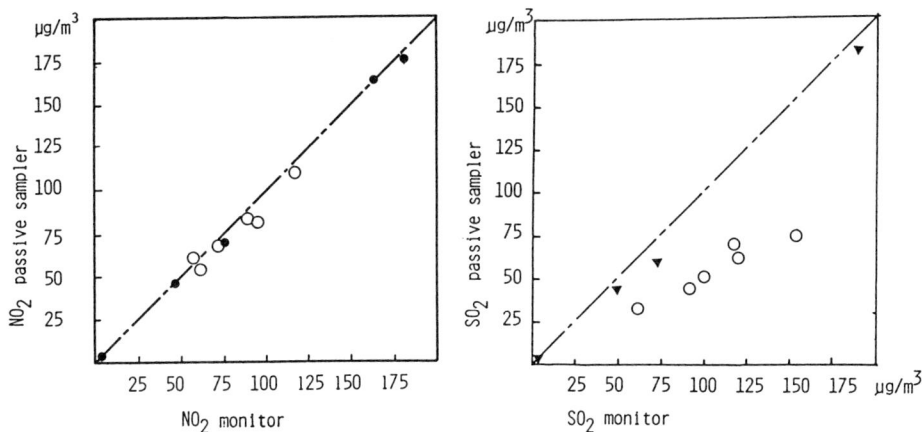

FIGURE 1: Comparison of values obtained by passive sampling and by continuous monitoring in laboratory (dots) and in field (circles)

As seen from the figure, there is a good agreement between passive sampler and monitor values for NO_2 under field conditions, but not in the case of SO_2. The sampling efficiency of the passive sampler for NO_2 does not seem to be affected by meteorological factors like wind, precipitation and temperature in the usual range of weather conditions (0.5 to 4 m/sec and -10 to 25°C). The storage time for the NO_2 samplers was at least 6 months before exposure and 4 months after exposure. They were stored at 4, 22 and 37°C and no changes in values for NO_2 samplers were observed. This was not the case with SO_2 samplers. Storage before exposure was not a problem, but storage after exposure longer than two weeks or at higher temperatures than 4°C resulted in considerable losses of SO_2.

In order to determine the lower detection limit the mean and standard deviation of blanks were taken into consideration.

The mean blank value for NO_2 was 0.0103 (u = 90) and for SO_2 was 0.090 (u = 144) absorbance units; their standard deviation being 0.00281 and 0.00147. This gave the lower detection limit of 0.022 for NO_2 and 0.0958 for SO_2 at 99.9% confidence level. These are equivalent to 0.8 nanomole NO_2 and 0.5 nanomole SO_2 which when converted to a 7-day exposure would give 4 $\mu g/m^3$ NO_2 and 5 $\mu g/m^3$ SO_2. The mean relative standard deviations were 3% for NO_2 and 12% for SO_2.

CONCLUSIONS

The NO_2 passive sampler described here could very well be used for determining the annual averages. The lower detection limit of 28 $\mu g/m^3$ for 24-hr exposure permits the surveillance of 24-hr average standard. The long storage time and the temperature stability up to 37^oC make them suitable for broad applications so that even mailing them is also possible. Because of integrated sampling the information on peak values could not be obtained and the comparison to short-term standards is not possible here.

It is felt that the samplers reported here might also be used for SO_2. But the sulfite ion is not stable enough to get adsorbed on triethanolamine and gets converted to sulfate when traces of oxidants are present; this is much more the case at higher temperatures. This shifts the lower detection limit nearer to 25 $\mu g/m^3$ and makes the method rather unsuitable for SO_2. There seem to be two solutions to this problem: either if the sulfite ion could be stabilised in the adsorbed phase or if it can be quantitatively oxidised to sulfate which then can be determined.

REFERENCES

(1) Palmes, E.D., Gunnison, A.F.: Personal Sampler for Nitrogen dioxide. Am. Ind. Hyg. Assoc. J. 37:570-577 (1976)
(2) Dockery D.W., Spengler J.D., Reed M.P., Ware J.: Relationships among personal, indoor and outdoor NO2 measurements. Environment International 1981; 5: 101-107.

(3) Seifert, B., K.-E. Prescher, D. Ullrich: Auftreten anorganischer und organischer Substanzen in der Luft von Küchen und anderen Wohnräumen. WaBoLu-Berichte 2/1984. Dietrich Reimer Verlag, Berlin 1984.

USE OF A PERMEATION DEVICE TO COLLECT VOLATILE

ORGANIC PRIORITY POLLUTANTS

J. K. Hardy and R. D. Blanchard

Knight Chemical Laboratory
The University of Akron
Akron, Ohio U.S.A.

ABSTRACT

The use of permeation in the collection of volatile organic priority pollutants is described. Permeation, in the collection of environmental samples, involves placing a synthetic membrane between a sampling environment and a collection medium. The analyte to be collected, permeates through the membrane from the sampling environment to the collection medium where it is stored for analysis. A permeation device has the advantages of providing time-weighted-average analyte concentration values, it's simple in construction and use, it's inexpensive, and it does not require a power supply. Because a permeation device is inexpensive and does not require a power supply, multilocation testing with such a device is feasible. A permeation device is described which was found to provide linear results for mass of analyte collected from a sampling environment versus the product of analyte concentration in the sampling environment and time of exposure of the permeation device to the sampling environment over a concentration range from the low parts-per-billion to the low parts-per-million.

INTRODUCTION

Federal water pollution regulations which include the Federal Water Pollution Control Act of 1972 (U.S. Statutes at Large, 1972), the Clean Water Act of 1977 (U.S. Statutes at Large, 1977), and a lawsuit termed the EPA Consent Decree (NRDC, et. al. versus Train, 1976) provided the impetus for the formation of a list classified as the Priority Pollutant List (U.S. Code of Federal Regulation, 1982). Contained in the Priority Pollutant List is a group of 31 species known as volatile or purgeable organics.

The following describes a method for the collection, isolation, concentration, and determination of as many as 23 volatile organic priority pollutants. The method is based on permeation methods developed for ambient air sampling (Hardy, et al., 1981). A synthetic membrane is used to affect collection of compounds from a sampling environment and the species are then trapped on the opposite surface of the membrane in a collection medium.

The method described utilizes a silicone polycarbonate membrane and a collection medium of activated charcoal. After exposure of the sampling device to an aqueous medium of interest, the charcoal is removed from the sampling device, the volatile components are desorbed with carbon disulfide, and their levels are determined by gas chromatography. The initial research on this technique was done with the following compounds: benzene, toluene, ethylbenzene, dichloromethane, chloroform, and carbon tetrachloride. This was followed by evaluation of the technique on 23 volatile organic priority pollutants.

EXPERIMENTAL

The sampling devices used to determine the relationship between mass of analyte collected in a permeation device, the analyte concentration in the sampling environment and the time of exposure of the permeation device to the sampling environment consisted of 17 mm i.d. glass tubes, 50 mm in length to which the silicone polycarbonate membrane (0.025 inches, General Electric) was affixed with silicone rubber cement. The calibration chambers were 4-L Erlenmeyer flasks in which the devices were suspended.

The gas chromatograph used for the analyses was a Hewlett-Packard Model 5730A unit with twin FID detectors. Six foot by 1/8 in. stainless steel columns packed with 1% SP-1000 on 60/80 Carbopack B (Supelco, Inc.) were used on all analyses prior to separation of 23 volatile organic components. Separation of 23 volatile organic priority pollutant was accomplished with an SPB-1 fused silica capillary column 60 m by 0.32 mm i.d. with a film thickness of 1.00 micrometers.

RESULTS

Previous work on ambient air monitors has shown that the mass of analyte collected in a permeation device is related to the concentration of the analyte in the sampling environment and the time of exposure of the permeation device to the sampling environment, and Equation (1) shows that relationship:

$$K = m/(Ct) \tag{1}$$

where C represents the time-weighted-average analyte concentration in parts-per-million, t is the exposure period in hours, m is the mass of analyte collected (micrograms), and K is the calibration constant for the analyte, and has units of micrograms/(ppm*hours). To determine if this relationship held for volatile organic priority pollutants, permeation devices were exposed to known levels of each of the six initial volatile organic priority pollutants for time periods ranging from 0.5 hours to 10 days. In each case, three devices were exposed to identical solutions in three separate exposure chambers. Each device was charged with 1 gram of

activated charcoal just prior to exposure. After an appropriate exposure
interval, the devices were withdrawn and the charcoal was removed. For
analysis, the charcoal was desorbed with 5 mL of carbon disulfide, the solution was allowed to stand for 30 minutes, and then 5 microliters of this
solution was assayed. It was found that a linear relationship was obtained between the mass of analyte collected and the product of its external
concentration and the time of exposure for each test compound. Initially
each compound was studied individually. After individual analyses of the
six test compounds, cumulative exposures were performed with the six test
compounds. Cumulative exposures did not adversely affect the collection
of any of the volatile organic compounds, and linear results were maintained. Table I lists the K value, ppm*hour values, concentration ranges,
and correlation coefficients for each compound at 22°C.

With exposure of calibrated devices for a known time period, it is
possible to determine the time-weighted-average concentration of each
species by

$$C=m/(K_o+sT)t \qquad (2)$$

where K is the constant for each species as defined in Equation 1,

Table I.   Calibration Results and Ranges Investigated.

Compound	K value[1] ($\mu g/ppm \cdot h$)	ppm·h range	Conc. range, ppm	Correlation coefficient
Dichloromethane	15.7±0.1	0.98-22.6	0.01-20.0	0.997
Chloroform	15.9±0.2	1.38-26.4	0.01-11.4	0.996
Carbon tetachloride	15.3±0.3	1.19-18.3	0.01- 4.0	0.984
Benzene	16.7±0.1	1.08-23.4	0.03- 6.5	0.998
Toluene	15.4±0.2	0.74-20.0	0.01- 5.0	0.994
Ethylbenzene	15.1±0.4	0.42-10.0	0.03- 3.0	0.959

[1]Average of 20 exposures, 95% confidence limits, t test.

It was found that the rate of permeation was dependent on the sampling
environment temperature. Increases in the permeation constant with temperature were observed for each compound investigated and approximates a
linear relationship. Variation in temperature can be accounted for by using a modified form of Equation 2 where the average temperature during exposure is known where K_o is the permeation constant at 0°C, T is the aver-

$$C=m/(K_o+sT)t \qquad (3)$$

age temperature during exposure (0°C), s is the slope of the temperature effect (dK/dt), and C, m, and t have been previously defined.

Results for 23 Volatile Organic Priority Pollutants

Results for permeation constants and permeation constants versus temperature were obtained for 23 volatile organic priority pollutants. Permeation constants for most compounds were similar to the values found in Table I. For the temperature study, it was found that in general, permeation constant values change between 3 and 4% per °C change in temperature.

To separate 23 volatile organic priority pollutants required an SPB-1 fused silica capillary column. With the permeation technique and the SPB-1 capillary column it is possible to analyze at least 23 volatile organic priority pollutants for one solution.

REFERENCES

Hardy, J. K.; Strecker, D. T.; Savariar, C. P.; West, P. W. 1981. A Method for the Personal Monitoring of Hydrogen Sulfide Utilizing Permeation Sampling. Am. Ind. Hyg. Assoc. J., 42, 283-286.
Natural Resources Defense Council, Inc., et al. versus Train, 8 ERC 2120 (D.D.C., 1976), modified 12 ERD 1833 (D.D.C., 1979).
Statutes at Large. 1977. Clean Water Act. GS4.111:91, 1599.
Statutes at Large. 1972. Federal Water Pollution Control Act Amendment, GS4.111:86, 819.
U.S. Code of Federal Regulations. 1982. Protection of Environment. 40, Part 122.

A NOVEL ADSORBENT FOR USE IN ORGANIC VAPOUR DIFFUSIVE SAMPLERS

M. Harper,   C.J. Purnell

T.U.C. Centenary Institute of Occupational Health,
London School of Hygiene and Tropical Medicine,
Keppel St. (Gower St.), London WC1E 7HT, U.K.

ABSTRACT

A novel solid adsorbent is described for application in diffusive sampling systems for the measurement of organic vapours in air. The adsorbent is derived by the insertion of organic molecules in a clay-mineral lattice structure. Experimentation has demonstrated that the adsorption isotherms resemble those of strong adsorbents and that the surface area is comparable to that of active charcoal. The adsorbent capacity, in terms of chromatographic retention volume, exceeds that of porous polymers. It is therefore suitable for sampling a wide range of organic vapours, especially hydrocarbons. The adsorbent is pale in colour and hence can be impregnated with chemicals to form an adsorbent/reagent substrate system eminently suited to direct or subsequent colour formation. In addition, it has a high thermal stability to permit its use in thermal desorption systems.

INTRODUCTION

Great concern has been expressed in recent years over occupational exposures to toxic gases and vapours. Recent legislation has contained a specific requirement for sampling as one of the criteria for assessing such exposure. Samples may be taken from static locations within the workplace, but personal samples which are more representative of the individual exposure are normally preferred. Sampling heads attached to clothing within the worker's breathing zone are used to obtain the best estimate. The sampling devices and analytical methods employed must be sufficiently sensitive and accurate to allow reasonable interpretation of the results which are usually presented as a time-weighted average exposure over a given period, either short (10 minute) or long (8 hour working day) term.

Traditional methods of sampling, the so-called 'dynamic' systems, utilise a pump to physically draw contaminated air through a collecting medium. Both liquid reagent absorbents (impingers) and solid adsorbents have been employed. The pumps are expensive, bulky and cumbersome to wear. A significant breakthrough in sampling technology was achieved by the use of active diffusion as a means of collecting the contaminants. Diffusive samplers are low-cost, light-weight and often re-usable, with no moving parts and inherently safe for use in flammable atmospheres. They therefore offer a unique opportunity for the collection of samples from large numbers of the work-force.

The measurement of inorganic species collected by either sampling technique has often been by means of a colourimetric method. Many specific and sensitive reactions are available to enable a choice to be made depending on the analyte concentration and the presence of interfering compounds. The precision and accuracy of such methods may be less than that obtainable with instrumental techniques but such high levels are only required in order to demonstrate compliance with legislation around the limit values. If there is a general practice of maintaining exposures to levels as low as reasonably attainable (the ALARA syndrome) then accuracy need not be the prime consideration in the choice of sampling method and other advantages such as cost or the ease with which results can be obtained on-site become equally important. In such circumstances the combination of diffusive sampling with colourimetric analysis must be seen as particularly attractive.

Collection of organic species has mainly been achieved through physical adsorption onto a solid phase with subsequent solvent or thermal desorption followed by instrumental analysis. Active carbon is usually regarded as the best general adsorbent although the black colour of carbon is not suitable for in-situ colourimetric reactions. Porous polymers are white in colour but their organic nature may be unsuitable as a substrate for colourimetric reactions.

DESCRIPTION OF THE ADSORBENT

The need to identify an adsorbent with advantages over both charcoal and porous polymers which would also lend itself to colourimetric reactions led to an investigation into organic- substituted clay lattice structures. Alkylammonium montmorillonite clays have been known since they were first evaluated by Barrer et. al. (1955) for use as adsorbents in gas chromatography. However the subsequent development of porous polymers prevented further interest in this application. The structure of montmorillonite clay consists of layers of a sandwich of aluminium ions in octahedral co-ordination between two sheets of linked silicate tetrahedra. Intralamellar water can be adsorbed between two such opposing layers as also can cations which fit into hexagonal holes on the silicate surface. The water molecules are strongly associated with these cations which are exchangable so that replacement by a more hydrophobic species reduces the intraporous affinity for water but increases the surface area on which organic vapours will be preferentially adsorbed. The alkylammonium cation may be chosen to give the optimum sorbent properties.

NITROGEN ADSORPTION ISOTHERMS

Experiments have demonstrated that the isotherms given by these materials are equivalent to Type II of Brunauer et. al. (1940) with H4 hysteresis characteristic of platey clay minerals with slit-like micropores (Sing et. al., 1985). In addition there is evidence of considerable microporosity and a long linear region so that over a wide range of relative pressures the isotherm may be considered equivalent to a Type I or Langmuirian isotherm similar to those obtained from 'strong' adsorbents such as charcoal. The nitrogen surface areas and monolayer volumes (an approximate measure of the microporosity) which we have determined are given in table 1 for four of the compounds examined. A large increase in both values is acheived by the organic substitution. Since the tetraethylammonium molecule covers the largest area, MM and TM have the highest surfaces available for adsorption. As the areas covered by the tetrahedral tetramethylammonium molecule and the linear monomethylammonium are similar the adsorption capacities are also similar. Closure of the hysteresis loop was not achieved with MM indicating a possible permanent lattice distortion during adsorption. TM was therefore felt to be the best candidate for further evaluation. Surface area results obtained by Gregory (1981) for charcoals taken from a Du-Pont Pro-Tek sampler (211 m^2/g) and from a 3M Gasbadge (193 m^2/g) are comparable.

TABLE 1   Adsorbent surface area and monolayer volume

Adsorbent	Surface area m^2/g	V_m cm^3/g
Acid-washed montmorillonite (AW)	82.8	19.0
Tetraethylammmonium montmorillonite (TE)	121.7	28.0
Monomethylammonium montmorillonite (MM)	159.7	36.7
Tetramethylammonium montmorillonite (TM)	170.1	39.1

THERMOGRAVIMMETRIC STUDIES

After heating to between 110 and 120°C both in air and in vacuo the percentage weight loss was related to the possible nature of water vapor adsorption. Using Barrer's assumption of the unit cell formula, TM was determined to have lost slightly less than three molecules of water per unit cell. This is in accordance with a theoretical adsorption by broken bonds at lattice edges. Since this will only involve around 20% of the available surface (Grim, 1968) it is unlikely to pose a problem. Further heating did not achieve significant structural breakdown until temperatures in excess of the $280^{\circ}C$ quoted by Barrer (1957), a temperature substantially higher than the working range of most porous polymers.

BREAKTHROUGH VOLUME STUDIES

Breakthrough volume is defined as the first appearance in the outlet stream of a vapour applied to the inlet of a column packed with adsorbent and flushed with a carrier gas. This characteristic is relatively simple to measure and provides a useful guide to the potential performance in sampling situations. A direct liquid injection technique as described by Russell (1975) was used initially to determine the comparative retention volumes for carbon disulphide of AW, MM and TM. Extrapolated to $20^{\circ}C$ the results were 0.4, 1.0 and 10.5 l/g respectively. Clearly TM was the most suitable material for further evaluation. Similar experiments on solutions of various organic compounds in carbon disulphide indicated that high results would be achieved with atmospheres of low concentration.

The method of Brown and Purnell (1979) was employed in producing the results in table 2. Aliquots of an atmosphere of known concentration generated by a static chamber method similar to that described by Nelson (1971) were introduced into the nitrogen flow of a modified Perkin-Elmer F-11 gas chromatograph using a flame ionization detector. The usual column was replaced by a Perkin-Elmer ATD 50 adsorbent tube packed with TM. The results obtained by Brown and Purnell using Tenax GC are presented for comparison in table 2, they apparently show that TM is the better adsorbent. Future work will include comparison with active carbon.

Table 2 Breakthrough volumes of organic vapours

Analyte	Breakthrough volumes (20°C)	
	Tenax GC (l/g)	TM (l/g)
Pentane	0.6	160
Hexane	4.3	3000
Heptane	23	40000
Octane	100	520000
Benzene	8.0	5200
Toluene	50	52000
Cumene	600	1750000
Tetrachloroethylene	63	720
1,1,1-Trichloroethane	5.0	42

CONCLUSIONS

The tetramethylammonium montmorillonite adsorbent offers the advantage of combining the qualities of a strong adsorbent with high thermal stability. In which case thermal desorption should be feasible. It has a high surface area and it is apparently unaffected by the presence of water vapour. In addition the inert silicate lattice and pale grey colour suggests a possible substrate for colourimetric reactions.

REFERENCES

Barrer, R.M. and Macleod, D.M. 1955 Preparation of Alkylammonium Montmorillonite Complexes, and Adsorption Properties Thereof. Trans. Faraday Soc., 51, 1290-1300.

Barrer, R.M. and Reay, J.S.S. 1957 Sorption and Intercalation by Methylammonium Montmorillonites. Trans. Faraday Soc., 53, 1253-1261.

Brown, R.H. and Purnell, C.J. 1979 Collection and Analysis of Trace Organic Vapour Pollutants in Ambient Atmospheres: The Performance of a Tenax-GC Adsorbent Tube. J. Chromatogr., 178, 79-80.

Brunauer, S., Deming, L.S., Deming, W.E. and Teller, E. 1940 On a Theory of the Van der Waals Adsorption of Gases. J. Am. Chem. Soc., 62, 1723-1732.

Gregory, E.V. 1981 An Investigation into Factors Affecting the Performance of Passive Samplers and Charcoal Tubes for Some Organic Vapours. Ph.D. Thesis (Dept. of Environmental Health, University of Cincinnati, Ohio).

Grim, R.E. (Ed.) 1968 Clay Mineralogy. 2nd Edition. (McGraw-Hill, New York).

Grimshaw, R.W. 1971 The Chemistry and Physics of Clay. (Wiley, New York).

Nelson, G.O. 1971 Controlled Test Atmospheres: Principles and Techniques (Ann Arbor, Michigan).

Russell, J.W. 1975 Analysis of Air Pollutants Using Sampling Tubes and Gas Chromatography. Environ. Sci. Technol., 13, 1175-1178.

Sing, K.S.W., Everett, D.H., Haul, R.A.W. et. al. 1985 Reporting Physisorption Data for Gas/solid Systems with Special Reference to the Determination of Surface Area and Porosity. Pure & Appl. Chem., 57, 603-619.

DOES DIFFUSIVE SAMPLING INFLUENCE THE STATUS
OF COMPLIANCE WITH A TLV?

E. Lehmann, J. Auffarth, J. Häger
Bundesanstalt für Arbeitsschutz
Vogelpothsweg 50 - 52
D-4600 Dortmund 17, F.R.G.

ABSTRACT
Workplace concentrations of various solvents were measured
in the adhesives and paint industries and in user shops. For
making decisions about compliance or non-compliance with given
TLVs, personal samples were taken in parallel by active and
diffusive sampling on activated charcoal using different types
of commercial devices. The findings are presented. To avoid
false compliance statements in monitoring workplace atmospheres
it is at present recommended to accept results from diffusive
sampling only if they are well below the exposure limits or if
they are verified by other methods.

INTRODUCTION
Threshold limit values (TLVs) refer to airborne concentrations of hazardous substances at the workplace and are
usually defined as the time weighted average concentration
over an 8-hour shift.

For two or more hazardous substances, which possibly may
act biologically additive,the sum of the particular concentrations of each component divided by its TLV has to be compared
with unity. This stands for the "TLV" of the mixture (ACGIH,
1984).

The validity of the statement of compliance with a TLV
strongly depends on the validity of the workplace measurements
and, ultimately, on the sampling method applied. We investigated the statements which are derived from different sampling
techniques on activated charcoal in the same work environment:
active (by pumps) and passive (by diffusion) sampling.

WORKPLACE MEASUREMENTS

Workplace concentrations of solvent vapor mixtures containing up to 13 components with a TLV were measured in the
adhesives and paint industries and in user shops. Personal
samples were taken in parallel on activated charcoal using
i) NIOSH- or G-type sampling tubes and ii) different types of
commercial diffusive samplers (ORSA 5 from Draeger, 3500 from
3 M, and VAPORGARD from MSA). Duration of sampling was between
2 hours and times corresponding to a calculated 50 % load of
the sampler. The analytical method is described elsewhere
(AUFFARTH et al. 1985).

MONITORING STRATEGY

Monitoring strategies help to minimize costs by using a
priori knowledge (BUNDESMINISTER FÜR ARBEIT UND SOZIALORDNUNG,
1984). A monitoring strategy for exposure limits has two basic
objectives. Firstly, the status of compliance with the exposure
limits has to be ascertained. In case of overexposures the status of compliance must be established by technical measures reducing exposure. Secondly, it must be controlled by regular
measurements, if the conditions once stated have changed. The
frequency of the control measurements depends on the ratio of
the concentration measured and TLV.

Measurements are designed to answer different questions.
For the initial survey, only the range of concentrations must
be evaluated to decide whether further action is necessary or
not. Exact concentration determinations are needed, if the concentrations are found to be in the TLV-range (See also Table 1).

RESULTS

50 sets of field samples taken in parallel were evaluated
with respect to the decision ranges of the monitoring strategy.

13 active samples were found to be beyond the level of
significance, 8 of the parallel diffusive samples were below,
the other 5 were greater than 1/10 TLV leading to additional
measurements.

Compliance with the TLVs on a high level of significance
(concentrations lower than 1/4 TLV) was showing by 22 active

TABLE 1   Decision ranges

C/TLV	statements
< 1/10	beyond level of significance, no further measurements necessary
≤ 1/4	compliance with TLV (high probability) next control measurement within 64 weeks
1/4 ... 1/2	compliance with TLV next control measurement within 32 weeks
1/2 ... 1	compliance with TLV (low probability), exposure reduction by technical measures recommendet, next control measurement within 16 weeks
>1	non-compliance with TLV, exposure reduction by technical measures necessary

and 16 diffusive samples taken in parallel. The other 6 diffusive samples again indicate higher concentrations. However, 3 diffusive samples are found in this range while the correspondig active samples state more unfavourable situations.

Between 1/4 and 1/2 of the limit 15 active and 8 diffusive samples were ranged. 5 diffusive samples again indicate a higher risk, while 2 diffusive samples understimate the risk that the TLVs will be exceeded.

Compliance statements on a low probability level (concentrations between 1/2 and 1 TLV) were possible in 7 cases with active sampling. From diffusive samples in parallel only 4 show the same result. 1 sample already indicates non-compliance with the TLVs. 2 diffusive samples give results beyond this level.

The TLVs were exceeded by 6 active and also 4 diffusive samples. In 2 cases, however, the results of the diffusive samplers were well below the limit, while the active samples clearly show non-compliance with the TLVs.

DISCUSSION

Generally, comparable statements concerning the occupational exposure to solvent vapor mixtures are derived from

active and diffusive personal sampling on activated charcoal.
The differences observed in field measurements might be due to
different sampling characteristics of standardized charcoal
tubes and diffusive samplers. For example, the polarity of the
substances is known to influence sampling. From the point of view
of prevention false compliance statements in monitoring workplace atmospheres cannot be tolerated, while false non-compliance statements "only" lead to possibly unnecessary safety measures. Within the frame of a measurement strategy results from
diffusive sampling should at present be accepted if they show
exposures well below the exposure limits or if they are verified
by other methods.

REFERENCES

ACGIH, 1984. American Conference of Governmental Industrial
    Hygienists: TLVs, Threshold Limit Values for Chemical
    Substances and Physical Agents in the Work Environment.
    (ACGIH, Cincinnati).
Auffarth et. al. 1985. J. Auffarth und J. Häger:
    Lösemittel am Arbeitsplatz, ein Meßverfahren für die
    Praxis (Schriftenreihe der Bundesanstalt für Arbeitsschutz
    Dortmund).
Bundesminister für Arbeit und Sozialordnung, 1984.
    Technische Regeln für gefährliche Arbeitsstoffe (TRgA) 402
    Bundesarbeitsblatt Heft 11 (Bonn).

FIELD EVALUATION OF A DIFFUSIVE SAMPLER FOR FORMALDEHYDE IN AIR USING
CHEMOSORPTION ON 2,4-DINITROPHENYLHYDRAZINE-COATED GLASS FIBER FILTERS

R. Lindahl, J.-O. Levin and K. Andersson
National Board of Occupational Safety and Health, Research
Department in Umeå, Box 6104, S-900 06 Umeå, Sweden

ABSTRACT
    A diffusive sampler for formaldehyde was field evaluated at three
different workplaces by compairing with active (pumped) sampling. The sampler
was evaluated with personal sampling in a hospital dissection room. The formaldehyde level was 0.1-0.5 mg/m^3 and the sampling time was 10-150 min. Good
agreement was found between the results obtained from diffusive and pumped
sampling, coefficient of correlation = 0.959. The diffusive sampler was evaluated with area sampling at a textile manufacturing industry and in a
conference room, with good agreement between the two sampling methods.

INTRODUCTION
    When measuring pollutant concentrations in air, it is important to know
the reliability and accuracy of the methods used. The uptake rate of a diffusive sampler can be a function of pollutant concentration, absorbent capacity,
exposure time, air velocity, humidity, etc. These factors are most easily controlled in the laboratory, but even if the laboratory experiments give a good
picture of the uptake rate the field evaluation is essential for a complete
diffusive sampler rate validation (Brown et al. 1984).
    Elsewhere in this document we have reported a diffusive sampler for sub-
parts-per-million levels of formaldehyde in air using chemosorption on 2,4-
dinitrophenylhydrazine-coated glass fiber filters (Levin et al.). The laboratory validation of this method showed that the sampling rate was not affected by changes in formaldehyde concentration, sampling time or relative humidity. The sampling rate decreased at an air velocity below 0.1 m/s and increased at an air velocity higher than 0.5 m/s. We now wish to report a field
evaluation of this diffusive sampler. The sampler was compared with active
(pumped) sampling at three different workplaces with both personal and area
sampling.

MATERIALS AND METHODS
Coated filters and passive sampler
    The coated filters for both diffusive and pumped sampling, as well as
the diffusive sampling device, were prepared as previously described elsewhere in this document.

Personal sampling

The sampler was evaluated with personal sampling in a dissection room where medical students were dissecting human organs which had been embalmed with formalin. Personal sampling with diffusive samplers was used in parallel with pumped sampling at 0.2 L/min.

Area sampling

Six diffusive samplers and six or twelve pumped samplers (0.1 L/min) were mounted closely together within an area of about 3 dm^2 on a grid. Sampling was performed in a textile manufacturing industry and in a conference room of an office building. In the conference room a fan was mounted about three meters from the sampling grid to provide a air velocity of 0.2 to 0.5 m/s.

Liquid chromatography

The liquid chromatography determination was performed as previously described (Levin, et al. 1985).

RESULTS AND DISCUSSION

The reference method used was active sampling with 2,4-dinitrophenyl-hydrazine-coated 13-mm glass fiber filters which permitted determination of formaldehyde down to approximately 1 $\mu g/m^3$ in a 50-L air sample.

Personal sampling in a hospital dissection room

The formaldehyde level was 0.1-0.5 mg/m^3 and the sampling time was 15-150 min. Good agreement was found between the results obtained from diffusive and pumped sampling. Figure 1 shows the correlation between the pumped and diffusive samples.

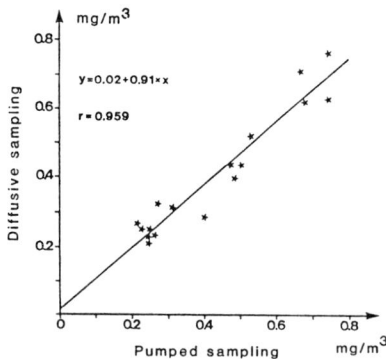

Fig. 1 Relationship between the results from diffusive and pumped personal sampling of formaldehyde in a hospital dissection room.

Area sampling a textile manufacturing industry

In the textile manufacturing industry three sets with six diffusive and twelve pumped samplers were set up at different locations. The sampling time was 6-8 hours and the formaldehyde levels were about 0.02-0.03 mg/m^3 (Table 1). The mean values obtained with diffusive and pumped sampling from each test site were compared with student's t-test. As can be seen in Table 1 there was no statistical difference between the two methods at the test sites A1, A2 and A3.

Area sampling in a conference room

In the conference room two sets with six diffusive and six pumped samplers were set up about one meter from each other. The sampling was performed during the night, with a sampling time of 16 hours. The formaldehyde level was about 0.025 mg/m^3.

The formaldehyde concentration found with the diffusive sampler at the test site B1 was statistically higher than the value obtained with the pumped method. The difference of 13% at B1 was observable due to very low standard deviations (see Table 1). At the test site B2 there was no statistical difference between the two methods.

TABLE 1. Results of area sampling. A1-A3 are three different sites in a textile manufacturing industry (8 hour sampling, RH 25%). B1-B2 are two locations in a conference room (16 hours, RH 33%).

Test site	Concentration found with pumped method ($\mu g/m^3$)				Concentration found with diffusive sampler ($\mu g/m^3$)				t-value
	m[a]	SD[b]	RSD[c]	n[d]	m	SD	RSD	n	
A1	20	2.59	13%	12	18	1.33	7%	6	2.09[e]
A2	24	1.40	6%	12	23	1.22	5%	5	1.47[e]
A3	27	2.77	16%	12	30	4.23	14%	6	1.58[e]
B1	23	0.85	4%	3	26	0.85	3%	5	4.83[f]
B2	25	0.34	5%	6	26	0.54	2%	6	1.70[e]

a) - mean value
b) - standard deviation
c) - relative standard deviation
d) - number of observations
e) - not significant at 5% level
f) - significant at 5% level

Conclusion

The field evaluations of the diffusive sampler for formaldehyde show a very good agreement with the pumped method, both in the personal and area sampling. The method is suitable for measuring formaldehyde concentrations at higher levels for comparing with the TLV as well as lower levels in indoor air.

REFERENCES

Brown, R.H., Harvey, R.P., Purnell, C.J and Saunders, K.J. 1984. A Diffusive Sampler Evaluation Protocol. Am. Ind. Hyg. Assoc. J., 45, (2), 67-75.

Levin, J.-O., Andersson, K., Lindahl, R. and Nilsson, C.-A. 1985. Determination of Sub-Part-per-Million Levels of Formaldehyde in Air Using Active or Passive Sampling on 2,4-Dinitrophenylhydrazine-Coated Glass Fiber Filters and High-Performance Liquid Chromatography. Anal. Chem., 57, 1032-1035.

USE OF DIFFUSION DEVICES
FOR ANALYSIS OF AMMONIA AND
HYDROGEN SULFIDE IN COMPLEX RESEARCH STUDIES

G. Michaylov, V. Velichlkova, R. Vasileva
Institute of Hygiene and Occupational Health
- Medical Academy, Sofia 1431, Bulgaria

ABSTRACT

The use of Bulgarian-made diffusion devices for analysis of NH3 and H2S in the work area for purposes of complex hygienic investigations are discussed. The strategy of sampling involves: personal passive dosimeters, assessment of dispersion of the studied chemicals by means of stationary passive monitors (map of the concentrations with curves of the "isonorms"), and simultaneous control by short-term sampling. An attempt is made to assess the relation between individual exposure, body levels of the studied chemicals and health state of the workers. Possibilities of increasing the application of diffusion devices in occupational hygienic analyses are discussed.

INTRODUCTION

The main advantage of diffusive sampling is the possibility for simultaneous measuring of the concentration in many places in the working environment. Personal exposure of workers is measured at the same time also using diffusive monitors. In this way we obtain information on the distribution of the chemical substances in all areas of the working environment.

STRATEGY OF SAMPLING

We used the following strategy:

1. A large number of diffusive monitors are exposed over the same period in the working area.

2. Personal monitoring is undertaken at the same time for the same substances.

3. Measurements are made of the same substances or their metabolites in blood or urine.

4. The workers health is monitored and any death evaluated.

5. Any correlation of the above points is examined.

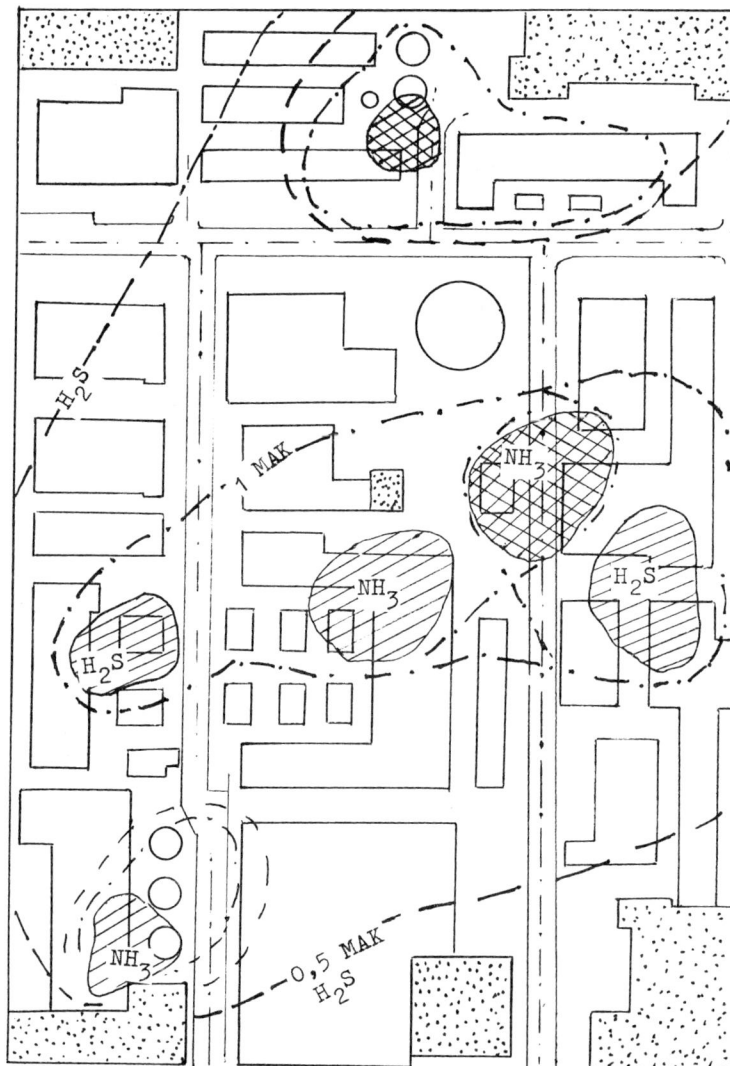

Fig.1   Map of concentration curves conecting   spots
having the same concentration   in the range of 0,5 ,
1,0   and 2 times MAK NH_3 and H_2S

- Technological instalations (uncovered)
- Buildings, control panels
- "isonorm area" with 2 and more MAK
- "hot spots" with "overlaped concentrations"

MATERIALS AND METHODS

The air monitoring study was done by using the Bulgarian-made passive monitors -"Hygitest" - PM 05020 for NH_3 and PM 1910 for H_2S. They are diffusive type direct-reading indicator tubes. The results are given on fig.1.

The same tubes were used for quantitative determination of the NH_3 in urine and of waste water The construction of the sampling device is given in Fig. 2.

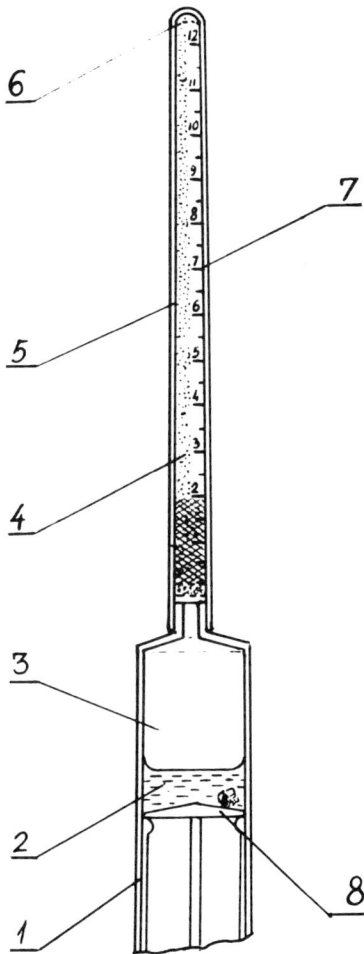

Fig. 2

Construction of the device -
gas/vapour in liquid or solid media

1-Polypropilene chamber
2-Analysed liquid
3-Diffusion space
4-Length of stain diffusion type
   indicator tube

5-Cylindrical transparent body -
   polypropilene
6-Conpensator
7-Scale in (mm)
8-Piston bottom

The liquid (0.1 ml) is added to the reaction chamber (3) and NH3 is liberated by Na2 CO3 addition (urine test). The NH3 diffusive indicator tube (specially shaped) is connected to the chamber. The length of stain at constant temperature and exposure time is proportional to the concentration of the NH3 in the urine. The calibration curve is given on Fig.3.

NH3 in urine is simultaneously analyzed by this method and by spectrophotometric method. The results are compared with Juden's method on Fig. 4. The correlation between the methods is R=0.975. The mean standard deviation is 5-8% for NH3 devices.

The examples given in this work show that passive type devices could be used widely in complex-hygienic investigations.

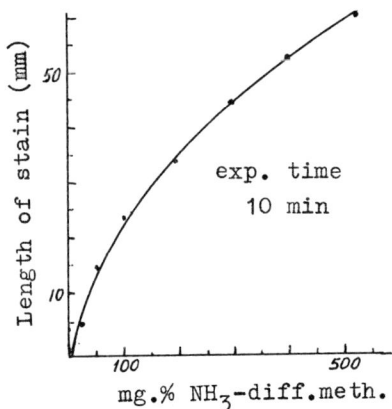

Fig. 3   Calibration
         curve for NH3 in urine

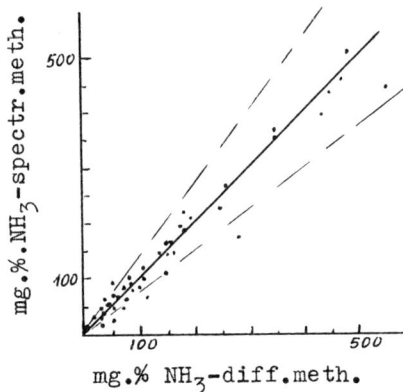

Fig. 4   Juden's comparative
         test

PRELIMINARY REPORT ON THE DEVELOPMENT OF A NEW
PASSIVE SAMPLER FOR ORGANIC VAPOURS

Zs. Schandl, J. Solt
National Research Institute
of Occupational Safety
H-1021 Budapest, Hungary

ABSTRACT

The effect of sampler geometry, sorbent material, barrier material, ambient air movement and pollutant concentration on the uptake was studied in order to develop a new personal monitor for organic vapours. On the basis of experimental findings it is believed that the use of several sorbents together with multiple sampling rate will result in an optimal passive sampler system.

INTRODUCTION

Recently, a recommendation has been outlined in Hungary for an occupational exposure monitoring strategy, giving priority to personal or breathing zone sampling. On the practical side, this made it worth trying to develop a personal monitor of our own /for organic vapours first/ considering import limitations and the lack of a Hungarian product. Because of the well-known advantages of passive samplers, the choice has fallen on the development of a new passive monitor.

EXPERIMENTAL SET-UP

The experimental apparatus consisted of
- an "exposure chamber" of stainless steel, suitable for the simultaneous exposure of 3 badge and 6 tube type samplers, and having 8 additional sampling points for independent test methods;
- a syringe injection system for generating a known test atmosphere. The true concentrations of the target component were calculated from the syringe injection rate;
- regulators for flow rate and temperature control.

Direct /"volumetric"/ sampling and conventional, pumped adsorption tubes were used as independent sampling methods. The test component in these experiments were toluene. The desorption and analysis was done by the common method of CS_2 elution followed by GC analysis. DuPont's method was used for the determination of the desorption efficiency.

SAMPLER GEOMETRY

In order to optimize the dimensions of the sampler, the inner diameter /D/ and the path length /L/ of the air-gap were varied as follows:

$$D = 1, 2, 4, 5, 6, 8, 20 \text{ mm}$$
$$L = 4, 8, 10, 20 \text{ mm}$$

The uptake rate was found to be a linear function of the cross section area/path length ratio from 1 to 4, clearly indicating the diffusion control in this range.

SORBENT MATERIAL

10 types of sorbents were examined, and three of them were found to be satisfactory: a charcoal cloth /made in the UK/, an activated charcoal /granulated, made in the USSR/ and TENAX GC.

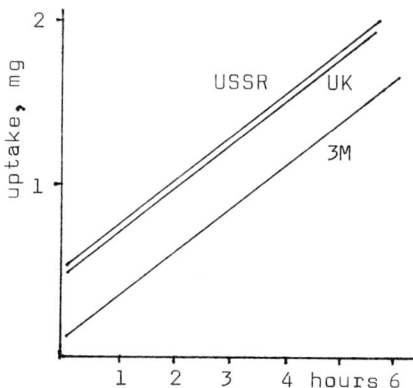

Fig. 1   Toluene uptake of different sorbent materials

The figure shows the uptake of the charcoal cloth and the activated charcoal compared to the 3M 3500 Organic Vapor Monitor. The slopes are similar, but there are marked differences in the y-axis intersections, indicating the unsatisfactory pretreatment of the charcoal cloth and the activated charcoal.

BARRIER MATERIAL

Fortunately the very first test material, a 4 mm thick mi-
niporous polypropylene made in Hungary proved to be quite satisfactory.

EFFECTS OF AMBIENT AIR MOVEMENT AND TOLUENE CONCENTRATION

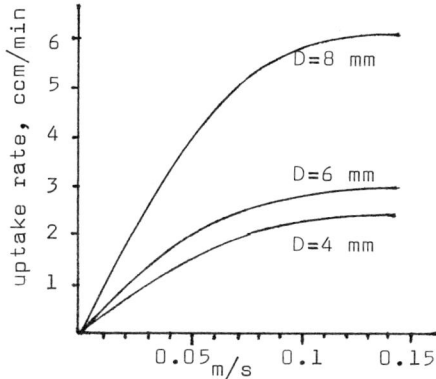

Fig. 2   The effect of ambient air movement on the
uptake rate

The figure demonstrates that the uptake rate is significantly
affected by ambient air movement below 0.1 m/s linear flow rate.
There is no indication of any dependence from the D/L ratio in
the investigated range.

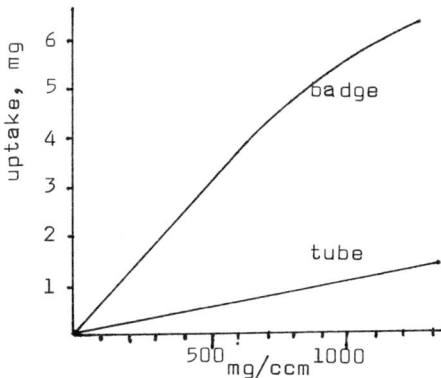

Fig. 3   The effect of toluene concentration on
the uptake

This figure shows the uptake rate of charcoal cloth in badge type and tube type samplers in 2 hours' exposure time and 0.10 m/s ambient air flow. Deviation from linearity occurred quite soon with the badge type sampler, while the tube type sampler was linear even at high concentrations.

CONCLUSIONS

Since the experimental results indicate that there is a definite effect of both ambient air movement and pollutant concentration on the uptake, a flexible device is required. We believe that the use of several sorbents together with multiple sampling rates /two different sorbents and two different diameters in the same device/ will result in the necessary flexibility, allowing the use of one optimal passive sampler system for a very wide range of conditions.

ORSA TUBES WORN INSIDE FACE MASKS :
A SIMPLE MEANS OF CHECKING THE EFFECTIVENESS
OF PROTECTIVE FILTER MASKS

Mr Karl Tchorz,

Chief Safety Engineer, Lufthansa, Hamburg

ABSTRACT

Workers using aircraft paints come into contact with hazardous substances. In addition to chromates (which are classified as carcinogenic), workplace air is especially affected by the various solvents. Measurements provided information on the concentration, spread and dispersal time of contaminant clouds. Parallel medical studies on selected workers produced unexpected results in some cases. Despite the use of full-face protective masks, biological tests revealed a high contaminant uptake. This cast doubt on the effectiveness of protective masks under the actual conditions of use. The supposition that workers are lax in wearing masks can only be confirmed or discounted by repeated test series under controlled conditions. An obstacle to such tests, however, is the unwillingness of subjects to provide blood and urine samples. Orsa tubes worn inside face masks offer an alternative. Test series confirm the reliability of the results obtained with such tubes for analyses to determine whether significant concentrations of dichlormethane or methyl ethyl ketone are present in inhaled air and hence taken up by the body. Orsa tubes are diffusive samplers filled with activated carbon. After sampling, the tubes undergo laboratory analysis. The method presented here is not intended to replace biological tests, but it can considerably reduce the number of tests found unpleasant by subjects. It thus provides those responsible for safety at work with a simple way of checking the level of safety by means of random sampling.

INTRODUCTION

Stripping old coats of paint from aircraft involves particular health hazards for workers - as does the application of new coats. The nature of the hazards, the protective measures employed to reduce the risk and the methods of these measures are explained below.

Removal of old coats of paint

Depending on its statem an old coat of paint needs to be sanded downm stripped or sprayed with plastic granulate before a fresh coat can be applied. In our case, the old coats are normally stripped. To do this, the painted surfaces are covered with a stripping paste by means of spray lances. After 1 to 2 hours the old paint structure is chemically broken down. It can then be removed with plastic scrapers.

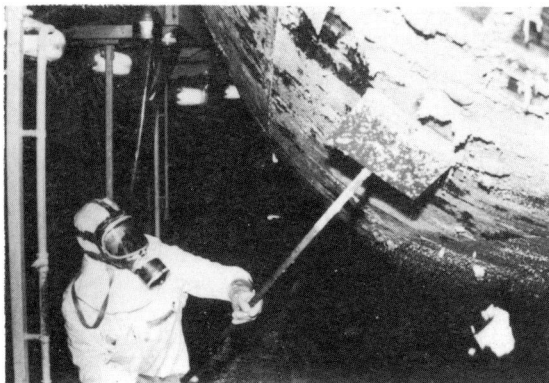

Fig. 1

Scraping off old coats of paint down to the underlying metal surface.

The hazardous substances used in stripping compounds cause health and environmental problems. For our (mostly routine) stripping work, a mixture of low-boiling dichlormethane, phenol and formic acid is used.

Protective measures

Some time ago, however, questions were raised as to whether the individual components of the overall protection provided met existing operational requirements and whether personal protective equipment was used properly. To examine these questions, measurement programmes were conducted under the auspices of the BIA (Berufsgenossenschaftliches Institut fuer Arbeitssicherheit - Industrial Safety Institute of the Industrial Accident Insurance Associations). At the same time as these measurements (e.g. using personal samplers), repeated tests were carried out on biological samples (blood and urine) from groups of workers. The initially unexpected finding was that some persons working directly with stripping agents were found to have absorbed unacceptably high levels of solvents and phenol.

The employer was then obliged to have measurements and tests carried out promptly until it could be demonstrated statistically that no danger to workers was likely. The design of the masks and laxity in wearing them were factors to be considered. Understandably, weaknesses in the mask were initially suspected, in addition to laxity. In exhaustive discussions with manufacturers, practical improvements and/or confirmation that safety was adequate were sought.

Testing methods

In accordance with a fixed programme, our medical service collected blood and urine samples from selected groups before and after a shift and submitted them for analysis. Initially disturbing solvent levels rapidly fell to acceptable values when wearing habits were improved, ABEK filters were introduced with a maximum limit of one hour in use and care was taken to clean masks thoroughly and ensure proper functioning of the valves. Very soon, however, we were faced with the problem that workers refused to be 'guinea-pigs', in particular for blood samples. In order to continue with the programme, we looked for a practicable way of monitoring the insides of masks for the presence of solvents. the use of Orsa sampling tubes (trade name) suggested itself. These relatively short, thin tubes are passive samplers able to bind solvents present in the ambient air - in a filling of activated carbon. Concentrations can then be determined in subsequent laboratory analyses.

This method of testing is quite workable. In our case, the aim is simply to establish the presence of solvents inside masks.

Fig. 2

Fitting of the Orsa sampling tube inside the panorama mask

The table below sets out the results of a test using this method.

Table:
Dichlormethane concentrations measured inside masks using Orsa tubes (in ppm)

Filter change	1	2	3	4	5	6	7	
Subject 1	<3	44	<3	28	<3	<3	<3	
Subject 2	7	53	<3	8	23	<3	12	time in
Subject 3	<3	40	9	10	22	5	5	use of
Subject 4	<3	41	5	13	51	<3	5	filter
Subject 5	11	41	8	16	7	6	8	0.5 h
Subject 6	11	52	34	8	–	–	–	time in
Subject 7	35	27	4	1	–	–	–	use of
Subject 8	5	27	5	2	–	–	–	filter
								1 h

Although the Orsa tubes did indicate some disturbing concentrations, the parallel biological test in the laboratory ultimately yielded negative results in all cases. This obvious discrepancy was explained by reconstructing the testing procedure. It turned out that the high ppm values particularly those in column 2, which are around half the MAC value, were clearly due to carelessness during the break, when masks contaminated by the stripping agent were removed in the rest room - with the sampling tubes left open. The high values in the other columns were likewise the result of similar procedural errors. However, the results of this test, which was selected after due consideration, show that this method does yield useful results provided the errors described are avoided.

CONCLUSIONS

In my view, there can be no doubt that Orsa sampling tubes inside masks are a useful aid for checking the effectiveness of such masks. This method can be employed if

- no or not enough workers are available for blood and urine sampling;

- care is taken to ensure that the correct procedure is followed during interruptions and breaks and when filters or tubes are replaced, in order to rule out accidental contamination of the Orsa tubes;

- simple answers (OK or not OK) are adequate;

- as and when necessary, the results are corroborated by other checks, for example biological tests.

DEVELOPMENT AND EVALUATION OF A DIFFUSIVE SAMPLER FOR
MEASUREMENTS OF ANAESTHETIC GASES

J. Kristensson*, M. Widén**

*Institute of Analytical Chemistry
University of Stockholm
S-106 91 Stockholm, Sweden
**Siemens-Elema AB
Ventilator Division
S-171 95 Solna, Sweden

ABSTRACT

Several epidemiological studies have been performed
to determine if occupational exposure to anaesthetic gases
increases the risks of spontaneous abortion, congenital mal-
functions in the offspring and neoplastic diseases in personnel.
The studies confirm a correlation between the spontanous
abortion and the exposure to nitrous oxide.
Exposure to trace amounts of anaesthetic gases might influence the performance of the anaesthesist. Other effects have
been reported such as difficulties in concentration and in
recall, headache, stomach upset, fatigue and other general
symptoms.
A diffusive sampler, Anaesthesia Gas Dosimeter 185, for
simultaneous measurement of nitrous oxide and halothane was
developed.
The sampler is tube formed and the adsorbent used is Molecular Sieve 5A.
The sampler is thermally desorbed and the analysis is performed by gas chromatography and electron capture detection.
The diffusive sampler was evaluated according to the HSE
protocol MDHS 27 and the NIOSH protocol for evaluation of a
diffusive sampler.
Anaesthesia Gas Dosimeter 185 can be used for nitrous
oxide in the concentration range of 1 to 1000 ppm and for
halothane in the concentration range of 0.01 to 100 ppm.
The sampler can be used both for full day measurements, 8
hours, and for short time measurements, 15 minutes.

ANAESTHESIA GAS DOSIMETER 185

Sampling in the dosimeter is based on diffusion according to Fick's law of diffusion.
The sampler, figure 1, is made of a stainless steel tube (1),
30 mm x 6 mm. The opening of the tube is covered with a
stainless steel gauze (2), 100 mesh. The length and cross
section area of the diffusion zone is defined by a glass
insertion (3). The glass insertion is follwed by a second
gauze (4), the adsorbent, Molecular Sieve 5A 60/80 mesh (5)
and a third gauze (6).
The sampler is inserted into a holder (7) which can be
carried close to the breathing zone by a chlip (8).
The sampler is closed by an end cap (9) when not used.

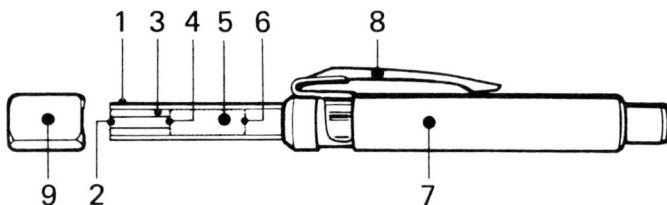

Figure 1.   Anaesthesia Gas Dosimeter 185.
             Explanation in text.

ANALYSIS

At the analysis the sampler is thermally desorbed. The
desorbed anaesthetic gases are cryogenic focused in a cold
trap. The cold trap is rapidly heated and the sample is
injected via a split system onto a micro packed column. The
split ratio is approximately 1:200 and the column used is
1.5 m x 0.7 mm id, pyrex, packed with Porapak QS 115/120
mesh.

A electron capture detector at 350°C is used for the
detection of both nitrous oxide and halothane.

EVALUATION

Anaesthesia Gas Dosimeter 185 has been evaluated according to the protocols from both NIOSH and HSE (MDHS 27).

Calibration and calculation of uptake rates were performed by gas mixtures with known concentrations. The uptake-
rates is not linear, it is a nonlinear function of the exposed time.

Analysis of known concentrations in the laboratory, six
by six according to HSE protocol, resulted in a stright line
with k close to one.

The concentrations in an operating theatre were simulated in the laboratory by changing the concentrations of
nitrous oxide in the range of 10 to 1000 ppm. The exposure
times at the different concentrations were changed and the
total exposure was 8 hours. The agreement between calculated
and measured results were good.

The effect of humidity was evaluated at the laboratory.
As long as the samplers are well conditioned and free from
water when the sampling begins, normal relative humidities
are not influencing the results.

Other evaluation parameters like storage stability and
exposure to zero concentration were controlled and the results were in accordance with the protocols.

Field evaluation was performed in operating theatres.
Three different methods were used. Samples were taken in gas
sampling bags and portable pumps. These samples were analyzed
by IR-technique and a direct gas chromatographic technique.
The obtained results were compared with the results from the
analysis by diffusive sampling.
The agreement between the two chromatographic techniques were
good but the agreement with the IR-technique was bad.
The disagreement obtained depended upon the calibration of
the IR-instrument. Cloosed-loop calibration of the IR-instrument is not recommended when sampling is performed by gas
sampling bags. Calibration should be done by the use of gas
sampling bags and gas mixtures with known concentrations.

RESULTS

Anaesthesia Gas Dosimeter 185, a diffusive sampler for simultaneous sampling and analysis of nitrous oxid and halothane, offers a simple and effective sampling method for occupational exposure measurements.

The evaluation of the sampler gave good agreements both according to reproducibility and accuracy at concentration normally obtained in operating theatres and at the dentists.

The sampler was well accepted by the personnel.

ASSORBIMENTO PASSIVO SU CARBONE

N. Zurlo*, G. Camesasca**, L. Metrico***

*Via Passo di Brizio, 4
20148 - Milano - Italia
**Divisione Ecologia ed Igiene Industriale
TEI s.p.a. - Via Paleocapa, 6
20121 - Milano - Italia
***Via Barbaro, 4
26013 - Crema - Italia

SOMMARIO

   Viene sviluppata la teoria dell'assorbimento passivo
di sostanze organiche volatili su substrato solido (carbone), basandosi sulla teoria della diffusione in condizioni
di equilibrio e della penetrazione nello strato solido assorbente, (diffusione interna con flusso costante in in-
gresso).
   Sono studiati i fattori fisici che influiscono sul fenomeno dell'assorbimento nel tempo ed in particolare il
rallentamento dell'assorbimento dovuto alla contropressione
generata dal solvente stesso adsorbito
   Si definisce poi la capacità di assorbimento del carbone, il suo grado di saturazione ed il tempo necessario
per raggiungere le condizioni massime oltre le quali il
prelievo risulterebbe soggetto ad errori significativi.

LA FASE DI ASSORBIMENTO

   Durante l'assorbimento di un solvente con carbone attivo, nell'aria sovrastante al carbone, la pressione Pc del
solvente in equilibrio con la quantità assorbita aumenta in
funzione della concentrazione C dell'adsorbito elevata
all'esponente n che è tanto più elevato quanto maggiore è
l'affinità del solvente per il carbone.

1) $Pc = cost\ C^n$

   Con aria satura il carbone è saturo quando Pa è eguale
alla tensione di vapore Ps del solvente alla temperatura
alla quale avviene l'assorbimento.

   A saturazione il volume del solvente assorbito (allo
stato liquido) è eguale allo spazio Wo disponibile nel

carbone, generalmente compreso tra 0.4 e 0.7 ml/gr di carbone.
Ne deriva che la quantità Qs contenuta è data da:

2) $Qs = \delta\ Wo$ gr/gr di carbone
dove δ è la densità del solvente liquido.
Normalmente nell'aria la pressione Pa del solvente è minore
di Ps; indicando con P.M. il peso molecolare del solvente a
25 gradi si ha:

3) $C = Ps\ \dfrac{PM}{18.5}$ $Ca = Pa\ \dfrac{PM}{18.5}$ mgr/litro

Con tensione di vapore   Pa nell'aria il carbone è saturo
quando il solvente occupa il volume Ws

4) $Ws = Wo \left(\dfrac{Pa}{Ps}\right)^{1/n}$

quando è stato assorbito il solvente contenuto nei Vs litri
di aria dati da:

5) $Vs = \dfrac{1000\ \delta\ Ws}{Ca} = \dfrac{1000\ \delta\ Ws}{Ca} = \dfrac{18.5}{1-1/n\ \ PM\ Ps}^{1/n}$

Indicando con Cc la concentrazione del momento nel
carbone, con Ccs quella di saturazione e con Pc la
pressione in equilibrio con Cc si ha:

6) $\dfrac{Pc}{Pa} = \left(\dfrac{Cc}{Ccs}\right)^n = A^n$ $A = \dfrac{Cc}{Ccs}$

dove A indica il grado di saturazione.
      A titolo orientativo nella tabella 1 sono riportati Ws
e Vs in funzione di n e Cc per alcuni solventi.
      Nel prelievo passivo il flusso è direttamente propor-
zionale a Ca in quanto si presuppone eguale a zero la C
nell'aria a livello dell'assorbitore. Indicando con H

l'altezza della camera di diffusione e con S la superficie
si ha:

7) $\quad dq = Ca \dfrac{D}{H} S\ dt \qquad dv = \dfrac{dq}{Ca} = \dfrac{D}{H} S dt$

Nell'assorbimento su carbone attivo la 7 è verificata
solo nella fase iniziale; in seguito il flusso tende a di-
minuire in funzione di Ao^n e cioè del rapporto, sulla superficie di contatto camera-assorbitore, della tensione Pc
del solvente nell'aria sovrastante il carbone e la tensione Pa nell'aria. D'altra parte il solvente penetra nello
strato accettore per diffusione attraverso all'aria degli
interstizi tra i granuli di carbone, penetrazione che
non può avvenire in assenza di gradiente di concentrazione.
Indicando con Ao^n il rapporto Pc/Ps sulla superficie
di contatto (X=0) e con Hc la penetrazione al momento nello
strato accettore, il gradiente di concentrazione è dato da:

8) $\quad \dfrac{dAo^n}{dX=o} = \dfrac{Ao^n}{Hc}$

Per la sezione unitaria S = 1, deve esistere la relazione

9) $\quad \dfrac{dq}{Ca} = dv = (1-Ao)^n \dfrac{D}{H} dt = Ao^n \dfrac{D}{Hc} Sp\ dt$

da cui

10) $\quad Hc = \dfrac{Ao^n}{1-Ao^n} H\ Sp$

dove Sp è il rapporto tra la sezione libera disponibile
nello strato assorbente e la sezione della camera di diffusione, rapporto generalmente inferiore a 1/10

l'altezza della camera di diffusione e con S la superficie
si ha:

7) $dq = Ca \dfrac{D}{H} S\, dt$ $dv = \dfrac{dq}{Ca} = \dfrac{D}{H} S\, dt$

Nell'assorbimento su carbone attivo la 7 è verificata
solo nella fase iniziale; in seguito il flusso tende a di-
minuire in funzione di Ao^n e cioè del rapporto, sulla superficie di contatto camera-assorbitore, della tensione Pc
del solvente nell'aria sovrastante   il carbone e la tensione Pa nell'aria. D'altra parte il solvente penetra nello
strato accettore per diffusione attraverso all'aria degli
interstizi tra i granuli   di carbone, penetrazione che
non può avvenire in assenza di gradiente di concentrazione.
        Indicando con Ao^n il rapporto Pc/Ps sulla superficie
di contatto (X=0) e con Hc la penetrazione al momento nello
strato accettore, il gradiente di concentrazione è dato da:

8) $\dfrac{dAo^n}{dX=o} = \dfrac{Ao^n}{Hc}$

Per la sezione unitaria S = 1, deve esistere la relazione

9) $\dfrac{dq}{Ca} = dv = (1-Ao^n) \dfrac{D}{H} dt = Ao^n \dfrac{D}{Hc} Sp\, dt$

da cui

10) $Hc = \dfrac{Ao^n}{1-Ao^n} H\, Sp$

dove Sp è il rapporto tra la sezione libera disponibile
nello strato assorbente e la sezione della camera di diffusione, rapporto generalmente inferiore a 1/10

Il prelievo è ancora accettabile con $Ao^n \le 0.05$ quando la perdita dovuta alla contropressione è inferiore al 5% con penetrazione Hc.

11) Hc $\le$ 0.05 H Sp   Hc $\le$ 0.005 H per Sp = 1/10

Con camera di diffusione di altezza H = 1 cm. la contropressione è già al limite ancora accettabile quando il solvente è penetrato per meno di 1/20 di mm e interessa meno di 1/20 dello strato accettore generalmente di 1 - 1.5 mm di spessore (40÷60 mg di carbone per cm2)

I FATTORI "n"- Wo-Sp

I fattori che regolano l'assorbimento, n, Wo, Sp cambiano sensibilmente da un carbone all'altro e, per lo stesso tipo di carbone, da una preparazione all'altra.

n

Aumenta con la temperatura di ebolizione Te dei solventi con differenze sensibili tra le diverse classi di solventi.

In generale:

Te ‹ 40 C   n ‹ 2.5 (incluso l'alcool etilico)
Te ‹ 80 C   n ‹ 5
Te › 100 C   n › 10

Per uno stesso solvente passando da un carbone all'altro "n" può risultare anche raddoppiato in funzione delle microporosità del carbone.

Wo

Generalmente è compreso tra 0.4 e 0.7 ml/gr e la Vs passando da un carbone all'altro può risultare quasi raddoppiata.

Sp

rapporto di superficie. Lo spazio vuoto tra i granuli del carbone attivo è circa 1/6 del volume (0.35 - 0.45

ml/gr di carbone il cui volume è di circa 2.5 ml/gr, δc = 0.4 circa). Se lo spazio vuoto fosse costituito da condotti verticali Sp risulterebbe circa 1/6. In effetti per penetrare nello strato il solvente deve compiere un percorso tortuoso. Per tale ragione i canali risultano più lunghi mentre diminuisce in proporzione la sezione di transito.

In definitiva la tortuosità del percorso ha un effetto quadratico; per una penetrazione verticale Hc e con percorso reale pari a due volte Hc occorre un Ao^n quattro volte più elevato per compensare il dimezzamento delle sezioni di transito ed il raddoppio del percorso da compiere.

Sp per lo stesso carbone cambia in funzione della compattazione dello strato e degli eventuali leganti che possono sottrarre spazio vuoto. A titolo orientativo in seguito verrà considerato Sp = 1/10, valore sicuramente superiore al reale per la maggior parte dei substrati in uso.

CORRELAZIONE TRA Ao - Ao^n

Nella Tabella 2 sono riportati i valori di Ao corrispondenti ad Ao^n in funzione di n.

Per assicurare il gradiente $\dfrac{d\,Ao^n}{dx} = 0$, indispensabile pe la penetrazione nel carbone, per i solventi ad affinità elevata lo strato superficiale è già quasi saturo non appena iniziato il prelievo. Per n = 20 Ao = 0.71 con Ao^n = 0.001, quando il solvente è penetrato per meno di un micron e interessa meno di 1/1000 dello strato assorbente.

In seguito, poiché Ao può aumentare solo di poco, l'accumulo è praticamente proporzionale alla penetrazione Hc ed il solvente si ferma sul primo carbone libero che trova comportandosi quasi come un liquido immesso dal basso in un tubo verticale.

E non potrebbe essere altrimenti; anche nel prelievo attivo nonostante la velocità di penetrazione assicurata dal flusso di aria, centinaia di volte superiore rispetto a quella della penetrazione per diffusione, il solvente si

ferma sul primo carbone libero che trova ed arriva sul
secondo strato di controllo della fiala solo quando il primo strato è quasi saturo.

La saturazione della superficie di contatto è meno
pronunciata per valori di n più bassi. Per n=1, di interesse puramente teorico, la concentrazione superficiale Ao e
la penetrazione Hc aumentano in parallelo.

DISTRIBUZIONE DEL SOLVENTE NELLO STRATO ACCETTORE

Dallo studio analitico del fenomeno la distribuzione
del solvente nello strato accettore è dato da:

$$12) \quad A = Ao^{1 - \left(\frac{X}{Z \, Hc}\right) z / n} \qquad Z = \left(\frac{1}{4/\pi - 1}\right) \quad n^{\frac{1}{4/\pi}}$$

dove X indica la penetrazione a partire dalla superficie di
contatto camera di diffusione-strato accettore.

Il volume Va, e il corrispondente qa, assorbito sono
dati da:

$$13) \quad Va = Ao \, Hc \, d'c \, Vs \qquad qa = Ca \, Va$$

dove d'c è la densità del carbone (generalmente 0.35÷0.45).

Nella Fig. 1 è riportata la distribuzione nello strato
accettore con Ao^n = 0.05 per solventi rispettivamente con
n=20; 10; 5; 2,5.

Come già detto, con H=I e Sp=0,1, Ao^n=0.05 già con una
penetrazione di 0.05 mm.

Con strato accettore di 1.5 mm di spessore (60 mg/cm2)
la contropressione 0.05 viene raggiunta dai solventi a
grande affinità (n = 20) con assorbimento pari al 3% circa
della saturazione; meno dell'1% con n = 2.5.

Queste percentuali possono risultare sostanzialmente
ridotte con carbone compatto (Sp < 0.1) o più elevate con
carbone supportato su rete a maglia larga.

La velocità di saturazione dello strato superficiale

varia essenzialmente con l'altezza H della camera di diffusione; con H = 0.5 le percentuali citate risulterebbero dimezzate e 5 volte più elevate con H=5

Con strato di 40/60 mgr/cm2 e H>5 l'assorbimento passivo su carbone ha una autonomia eguale o anche superiore a quelle delle normali fiale a doppio strato.

PERDITE DI PRELIEVO A CAUSA DELLA CONTROPRESSIONE Ao^n

Dalla 8) e dalla 10)

14) $Va = \dfrac{\dfrac{Ao^{n+1}}{n}}{1-An}$ H Sp δc Vs

per la 9)

15) $dVa = HSp\ \delta c\ \cdot Vs\ d\ \dfrac{Ao^{n+1}}{1-Ao^n} = (1-Ao)\ \dfrac{D}{H}\ dt$

Integrando, nel campo di interesse pratico, con approssimazione +/-1% si può porre:

16) $\dfrac{\dfrac{Ao^{n+1}}{}}{\left(1 - Ao^n\right)^{5/3}} = \dfrac{t}{k}$

dove , per t in ore

17) $K = \dfrac{Sp\_H\_\_\delta c\_Vs^2}{3,6\ D}$

Per la 13) la quantità assorbita Va è direttamente

proporzionale a $\dfrac{Ao^{n+1}}{1-Ao^n}$ indipendentemente dalle

caratteristiche del carbone e dalla Vs da cui dipende K
Confrontando la 15) con la 13) il rapporto R tra la durata
te del prelievo e il tempo t_t teorico corrispondente al
prelievo effettivamente eseguito è dato da:

$$R = \frac{1}{\left(\dfrac{n}{1-Ao}\right)^{2/3}}$$

Con $Ao^n = 0.05$; 0.1; 0.2 R risulta rispettivamente 1.035;
1.073; 1.16 con un errore per difetto nel prelievo rispettivamente del 3.5; 7.3; 16%

VARIAZIONE DI Ao^n NEL TEMPO
Ponendo H = 1 D = 0.1 Sp = 0.1 δc=0.4 dalla 15) e 16) si
ha

$$18)\quad t = 0.11\ V_s\ \frac{Ao^{n+1}}{\left(\dfrac{n}{1-Ao}\right)^{5/3}} \qquad t\ in\ ore$$

Con la 18), a titolo orientativo, si può calcolare la
variazione nel tempo delle contropressioni Ao^n per un carbone con le caratteristiche indicate.
Nella Fig. 2 sono riportate le variazioni nel tempo
di A^n per Xilolo (n=20), Etilacetato (n=10), Alcool
Isopropilico (n=5) e Alcool Etilico (n=2.5) alla concentrazione nell'aria di 1 e 0.1 gr/m3 in base alla Vs della tabella 1.
Con Ca = 1 gr/m3 Ao^n risulta eguale a 0.05 dopo 2-3
ore anche per i solventi con grande affinità per il carbone
(n=20); questo tempo può risultare sensibilmente ridotto o
aumentato, in funzione delle caratteristiche del carbone
(Sp, n, Wo, ecc). I solventi di media affinità raggiungono
Ao^n=0.05 nel giro di qualche ora anche con Ca=0.1 gr/m3.
Praticamente impossibile il prelievo con n<2.5 (Alcool
Etilico, Diclorometano ecc)
Di importanza fondamentale risulta l'altezza H della

camera di diffusione; con H=0.5 i tempi risultano ridotti
di 4 volte rispetto ad H=1 e con H=7 aumentano di 50 volte
e risultano possibili prelievi prolungati per tutto il turno di lavoro anche per solventi con scarsa affinità (n=2.5)
alla Ca = 1 gr/m3.

Aumentando opportunamente H si possono prelevare dalla
atmosfera anche i bassobollenti difficili da captare con le
fiale a carbone. In base ai dati relativi ai campionatori
passivi in uso si può ritenere che con H di 3÷4 cm si pos-
sano eseguire prelievi corretti nella maggior parte degli
ambienti di lavoro.

MISCELE DI PIU' SOLVENTI

Con una miscela di solventi ciascuno di essi occupa
una parte del Wo disponibile nel carbone sottraendolo agli
altri . A parità di tempo il grado di saturazione per i
singoli solventi risulta più elevato in quanto, disponendo
di un Wo ridotto, diminuisce la loro Vs.

L'effetto competitivo varia in funzione delle affinità
dei singoli solventi per il carbone, della loro concentrazione nell'aria ecc.. Uno o più solventi con grande affinità per il carbone, come ad es. Xilolo + Isobutilacetato, a
concentrazioni atmosferiche eguali, o meglio, con eguale
Vs, occupano il medesimo strato iniziale del carbone; con
2/3/4 solventi contemporaneamente presenti le singole Vs
risultano rispettivamente 2/3/4 ecc. volte inferiori.

Con solventi di diversa affinità risulta accellerata
specialmente la saturazione del solvente con n inferiore.

UMIDITA' ATMOSFERICA

Con umidità relativa (U.R) inferiore al 50% la concentrazione dell'acqua nel carbone è molto bassa quando Ws <
0.1 Wo; con umidità relativa 80-90% la Ws è rispettivamente
superiore a 0.5 - 0.7 Wo (valore cui corrisponde n=1/3 circa).

Con U.R. = 80% a 22°C (16 gr di acqua per m3 di aria)
con Wo=0.5 ml/gr la Wo corrisponde a 250 mgr di acqua per
gr di carbone con Vs = 16 l/gr

Quando il solvente raggiunge Ao=0.05 nel tratto iniziale dello strato accettore il vapore d'acqua occupa già più di 1/3 di Wo con conseguente diminuizione del Vs del solvente. Con U.R. del 90% risulterebbe già occupato più del 40% di Wo.

Questo effetto massa è dovuto alla notevole differenza tra i flussi per diffusione che con H=1 per il solvente e con Ca=1 gr/m3, è di 0.2-0.3 mg/cm^{2}h mentre per l'acqua è di 16 mg/cm^{2}h e cioè da 50 a 80 volte più elevato.

ANALISI DI CONTROLLO

Per le numerose variabili in gioco la contropressione può risultare anche 3/4 volte più elevata rispetto a quella corrispondente a condizioni ideali; ciò specialmente in presenza di miscele, umidità atmosferica relativa superiore al 70%, concentrazioni atmosferiche globali dell'ordine del gr/m^3.

Per contenere la contropressione entro i limiti accettabili occorre aumentare l'altezza H della camera di diffusione o, per evitare camere di altezza eccessiva, aumentare l'altezza equivalente (He) riducendo per esempio la sezione di una parte della camera.

Con atmosfere difficili e per ambienti non controllati in precedenza conviene comunque confermare i risultati con analisi di controllo.

La contropressione, e il difetto di prelievo conseguente, sono direttamente proporzionali al volume d'aria campionato.

Eseguendo 2 prelievi passivi in parallelo con flussi diversi su substrati accettori eguali la perdita conseguente alla contropressione è maggiore per il prelievo a flusso più elevato e confrontando tra loro i risultati si può risalire al valore esatto.

Con campionatore a cono ZURLO (altezza reale della camera 1,7 cm) eseguendo in parallelo due prelievi di cui uno con cono normale (He equivalente 7 cm) ed il secondo con il cono riduttore (He = 14 cm) che preleva la metà rispetto al cono normale, indicando rispettivamente con Qn e Qr i risultati ottenuti e con Qe il valore corretto si ha, con approssimazione $\pm$ 1%

$$Qe = Qr \left(\frac{2Qr}{Qn}\right)^{2/3} \qquad Qe = Qn \left(\frac{2 \cdot Qr}{Qn}\right)^{5/3}$$

CONCLUSIONI

Nel campionamento passivo con carbone attivo l'analita penetra nello strato accettore per diffusione attraverso l'aria che circonda i granuli di carbone. Proseguendo il prelievo aumenta il grado di saturazione A nel carbone.

$$A = \frac{Ca}{Csa} = \frac{\text{concentrazione del carbone}}{\text{saturazione corrispondente alla Ca dell'aria}}$$

e il rapporto tra la tensione di vapore An dell'analita nell'aria sovrastante il carbone e la sua pressione Pa nell'aria.

$$Ca = Pa \ \frac{PM}{18.5} \ \text{mg/l}$$

L'esponente n è > 10 per i solventi con grande affinità, 2.5 per i solventi difficili da captare con il carbone attivo. La penetrazione nello strato accettore può avvenire per diffusione solo in presenza di un gradiente di concentrazione.

$$\frac{dA}{dx}^n \qquad \frac{dAo}{dx=0}^n = \frac{Ao^n}{Hc}$$

che sulla superficie di contatto aria-carbone (X=0) deve consentire la penetrazione del flusso in arrivo.

$$dv = \frac{dq}{Ca} = (1-Ao^n) \ \frac{D}{H} \ dt = Ao^n \ \frac{A}{H_c} \ Sp \ dt$$

Il rapporto Sp tra la superficie disponibile per la penetrazione e la superficie di contatto aria-carbone è normalmente inferiore a 0.1. Ao^n deve aumentare via via con l'aumento della larghezza Hc dello strato accettore interessato dall'analita:

$$Hc = \frac{Ao \_\_ H \_ Sp}{1 - Ao} \quad n$$

Per assicurare il gradiente indispensabile alla penetrazione lo strato superficiale è già quasi saturo (specie con n elevato) non appena iniziato l'assorbimento.

Il solvente in arrivo, per assicurare il gradiente, deve occupare il primo carbone libero che trova e penetrare nello strato solo dopo aver quasi saturato il settore a monte.

Mano a Mano che aumenta Ao^n diminuisce il gradiente con l'esterno e di conseguenza il flusso in entrata.

$$R = \frac{prelievo\_effettivo}{prelievo\ teorico\ Ao} \quad n = 1 - Ao^{n\ 2/3}$$

Per evitare un errore eccessivo è consigliabile sospendere il prelievo quando $Ao^n \le 0.05$, con penetrazione Hc di circa 0.05 mm (con $Sp=0.1$). Lo spessore dello strato accettore è dell'ordine di $1-1.5$ mm ($40/60$ mg di carbone cm2); con $H=1$ si sospenderebbe il prelievo con assorbimento inferiore al 3% della saturazione anche per i solventi con grande affinità.

Aumentando H aumenta in proporzione diretta la possibilità di sfruttamento del carbone; sempre per strato con spessore di 1.5 mm e $Ao^n = 0.05$ ad $H=5$, 10, 20 corrispondono saturazioni rispettivamente del 15, 30, 60% .

Il flusso è inversamente proporzionale ad H ed in definitiva il tempo di saturazione è direttamente proporzionale ad H^2.

Con H=1 Ro^n = 0.05 in meno di un'ora per i meno affi-
ni; in 2/3 ore con affinità elevata; con H = 5 o 10 questi
tempi aumentano rispettivamente di 25 e 100 volte.

Con H di 5÷10 cm si possono eseguire prelievi prolungati per tutto il turno di lavoro anche in condizioni difficili o tali da rendere aleatorio il prelievo tradizionale
con fiale.

Come per il prelievo attivo la saturazione del carbone
è più rapida, in misura non valutabile a priori, con concentrazioni elevate di miscele di solvente e/o con umidità
atmosferica relativa del 70%.

I parametri in gioco n, Sp, ecc. cambiano sostanzialmente da un carbone all'altro e per lo stesso tipo di carbone da un lotto all'altro; con atmosfere difficili, o comunque non controllate in precedenza, è consigliabile confermare i risultati ottenuti.

A parità di tempo la saturazione A e la perdita R sono
proporzionali al flusso; eseguendo due prelievi in parallelo a flusso diverso si può risalire al valore corretto confrontando tra loro i risultati ottenuti.

Con il campionatore a cono ZURLO (altezza della camera
di diffusione 1.7 cm) eseguendo in parallelo un prelievo
con cono normale (altezza relativa H rispetto allo strato 7
cm) con cui si determina Qn, ed uno con cono riduttore (He
= 14) con flusso dimezzato, con cui si determina Qr, la Qe
corretta è data da:

$$Qe = Qr\left(\frac{2Qr}{Qn}\right)^{2/3} = Qn\left(\frac{2\,Qr}{Qn}\right)^{5/3}$$

Il prelievo in parallelo corrisponde all'analisi di
controllo sul gorgogliatore di coda o sul secondo strato
della fiala adsorbente del prelievo attivo; il risultato

però rappresenta la media di due determinazioni.

SIMBOLOGIA

A   = Grado di saturazione
Ao   = Grado di saturazione dello strato X =o
C   = Concentrazione del solvente in aria
C_c = Concentrazione del solvente nel carbone
Ccs = Concentrazione del solvente nel carbone a
       saturazione
δ = Densità del solvente
δc = densità del carbone
D   = Coefficiente di diffusione
H   = Altezza della camera di diffusione
Hc   = Altezza di penetrazione
He   = Altezza equivalente della camera di diffusione
n   = Esponenziale dell'isoterma di Freundlich
Pa   = Pressione parziale del solvente in aria
Pc   = Pressione parziale del solvente sullo strato
       adsorbito (contropressione)
Ps   = Tensione di vapore del solvente
PM   = Peso molecolare
q   = Quantità di solvente adsorbito
qs   = Quantità di solvente adsorbito alla saturazione
R   = Rapporto te/t_t
S   = Superficie della camera di diffusione
Sp   = Rapporto sezione libera /sezione della camera di
       diffusione
t   = tempo
te   = tempo effettivo di prelievo
t_t = tempo teorico corrispondente alla quantità
       effettivamente adsorbita
T   = Temperatura
Te   = Temperatura di ebollizione del solvente
V_a = Volume di aria assorbita
Vs   = Volume di aria corrispondente alla quantità qs
       di solvente
Wo   = Spazio disponibile nel carbone

Ws   = Volume occupato alla saturazione del solvente
        nel carbone

X   = Strato di carbone

Xo   = Strato superficiale del carbone

REFERENCES
Gregory E. D. and Elia V.J. 1983. Sample retentivity
    properties of passive organic vapor samplers and
    charcoal tubes under various conditions of sample
    loading, relative humidity, zero exposure level
    periods and a competitive solvent. Am. Ind. Hyg.
    Assoc. J., 44, 88-96.
Nelson G.O. and C. A. Harder 1976. Respirator cartridge
    efficiency studies: VI; effect of concentration. Am.
    Ind. Hyg. Assoc. J., 37, 205.
Nelson G. O. , A.N. Correia and C.A. Harder 1978.
    Respirator cartridge efficiency studies: VII; effect
    of relative humidity and temperature. Am. Ind. Hyg.
    Assoc. J., 39, 240.
Dubinin M.M. 1975. Physical adsorption of gases and vapors
    in micropores. Prog. Surf. Membr. SRI, 9, 1-70
Reucroft P.J., Simpson W.J. and Jonas L.A. 1978. Sorption
    properties of activated carbon. J. Phys. Chem. 75,
    3526-3531.
Moore G., Steinze S. and Lefrevre H. 1984. Theory and
    pratice in the development of a multisorbent passive
    dosimeter system. Am. Ind. Hyg. Assoc. J. 45, 145.
Melcher R.G., Langner and R.O. Kagel 1978. Criteria for
    the evaluation of methods for the collection of
    organic pollutants in air using solid sorbents. Am.
    Ind. Hyg. Assoc. J. 39, 349.
Underhill D.W. 1984. Efficiency passive sampling by
    adsorbents. Am. Ind. Hyg. Assoc. J. 45, 306.
Sansone E. B., Jonas L.A. 1981. Prediction of activated
    carbon performance for carcinogenic vapors. Am. Ind.
    Hyg. Assoc. J. 42, 688.

Tab. 1   Wo e Ws IN FUNZIONE DI n e Ca - Wo = 0.5 ml/gr

Solvente	n	Ws/Wo mgr/ m^3		Ws (1/gr) mgr/m^3	
		1.000	100	1.000	100
Xilolo	20	0.825	0.74	355	3160
I.butilacetato	20	0.785	0.70	342	3040
Tetracloruro di carbonio	10	0.52	0.41	410	3270
Etilacetato	10	0.55	0.44	250	1960
Cloroformio	5	0.28	0.18	209	1320
Alcool isopropilico	5	0.39	0.25	154	970
Diclorometano	2.5	0.053	0.02	35	140
Alcool etilico	2.5	0.16	0.06	64	250

Tab. 2   CORRELAZIONE Ao - Ao^n IN FUNZIONE DI n

n	Ao^n				
	0.001	0.01	0.05	0.10	0.20
1	0.001	0.01	0.05	0.10	0.20
2.5	0.063	0.16	0.30	0.40	0.52
5	0.25	0.40	0.55	0.63	0.72
10	0.50	0.63	0.74	0.79	0.85
20	0.71	0.79	0.86	0.09	0.92

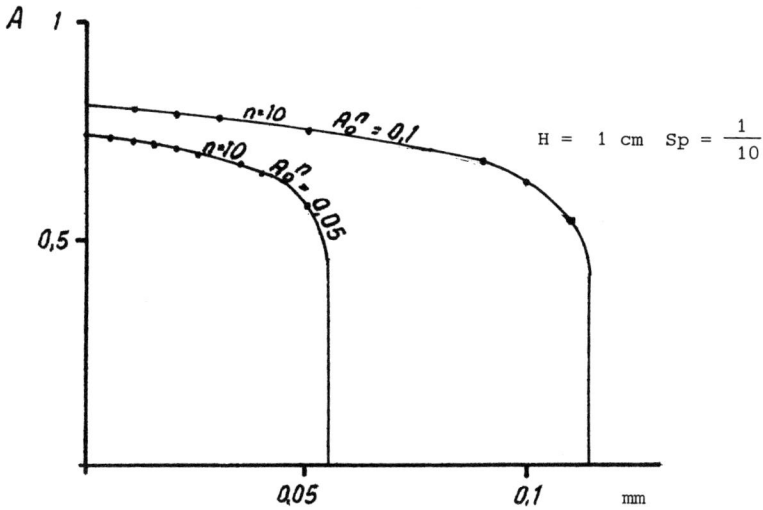

Fig. 1   PENETRAZIONE NEL CARBONE E PER Ao^n = 0.05 e 0.1

CON n = 10

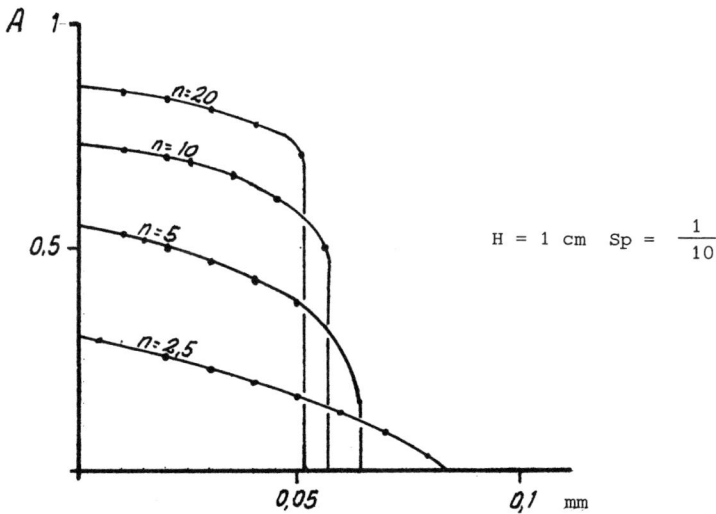

Fig. 1   PENETRAZIONE NEL CARBONE IN FUNZIONE DI n   CON Ao^n = 0.05

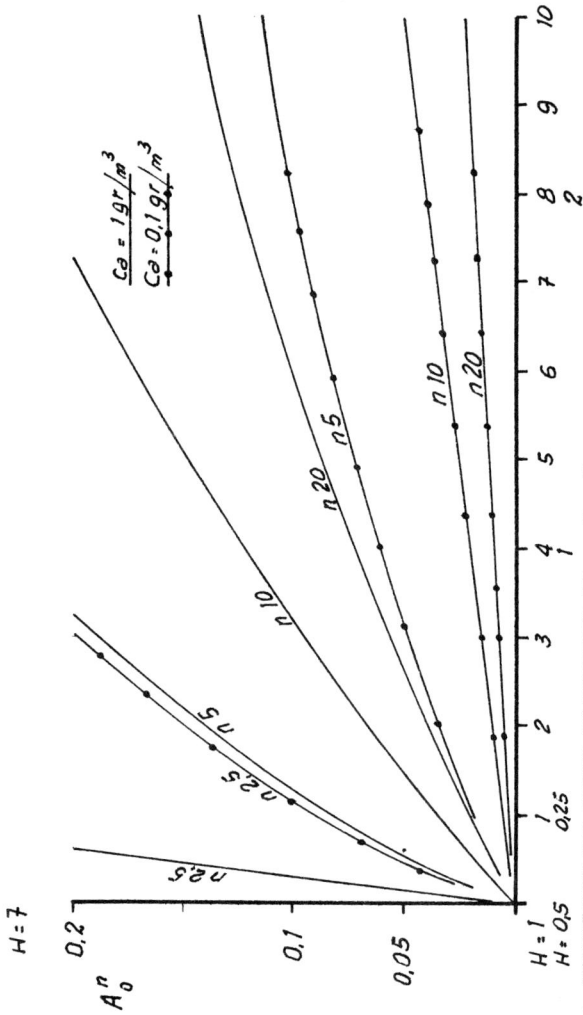

Fig. 2 A_o^n = VARIAZIONE NEL TEMPO IN FUNZIONE DI n, Ca, H

$(Sp = \frac{1}{10}$ Wo = 0.5 ml/gr$)$

PASSIVE CONE SAMPLER WITH LIQUID OR SOLID SUBSTRATE

N. Zurlo

Clinica del Lavoro, University of Milan,
Via S. Barnaba 8, I-20122 Milan, Italy

ABSTRACT

Following theoretical consideration of diffusion processes, laboratory experiments and a large number of recently published works, a passive cone sampler was developed which is suitable for investigations at the workplace and for the measurement of the main pollutant gases in the atmosphere. The parameters which affect diffusion are examined.

SIZE OF THE SAMPLER

In a passive sampler the flow of an analyte with diffusion coefficient D through the unit surface is expressed as:

$$dV = \frac{D}{He} \, dt \qquad (1)$$

where He (equivalent height) is the height of a diffusion chamber with the same flow.

Equation (1) is true if the air is renewed almost instantly on the air-chamber contact surface. In fact, the initial concentration of the analyte Ca is depleted along the contact surface with the result that the level of Ca = const. gradually rises above the contact surface (*1, *2) and diffusion occurs through a chamber with a depth of:

$$Hc = H + h \qquad (2)$$

where h is the distance from the contact surface on which the concentration remains Ca. If u indicates the air velocity and L the length of the contact surface in the direction of u, we have:

$$h = (\frac{2 \, DL}{a^2 u})^{1/2} \qquad (3)$$

where a2 is the ratio between the air contact surface and the section of passage through the diffusion chamber (*1). The sample can be regarded as still being acceptable when:

$$\frac{h}{a.H} < 0,05 \qquad (4)$$

A height H of 5-10 cm is required to sample a pollutant at the heart of the environment with u=5-10 cm/s, with L=2-3 cm and with a diffusion chamber of constant section equivalent to that of the contact surface.

SPHERICAL DIFFUSION CHAMBER

The depletion of Ca is reduced on the surface of contact with the atmosphere when the diffusion chamber is spherical. The sections of passage inside the chamber are directly proportional to the square of radius r. In a state of equilibrium it should be:

$$\frac{dC}{dr} \, r^2 = \text{cost} \qquad (5)$$

If the concentration of the absorber of radius X_o is zero and if X_L expresses the external radius, we have:

$$C = Ca \left(\frac{X_L}{X_L - X_o} \right) \left(1 - \frac{X_o}{X} \right) \qquad (6)$$

$$dC = -Ca \, \frac{X_L}{X_L - X_o} \cdot \frac{X_o}{X^2} \qquad (7)$$

and thus

$$dV = \frac{X_L}{X_L - X_o} \, 4 \, \Pi \, X_o \, dT \qquad (8)$$

and for $X_L \longrightarrow \infty$

$$dV \longrightarrow D \, 4 \, \Pi \, X_o \, dT$$

CONICAL DIFFUSION CHAMBER

In a spherical diffusion chamber the air tends to flow through the sphere with internal currents which increase the flow, with the result that the specific flow is high ($He<1$).

These drawbacks are eliminated if the diffusion chamber is restricted to a conical sector with an angle of aperture $\leqslant 60o$ C. For ease of design, it consists of a truncated cone resting on an external beaker containing at the bottom the liquid or solid absorbent. For practical reasons, a cone with an aperture of $60o$ C and with $X_L = 3$ cm was chosen. There is no leakage of liquid when the beaker is shaken or turned upside down. The air chamber between the cone and the beaker ensures that the diffusion chamber is adequately insulated and prevents internal convective movements. The passage from the smaller base of the cone to the collection layer involves additional resistance and the flow is slightly below the theoretical value for the cone and is equal to:

$$dV = D \, dt \qquad (9)$$

corresponding to $X_o = 0.9$

The distance between the smaller base of the cone and the bottom of the beaker is such that the flow is not affected by the inclination of the beaker. The flow can be reduced by inserting extra reducer cones. The actual height of the diffusion chamber is 1.73 cm, while the equivalent height He is equal to 7 cm. Experiments proved the theory concerning the effect of turbulence on the rise of the chamber h:

$$h = (-\frac{2\ D\ X_L}{3 + u})^{1/2} = (\frac{6\ D}{3 + u})^{1/2} \qquad (10)$$

with a corresponding empirical correction factor:

$$Ft = 0.97 + 0.4 (\frac{D}{3 + u})^{1/2} \qquad (11)$$

During the sampling of solvents in liquid (*3) the return pressure of the absorbed analyte (if vapour tension exceeds 20 mm Hg) slows down diffusion and the precise volume sampled is expressed:

$$Va = V (1 - \frac{V}{2\ Vs}) \qquad (12)$$

$$\text{where} \qquad Vs = \frac{Q\ .\ 18.5}{Ps\ .\ PM_L}$$

CONCLUSIONS

The passive cone sampler can be used at the workplace to monitor the heart of the environment and as a personal sampler and also to monitor external atmospheric pollution. In all cases the same collection medium can be used as in equivalent active samplers. The accuracy and reproducibility of the results are the same as or better than those achieved using active samplers.

BIBILIOGRAPHY

*1 F. Andreoletti, 1986, Incidenza della turbolenza dell'aria sul campionamento per diffusione; International symposium Workplace Air Monitoring; Diffusive sampling - an alternative approach.

*2 Fowler, W.K., 1982, Fundamentals of passive vapor sampling, American Laboratory 31, 80-87

*3 Zurlo, N., Metrico, L., Camesasca, G., 1985, Uso del campionatore passive Zurlo mediante assorbimento su substrato liquido, Igienisti Industriali X, 85-92

PANEL DISCUSSION

BERLIN (CEC)

I would now like to ask the panel members, who represent the regulatory authorities and social partners, to give their personal comment on how they view the acceptability of diffusive sampling as a monitoring technique.

FIRTH (UK)

From the point of view of HSE, the regulatory body looking at results coming from measurements, then we of course have the usual government phraseology that we will accept results which are satisfactory for the purpose for which they're taken. This means that the accuracy requirements which we want will be set by the use to which those data are going to be put, whether it's going to assess whether a firm is in compliance with some limit or whether the data is collected for archiving purposes, for looking at worker exposure for extended periods of time, which may be used at a future data for the purpose of setting limits. Now, many of us seem to be moving towards the position of setting exposure limits. In the UK we have two types of exposure limits: control limits and recommended limits. For control limits one is in breach of regulations if these are exceeded. One has to define what is meant by exceeded and I think that we have to be 95% confident that the limit is being exceeded. This figure of 95% confidence then sets the accuracy requirements of the measuring techniques. For our archiving purposes and for the surveys of exposure for general information, then we would accept a lower accuracy level. I think for that compliance testing we are looking for techniques of a relative standard deviation of about 10-15% and accuracies of maybe plus or minus 25%. I think that diffusive samplers which have been validated do meet this requirement. Yes we will accept diffusive samplers if they have been validated. We also have as a government body, a second consideration for samplers. We are users of them as well as looking at the results coming from them and we would use samplers for the same reason that industry would use them. In our opinion they are a cost-effective use of our resources. As Mr. Bord said earlier it releases highly qualified staff, such as inspectors, for doing more constructive work than pinning badges and pumps onto workers when large scale surveys are being carried out. The second use that we would put them to follows from the fact that we can get far more data if we use diffusive samplers. We'll start doing much more intensive surveys in the industry for the purpose of setting control limits. Control limits are set in the UK as a balance between what is medically required and what can be practically achieved in industry and the exposure levels which pertain to industry at a particular time are part of the considerations as to what levels can be accepted.

GROSJEAN (Belgium)

On the acceptability of diffusive sampling, I would like to say the following: as a general principle any method is acceptable as long as it is used according to its possibilities and its limitations by a skilled person. Talking about limitations I see that the most important limitations now are as follows: interfering compounds, face velocity dependence, uptake rate as a function of sampling time and the fact that most samples do not allow us to see if the sorption capacity has been exceeded. An argument which is often used is the argument of simplicity. Now I would like to say that everyone working in the field will welcome a simple method. However simplicity is only reached after a huge amount of work on the validation in the laboratory, then the validation in the field, and then a whole lot of background information has to be provided in order to be able to make a correct decision. This background information, the fact that it has to be provided is in contradiction with the practices as I see it now in my country where people buy a box of samplers, put the samplers on the lapel of the worker, record the end and the beginning of exposure and send the samples to the laboratory without any further background information. A third point are the costs and as Mr. Coker said, and I think some other people agree, in the context of the whole budget spent on industrial hygiene the costs for acquisition, maintenance and use of pumps are marginal. Together with costs it is often said that supervision can be reduced when using diffusive samplers. But if one really has to supervise conditions in which diffusive samples are taken, one person cannot take charge of more diffusive samplers than active samplers. So as a conclusion I would say that diffusive sampling is a powerful complement to already existing practices but not yet as a full alternative.

JOURDAN (CEFIC)

I'm giving my personal point of view here but it will be tinged with twenty years of experience in industry. Any method, any device, which allows us a good degree of reliability and an acceptalbe level of cost to assess workers exposure to chemical substances in the workplace is acceptable as compared to measurement in a fixed place. What should be measured is not the concentration of such and such a pollutant substance at such and such a point in the workplace. I think that everybody concerned knows to what extent concentration can vary from one point to another in the workplace in a plant because of the physical conditions. So it is important to measure personal exposure but this is not something that is widely enough recognized yet even in the European Community. For example, in the Draft Directive on Benzene in the Economic and Social Committee a proposal to amend it in support of personal monitoring was defeated. The results of the vote in the plenary session on the Economic and Social Committee was 31 in favour, 86 against, and 8 abstentions. This means that in a group of people who participate in decisions concerning industrial health and safety, three-quarters were not aware of the advantage of a personal monitoring system. The badge, the passive sampler, in so far as it allows these techniques for personal monitoring to be accepted more widely, is of course very strongly supported by industry.

SAPIR (ETUC)

I have to say first of all that the aim of the European Trade Union organization such as the Confederation is to talk about acceptability. Our two objectives in this field are: prevention of course and an active role of workers in prevention policies. As Mr. Jourdan said, we are working within European rules which are being discussed at the moment or which have already been promulgated. Measurement stategy and threshold values are still being discussed within the Community. Measurement strategies are very important, it is clear that there is a need to monitor the level of exposure and also we have to know the varieties of situations; how these situations change. In certain cases fixed detector systems are very important. We are convinced that all sources of information should be used, not just personal monitoring systems, but all sources of information that exist about what's happening in the plant. We should fall back on the experiences of workers, medical facts, and so on. I wouldn't like us just to be restricted to one particular measurement technique. Any technique should allow for the workers and their representatives in the various safety organizations to be able to play their role. there is a question here of how the data can be presented to the workers, how we can explain things to the workers, I can see that these new techniques certainly have major advantages, but I also wonder whether in an analysis and assessment strategy these techniques are actually going to allow us to assess the places and the sources of contamination and allow us to tackle the abnormal situations. I think that for these techniques to be developed, it is absolutely indispensible for validation protocols to be recognized not only at national level, but also European level.

EYRES

I was very attracted to personal samplers right from the very start. In my first years as a hygienist I had to survey all my company's plants in Europe and I spent my time struggling with a very heavy suitcase full of sampling pumps. So I think the light weight of passive samplers is a real attraction. However there are other useful advantages. They are more acceptable to the worker and you are more likely to get a typical exposure of the normal job rather than the wrong job, I think it's important. Passive samplers are also much easier to use on the site. You don't have to worry about recharging batteries, calibrating air-flows, etc. and therefore on-site technicians can be trained to collect samples. In my view hygienist are not the only people who are able to or should collect samples and generally it is not cost-effective for them to do so. They should be doing much more useful work and defining the monitoring programmes required, presenting and interpreting the results to the people who need to have them, and implementing controls where necessary. Costs of monitoring are becoming more important and I believe that passive samplers present possibilities for minimizing these, but it would be interesting to see some more actual case history comparisons. I'd also like to make a few comments on accuracy. Personally I believe the need for accuracy is often over-emphasized and that may be a little provocative in an audience of mostly analysts who are trained to seek accuracy. However from the viewpoint of protecting workers rather than complying with legislation, accuracy is only important if it changes the decision which you would have made. There are a number of situations which I can believe you can tolerate quite a considerable loss in accuracy without changing that decision. For example if mean exposures are well below the exposure limit or well above the exposure limit, accuracy is not that important. If you are getting mean results, around the exposure limit, then I believe than normal good industrial hygien practice is to seek to reduce those exposures to a much lower level. Another important aspect related to accuracy in my view is that exposure limits themselves are not accurate values. Most of these are based on rather questionalbe data, old data, and they also include some rule-of-thumb safety factors, so that if you occasionally exceed an exposure limit then I don't believe that has any great significance for worker protection. Finally I'd just like to comment on what was said by Mr. Griepink yesterday that accuracy is vital and also that it is better to have no value than the wrong value. I would suggest that it is better to have a wrong value which is sufficiently accurate to lead to the correct decision to protect worker health than to have no value. In conclusion I believe that the very impressive developments in passive samplers and their analyses which have been discussed in the Symposium have reached the stage where in many cases they will lead to the right decisions for protecting worker health and that more advantage can be taken of their potential benefits.

LEHMANN (FRG)

In the Federal Republic of Germany, as of October 1st, 1986 there will be a new regulation concerning dangerous substances at the place of work. In this regulation the employer is obliged to monitor the TLVs at the workplace. At the same time, there have been technical rules drawn up and these give advice on how to plan a measurement strategy, how to implement it, and how to evaluate the results. Our task is to decide to what extent the measurement methods that do exist can satisfy these requirements and we have done this in connection with the diffusive sampling method. I want to give my own personal opinion now. There is not one diffusive sampler but rather there are several sampling methods that are based on the diffusion. Given the present state of the art we cannot talk about the diffusive sampler generally as an alternative method, but rather we have to indicate that there are substances which can be collected with these diffusive samplers; hence, within the framework of the regulations that apply in Germany we can work with these. In this whole area there is a great deal of development to be done. We can assume that a lot of data exists that have not yet been made available to all. I see a great need to have all data published and I think that it's necessary for research to be initiated in order to fill the gaps that exist. In order to do this we should take stock of the knowledge that we have to date. I think a validation report is essential and this should be based on a protocol given to the manufacturer so that before marketing his product he can decide whether or not this product is acceptable. We can't leave development in this field to chance as we have done in the past. I think that it is now time for us to actively participate in the development and to influence it consciously. Lastly, the government authorities should lay down criteria for the user so that he has standards by which to make his decisions.

CRABLE (USA)

NIOSH (the National Institute for Occupation Safety and Health) is not a regulatory agency. We are a research institute. We do research in chemistry, engineering control, toxicology, ergonomics and other occupationally related areas. The regulatory agency is OSHA (the Occupational Safety and Health Administration). They are the ones who establish the standards, although we do recommend methods to them. The final step in the regulatory legal process is the promulgation of the standard in the Federal Register. With the exception of one or two, and the one that immediately comn to my mind is asbestos, these regulations do not specify either the sampling or the analytical procedure. WWhat they do state is that the sampling and analytical procedure must meet accuracy and precision criteria. If you have a passive monitor which meets those criteria then there is no legal impediment to the use in the United States. I might add it should have sufficient validation data and field use so that it can meet any strenuous legal challenge. The problem is that some of the devices that we have looked so far have not met the criteria which would make them legally acceptable in the United States. If you can sample and analyze within plus or minus 25%, then whether you are using a passive monitor, an impinger, or whatever then that is legally acceptable. But I do say that you need to be prepared to have sufficient back-up data to meet any legal challenge.

BERLIN (CEC)

We how have a first feeling about what is going on in many countries in industry and from the trade union point of view.

We should keep in mind that these samplers are not only to be used to check compliance with limit values but also to assess exposure levels with respect to action levels. Can we request from the panel members a fisrt reaction to this overall feeling.

FIRTH (UK)

I think that I would really go along with what Dr. Crable said, that diffusive samplers are another means of making measurements, as long as they meed the requirements of accuracy, for assessing against levels whether they are control levels or action levels then they're acceptable and like any other measuring technique, they have to have proper validation and data to support this. I think coming through as a flavour from the various comments has been that we ought to perhaps be looking at validation procedure and we ought to be considering common levels of accuracy and precision But these will apply to any measuring technique. So I think that we are from a regulatory point of view primarily looking at the performance levels of these devices, not the operational acceptability of these of use. So as long as this particular type are acceptable then I would of course reinforce what Mr. Lehmann said that we should not be regarding diffusive samplers as a single species. There are a number of diffusive samplers, as there are a number of instruments and it's my feeling that groups of samplers will be better than others for specific purposes and that's the purpose of proper validation.

GROSJEAN (Belgium)

I would like in the first place to react to something that you said about action levels. Everyone agrees that there should be some level below which action is necessary. However, there is a big danger here that when a first assessment is made, employers will try their best to arrange things so that no further measurements need be taken.

JOURDAN (CEFIC)

We shouldn't be afraid of the cost aspect. We are living in a word which is governed by economic rules and if the technique will allow us to achieve the same result with less outlay then it means that the money saved could be used for other areas, to satisfy other needs in the society.

SAPIR (ETUC)

There is still this question as to how this technique of passive sampling can be fitted in with other existing techniques; some people talked about it being a complement and some people talked about it being an alternative. I would like to see how we can use the techniques and procedures to assess the situation, to improve it, and to agree on what should be done because I think everybody has pointed out that once a measurement has been taken than we need to have action afterwards. So I'd like to ask whether the passive sampler technique allows precise enough measurements and whether it allows us to specify where the problem originates.

BERLIN (CEC)

I would like to have a reaction from the floor regarding the possibilities of passive sampling techniques and as an alternative or as a complement to static sampling regarding the detection of exposure sources. Since it's very easy to have a lot of diffusive samplers, perhaps they would allow us to have better knowledge of the distribution within the atmosphere?

EYRES

How do you define compliance with exposure limits and action limits? If an exposure limit for example is defined as a limit never to be exceeded, then I think you require something considerably more accurate than if it is defined in some way as not normally to be exceeded. I think this is an important aspect in relation to the whole question of accuracy and costs.

LEHMANN (FRG)

I get the impression that the performance of a diffusive sampler depends on the concentration level. This means that if we wish to use this method in the workplace we have to look at not only the characteristics of the substance but also the exposure limits. Now my second point. This refers to the difference between the accuracy of the measurements and what can be said in compliance with TLVs. We're now speaking about two different statistics; on one hand we have statistics concerning the whole method, that's the statistics of the measurement results and sampling results, here we have the errors that have been made when we tried to get a measurement results, but there is a second statistic here and that is concerning the distribution of measurement results relative to the time duration, be it one month, one year of 8 hours if we carry out more random samples. So we really have two different groups of data that have to be dealt with, with two different types of statistics.

CRABLE (USA)

Legally, in the United States, we are required to take a breath zone. That's what the standard is. But in any occupational hygiene survey the industrial hygienist will take area samples and in and ever increasing number of cases where there is biological monitoring, they will take biological samples. Having done all of that, one of the things that we're not discussing, and it may not really be a part of this

Symposium, is how do you use this data and that's the next stage. There are really three parts to all of this the third being engineering control. By proper sampling strategy you can identify where some of the emissions are and with intense sampling you can identify much more of these. So I think we have to identify good sampling stategy. I see some of these passive monitors as another way of sampling. Not a replacement for anything.
BERLIN (CEC)

The panel discussion will now be broadened to comments fom all participants.

BLOME (FRG)

I welcome the fact that in the panel discussion, Mr. Crable again mentioned the fact that a series of diffusion samplers actually failed the protocol tests and I think that we have to make sure that this information also gets through to the user.

SQUIRELL (UK)

I would just like to say a little on quality and costs. I think that the panel discussion has shown that we must demand quality in materials and products. Much of the rest of the discussion is not specific to diffusive sampling but to any analytical investigation which includes sampling and analysis. The whole business must be cost-effective. This includes selection of the method, the communication of the results upwards and downwards to management and the workforce. It also concerns value for money. You get what you pay for for the effort you put into the administration and the training of people to take samples, etc. You should not change the worker patterns; they should follow their normal activities. One of the major objectives that were drawn up right at the beginnng of research into passive sampling was that they should be designed for painless, unobtrusive, efficent sampling of personal worker exposure over about a four to eight hour period, responding certainly to short term high exposures. This was particularly needed for large volume sampling, when large numbers of samples might be required. I think we must be careful not to confuse the issue too much by the real bonuses that I see have come out as a result of all our work in that they can also be used with discretion for short time sampling. I want to make the point that it's value for money we're talking about, not just cost, and many analysts will agree that the best analysis is done by a single means and a simple means of sampling.

RHEIN (FRG)

I'd like now to reflect my interests as a representative of the workers and as a trade unionist. I am surprised that Mr. Eyres feels that the weight of the device is of more importance that the accuracy of the results. For myself as a workers' representative, accuracy of the measurement results is decisive because it is on the basis of those results that the health of the worker is dependent. I would be very keen therefore to see the development of diffusive samplers so that perhaps in the intermediate term, they can become an alternative to active sampling.

LEICHNITZ (FRG)

We have heard that the United States has acceptance levels of 25% and I know that in Germany, in the technical rule for toxic substances we have 30%. I have the feeling that many experts stick too closely to these 25 or 30% level, when every method that gives results above those levels fails validation tests or certification tests. I would like to inform you that other organizations such as the International Union of Pure and Applied chemistry came to the conclusion that one should distinguish between at least two different levels; a method with an overall error in the range of up 25% is a so-called excellent method and a method with an error of 25 to 50% is still acceptable, but that these levels should be published and also confirmed by several experts.

CHOO YIN (UK)

We have looked into passive sampling for a particular application with regard to the proposed legislation action levels which we spoke about earlier. The thing about these action levels is that they seem to be getting lower and lower. What is important to me is that I have a reasonable estimate of the accuracy for a particular sampler at a particular level in question and how much money I need to spend to measure this level.

KENNEDY (UK)

The conference has talked a great deal about the accuracy of diffusive samplers, vis-a-vis pump samplers both in the laboratory and in field conditions. In my opinion the accuracy of both methods is small in comparison to the day to day variations which take place in the workplace. In order to obtain a meaningful measure of whar actually happens in the workplace the greatest errors that we face are the statistical errors that are inevitably associated in dealing with small amounts of data. I welcome the advent of passive samplers because I believe that they will enable us to obtain much more data and more easily from the same resources and by looking at greater numbers of results we will obtain a far more useful assessment of the conditions in the workplace than would have been possible with a smaller number of results which may be more accurate in an academic sense.

BERLIN (CEC)

I think this is a very important comment in a number of ways. Sampling strategies in a number of countries have been devised to try to reduce as much as possible the number of measurements taken while still maintaining the accuracy. Maybe with the advent of diffusive samplers, one might reconsider how these strategies could be affected by their wider availabity.

MEYER (The Netherlands)

There is a need for a common protocol and that in the future manufacturers should be responsible for the products. I would recommend that there is a need for a protocol which is valid at least for the countries within the European Communities and that it is a task for the Commission to see that such a protocol will come into being.

BORD (UK)

One of the problems that we seem to have is that there is very little science available for what I would call the front end of this process of sampling and analysis. I wonder has any group, anywhere, carried out any research into worker responses in a scientific way to sampling methods. there are behaviour scietinsts who probably could help us in this. I think these ergonomic considerations should be given much more consideration. One of the reasons is that the nature of work is changing and some of the stereotypes that we have of the industrial worker I think are changing and the sort of people we have to sample may be different in the future. I jotted down a few examples of people who we have been concerned with recently; helicopter pilots, cleaning ladies, workers working inside aeroplane wings, teenagers in prison workshops, even patients in dental hospitals. I think that we need some good quality research into how people behave when they're sampled and I don't think this problem has been addressed and that it is relevant to the kind of sampling that is done.

TINDLE (UK)

We need a validation. We need a criteria of acceptance. I would like to know who is going to pay for the validation and who is going to direct it?

ROBERTSON (UK)

It seems to me that everybody is paying for validation at the moment and one of the most effective ways to reduce the costs of validations would be an effective system of collecting and dispersing validation information. So rather than ten people, or ten companies paying for validating exactly the same sampler and method for each solvent, one person could do it and the information could be shared. It would save an awful lot of money.

BROWN (UK)

The reason that the HSE committee of analytical requirements working group 5 was set up was precisely this. We realized that half a dozen of us were all doing the same thing in different laboratories at vast expense so we attempted to try and share the data that we had one with the other. In my own experience, there is no evidence than the diffusive system is more accurate or the pump system is more accurate. Statistically if you look at a large number of different systems there is really nothing to choose between them and I think that this idea that the diffusive is less accurate that the pumped is a myth. I think it's worth pointing out that the NIOSH standards completion programme that the pump error is 5%. Now this very often is far from the truth and if you put a realistic error into the pump calculations I would guess that a good number of the methods that we all accept, the charcoal tube and so on, would fail the NIOSH standards completion programme plus or minus 25% miserably.

NORBACK (Sweden)

It would be a good idea if somebody were to create a data base internationally where you could put in information about what you have done. I feel that many people are doing experiments and for some reason are not publishing them.

BERLIN (CEC)

The exchange of information is fundamental, and ways and means to develop it will have to be considered. I think this very lively panel discussion has been a first step in this direction.

INTERNATIONAL SYMPOSIUM ON WORKPLACE AIR MONITORING

DIFFUSIVE SAMPLING - AN ALTERNATIVE APPROACH

LUXEMBOURG, 22-26 SEPTEMBER 1986

SUMMARY REPORT

A. BERLIN   Commission of the European Communities
       Health and Safety Directorate
       Luxembourg
       GD Luxembourg

R.H. BROWN   Occupational Medicine and Hygiene Laboratories
       Health and Safety Executive
       403 Edgware Road
       London NW2 6LN
       England

K. LEICHNITZ Drägerwerk AG
       Moislinger Allee 53-55
       2400 Luebeck 1
       West Germany

B. MILLER   Imperial Chemical Industries PLC
       Chemical and Polymers Group
       P.O. Box 8
       The Heath
       Runcorn
       Cheshire WA7 4YD
       England

K.J. SAUNDERS BP Research Centre
       Chertsey Road
       Sundbury-on-Thames
       Middlesex TW16 7LN
       England

B. STRIEFLER Niedersächsisches Landesamt für
        Immissionsschutz
       Davenstedter Strasse 109
       3000 Hannover 91
       West Germany

WORKPLACE AIR MONITORING

DIFFUSIVE SAMPLING - AN ALTERNATIVE APPROACH

SUMMARY REPORT

A.   BERLIN,   R.H.   BROWN, K. LEICHNITZ,   B.   MILLER,

K.J. SAUNDERS AND B. STRIEFLER

1. INTRODUCTION

1.1 Organisation and aims

1.2 Philosophy of diffusive sampling

2. CURRENT FIELD APPLICATIONS

2.1 Overview of sampler types and applications

2.2 Comparison of diffusive and active sampling

2.3 Parameters affecting performance

2.4 Evaluation protocols

3. ROLES OF DIFFUSIVE SAMPLING IN MONITORING

3.1 Influence of diffusive sampling on sampling strategy

3.2 Cost-benefit analysis

3.3 Quality of monitoring data

3.4 Roles of workplace air and biological monitoring

3.5 Education and training

4. CURRENT TRENDS IN DEVELOPMENT OF DIFFUSIVE SYSTEMS

5. ACCEPTABILITY OF MONITORING DATA BASED ON DIFFUSIVE
    SAMPLING

6. CONCLUSIONS AND RECOMMENDATIONS

1. INTRODUCTION

1.1 ORGANISATION AND AIMS

This Symposium was organized jointly by the Commission of the European Communities and the United Kingdom Health and Safety Executive in cooperation with the World Health Organization and the Royal Society of Chemistry of the United Kingdom.

The 1984 Action Programme of the European communities on Safety and Health at Work stresses the need of a common methology for monitoring pollutant concentrations at places of work; to carry out intercomparison programmes and to establish reference methods for the determination of the most important pollutants.

A diffusive sampler is a device which is capable of taking samples of gas or vapour pollutants from the atmosphere at a rate controlled by a physical process such as diffusion through a static air layer or permeation through a membrane, but which does not involve the active movement of the air through the sampler.

The objectives of this Symposium were:

- to review the state of the art of diffusive sampler techniques

- to stimulate the exchange of technical information

- to assess the suitability and range of applications for workplace monitoring

- to promote the further development of this technique and its wider use.

The Symposium was attended by more than 200 participants (industrial hygienists, occupational physicians, analysts, officials from regulatory agencies) from all Member States of the European Community and a number of other countries such as Austria, Canada, Finland, Japan, Norway, Sweden, Switzerland and the United States of America.

1.2 PHILOSOPHY OF DIFFUSIVE SAMPLING

Diffusive sampling in the occupational environment dates back at least to the 1930s, when qualitative devices were described, but the first serious attempt to apply science to quantitative diffusive sampling was in 1973, when Palmes described a tube-form sampler for nitrogen dioxide. Since then, a wide variety of samplers have been described, some relying on diffusion through an air-gap, some relying on permeation through a membrane, and some using both techniques, for the rate-controlling process in sampling. Many of these devices are commercially available.

The theoretical basis for diffusive sampling is now well-established. Diffusion and permeation processes can both be described in derivations of Fick's first law of diffusion, which result in expressions relating the mass uptake by the sampler to the concentration gradient, the time of exposure, and the sampler area exposed to the pollutant atmosphere. Expressions have also been derived for the application of Fick's law to diffusive sampling in the "real" world, i.e. taking into account non-steady state sampling, the effects of fluctuating concentrations, sorbent saturation, wind velocity and turbulence at the sampler surface, temperature, pressure and so on. Except for sorbent saturation, which may lead to reduced (although sometimes predictable) uptake rates, these modifications to the basic Fick's law expression do not lead to significant errors for well-designed samplers.

Descriptions of diffusive samplers based on sorption and subsequent desorption, in both badge and tube forms have been numerous. Desorption of the pollutant has been mainly by solvent desorption, but the use of thermal desorption is gaining favour, mainly because of the ease of automation of this stage and the possibility of direct coupling with gas chromatography, particularly with samplers in the tube format. In addition, there is no interference from a solvent peak.

Most diffusive samplers have been designed for wearing on the lapel, i.e. for personal monitoring. Some, however, may also be used in static positions or in non-occupational environ-ments e.g. for ambient air monitoring. Similarly, most samplers have been designed for a collection period of 8 hours, but some particularly thermally desorbable types, have sufficient sensitivity for monitoring over shorter time periods or at very low levels of pollutant concentration. Some samplers may also be appropriate for environmental monitoring over extended periods.

2. CURRENT FIELD APPLICATIONS

2.1 OVERVIEW OF SAMPLER TYPES AND APPLICATIONS

Direct-reading reagent-type diffusive samplers rely on the pollutant reacting chemically with a reagent system dispersed on a suitable support to produce a colour change. Samplers may be of a tube-form, similar to a conventional gas detector tube but with a defined air-gap at the sampling end, the length of stain indicating the extent of exposure in ppm x hours. The support may be granular or a paper strip. Alternatively, samplers may be in badge form, similar in size and shape to a radiation badge, in which case exposure is generally indicated by the intensity of colouration.

Direct-reading devices have been mostly designed for inorganic gases such as ammonia, hydrogen sulphide, sulphur dioxide, nitrous fumes and carbon dioxide. A few applications for organic vapours have also been reported, particularly for gas detector tubes, for which the technology already exists for the pumped versions. A specific badge for phosgene has also been reported.

Such samplers, having a direct read-out of exposure dose, may warn the wearer of short-term high exposure, but only by visual examination.

Similar reagent-type diffusive samplers are also available for pollutants which do not lend themselves to direct colourimetry; in such cases, indirect read-out is accomplished by a measurement of conductivity, X-ray fluorescence, or by chemical analysis. Indirect devices are mostly of badge type and may use a solid or liquid reagent medium. Most of the chemical reagent-type samplers can only be used for one compound or group of compounds. These types have also concentrated on inorganic gas applications but can be applied to other specific analytes, such as formaldehyde, mercury and halogenated hydrocarbons.

The reversible sorption systems, particularly for organic compounds, are usually capable of addressing a wide range of compounds either singly or simultaneously, the analysis being completed by gas chromatography. Current applications have included organic solvents, anaesthetic gases, polymer precursors (vinyl chloride, acrylonitrile, styrene, etc.) and gasoline components.

The range of applications of diffusive samplers is broadly similar to that of conventional pumped samplers with regard to compound specificity. Indeed the same or very similar device is often used in either the pumped or diffusive mode. However, because of its convenience and simplicity, the diffusive sampler is ideally suited to large-scale surveys and screening procedures.

2.2 COMPARISON OF DIFFUSIVE AND ACTIVE SYSTEMS

Most described applications of diffusive samplers emphasize the value of comparisons with active systems. It is generally accepted that pumped methods can be regarded as established methods, both in regard to the "true" result and the achievable precision. The diffusive system is then tested, ideally both in the laboratory and the field (i.e. the occupational environment) using the pumped method as an independent check method. Generally, laboratory experiments establish the sampling rate. This sampling rate is then used to determine measured concentrations in field experiments, to check bias and precision under field conditions.

The comparison is usually expressed as a plot of diffusive result against pumped result for a series of paired personal samplers. The interpretation of such plots, however, is not straightforward and reported results and conclusions vary very widely from excellent correlation between pumped and diffusive results to very poor agreement.

Much of the Symposium was concerned with possible reasons for the poor agreement sometimes observed and whether there were, as a consequence, any inherent problems associated with diffusive sampling.

Poor agreement of one particular design of diffusive sampler with an independent method may simply be due to bad design. Not all diffusive samplers are the same; some may well have been better designed than others. It is known, for example, that some early designs were particularly susceptible to wind velocity/air turbulence on the sampler surface and adsorbent saturation effects.

Differences between individual pairs of results do not necessarily imply a deficiency in the diffusive approach. Pumped methods have also been observed to be in error. For example in a TNO/ISO exercise solely on pumped charcoal tubes, and involving experienced analysts, a significant proportion of the samples loaded from a standard atmosphere of chlorinated hydrocarbons were found to contain no recoverable analyte, even though the pumps were apparently operating normally.

Evidence was given that real concentration variations can occur within the sampling area (the breathing zone for personal monotoring). These will lead to differences in paired results even from identical samplers within this area.

In some examples given, the lack of agreement between individual pairs of results (pumped and diffusive) was sufficiently important to warrant special consideration.

The differences between pumped and diffusive systems might be real. As noted above, the actual concentrations measured in the sampling area might be different. Also, because of the large difference between the diffusion coefficients of gases and aerosols in air, aerosol is essentially not collected in a diffusive device, but may be collected to a greater or lesser extent (depending on flow rate) by a pumped system. Conversely, in a very dusty environment, some particulate may fall onto a diffusive sampler and give a "false positive" if, for example, the dust contained occluded analyte. The latter problem can be avoided by employing a diffusive sampler face-downwards, but this procedure leaves open the question (asked many times) or what is the "true" exposure value - with or without dust?

The sampling mechanisms and sampler geometry might also affect the results obtained. Because of the different effective flow-rates of the two types of samplers, the size/area of the sampling zone might be different, and the samplers might move about in different ways during personal monitoring. Furthermore, pumped sampling systems might be easily interfered with by the wearer.

2.3 PARAMETERS AFFECTING PERFORMANCE

Particular concern was expressed on the performance of diffusive samplers in measuring very low pollutant concentration, or short-term exposures. Since the effective sampling rate is much lower than for a similar pumped system, a smaller quantity of analyte would be collected. This might be offset by using thermal desorption instead of solvent desorption, but it means particular care needs to be taken in ensuring sampler cleanliness before, during and after sampling. Evidence was presented to show that at least in some cases good agreement could be obtained between diffusive and pumped methods down to ppb levels. There would seem to be no theoretical limitation with respect to concentration (although there may be with exposure time). In some cases, for example, some porous polymers used to sample acrylonitrile, the adsorbent itself might contain traces of analyte, or be subject to degradation to the analyte if thermally desorbed and hence cause blank problems.

Concern was also expressed on the effects of wind face velocity, particularly with samplers or large diffusive area and possibly ineffective draught shields. This problem appeared to be very dependent on sampler design and geometry.

For better assessing the performance of diffusive samplers at low pollutant concentrations and short exposure times, it was concluded that

- correlation plots of pumped against diffusive samples should not obscure behaviour at low concentrations by compression of points at the origin; log or other transformations were suggested to spread the points;

- evaluation protocols should include experiments to determine any limitations on sample performance at low concentrations or short sampling times.

2.4 EVALUATION PROTOCOLS

The general pattern of diffusive sampler evaluation - laboratory experiments to determine the sampling rate and field experiments to determine bias (systematic difference) and precision - was followed by most contributions. Such evaluations had been formalised, in particular, by the UK Health and Safety Executive (HSE) and the US National Institute for Occupational Safety and Health (NIOSH).

The general approach of both the HSE and NIOSH protocols was the same. Diffusive samplers are exposed to independently-calibrated standard atmospheres of pollutant vapours and the sampling rate (or the apparent concentration, if the sampling rate is known) determined. Such exposures are conducted over a variety of regimes in order to determine the effects, if any, on the sampling rate of exposure concentration, time, temperature, humidity, interferents, air movement and monitor orientation.

The two protocols differ in the parameters they consider most important, partly because the protocols are modelled on different monitor designs. The HSE protocol essentially addresses thermal desorption tube samplers, where non-ideal adsorption is important, whilst the NIOSH protocol essentially addresses charcoal badges, where air movement sampler orientation, sample recovery and adsorbent saturation are important. Other significant differences arise in the protocols' treatment of field experiments and sampler accuracy. HSE lays much more emphasis on field trials than NIOSH to counterbalance the absolute minimum of laboratory work. NIOSH, on the other hand, have many more laboratory experiments and ask for a larger number of replicates in each experiment. With regard to accuracy, NIOSH include a specific acceptability criterion, although this criterion relates only to the laboratory trials. The HSE protocol carries no explicit criterion, but insists on a statement of accuracy in both laboratory and field determinations.

There was a clear call from the Symposium for the elaboration of a unique, simple and adaptable validation protocol which would combine elements from the existing protocols. Such a protocol would have to address all types of diffusive sampler.

The cost of validation, however, was recognized to be high, and in the short term might alter the balance of the cost-benefit analysis. It was important to share both effort and results and there was a strong feeling from the Symposium that some form of international data base of samplers/ analytes/ uptake rates/ validation ranges should be set up.

3. ROLE OF DIFFUSIVE SAMPLING IN WORKPLACE AIR MONITORING

3.1 INFLUENCE OF DIFFUSIVE SAMPLING ON SAMPLING STRATEGY

The relative simplicity and convenience of diffusive sampling are likely to introduce flexibility into a sampling programme, allowing much larger numbers of samplers to be handled. Such samplers could be operated by relatively unskilled personnel, although this approach would have to be considered with care. The role of the occupational hygienist would change, as it will no longer be necessary for him to be present purely to operate the samplers. There is thus a danger that important information regarding work-practices may be missed or not related to the specific sample.

It was generally accepted that the accuracies of diffusive samplers were, in most cases, very similar to those of conventional pumped methods. There might be some cases, however, where accuracy did not meet target values e.g. $\pm$ 25%. Such might be the case, for example, with some colourimetric devices with visual read-out. This would have obvious importance in determining compliance with legal exposure limits, although in practice quite large errors might be acceptable for measurements much higher or much lower than the exposure limit. It was relatively straightforward, however, to improve precision by the simple step of taking replicate samples.

3.2 COST-BENEFIT ANALYSIS

Accurate cost-benefit analysis is usually complex and the conclusions reached are often very dependent on the assumptions made. The main factors affecting the balance between diffusive and active systems are the costs of buying and maintaining sampling pumps and the costs associated with the personnel needed to operate the samplers.

It may be that pumps are needed anyway for applications for which diffusive sampling is not appropriate, for example measuring dusts. The presence of the occupational hygienist for hygiene assessments and work practices evaluation will be required in most instances.

It has been estimated, however, that a hygienist can supervise approximately 10 active samplers or approximately 30 diffusive samplers per man-day. The cost-benefit balance, therefore, is markedly affected by the number of samples taken; or conversely, the potential cost-effectiveness of diffusive sampling may stimulate more frequent and more representative sampling and hence development of improved hygiene conditions.

3.3 QUALITY OF MONITORING DATA

Quality control and good monitoring practice are essential at all stages in the manufacture, use and interpretation of results from diffusive samplers as for any monitoring system. Samplers must be well-designed, in accordance with theoretical considerations and must be well-engineered to exacting tolerances. Unless samplers are to be individually calibrated, which is unrealistic, samplers should be identical, or nearly so, particularly with regard to the diffusional resistance.

Quality assurance at the analytical stage is also of importance and the use of certified reference materials, such as are being investigated by the Community Bureau of Reference, was mentioned. The existence of quality assurance schemes such as PAT (organized by NIOSH) and AQUA (organized by HSE) were also noted.

3.4 ROLES OF WORKPLACE AIR AND BIOLOGICAL MONITORING

Much of the discussion during the Symposium related to a comparison of diffusive sampling results with parallel pumped sampling results obtained in the field. In several instances biological monitoring had been employed as an alternative independent check method, since this is generally recognized to be an indicator of occupational exposure. However, personal monitoring (whether pumped or diffusive) and biological monitoring are complementary rather than parallel techniques.

The former indicates the airborne exposure hazard, the latter the individual's actual uptake. Personal sampling does not take account of individual biological variability, or absorption by routes other than inhalation.

Several reports showed that diffusive samplers, which could be used in large numbers, allowed significantly better correlations to be established between ambient sampling and biological monitoring. Such correlations did not prove the validity of diffusive sampling: only that a (approximately) linear relationship existed between the two methods. Nevertheless, the many examples given reinforced the potential of diffusive sampling for monitoring large numbers of workers with more indirect supervision.

3.5 EDUCATION AND TRAINING

As with the introduction of any new technique, it is important that adequate instructions are available for the selection and operation of a suitable sampler. Even if the operation of the sampler is comparatively simple, care must still be taken to avoid contamination, to position the sampler correctly and to package it correctly if the sampler needs to be transported to a central laboratory for analysis.

On a more general note, it may be that the introduction of diffusive samplers, particularly direct-reading ones, will increase worker awareness of occupational hazards and will enable workers to anticipate and avoid such hazards.

4. CURRENT TRENDS IN DEVELOPMENT OF DIFFUSIVE SYSTEMS

New types of diffusive sampler and new applications for existing ones are continually being developed and many were described at the Symposium. Major advances are also being made on the analytical front. Thermal desorption of both diffusive and pumped samplers is becoming increasingly popular, both from porous polymers and graphitized carbon because of the high sensitivity and potential for automation of this method. Because analysis by thermal desorption is usually possible only once, capillary columns are often used to improve resolution and aid confirmation of peak identity when complex mixtures of analytes are involved. For very large sample throughput, a mass selective detector may be justified for both qualitative measurement. Even for modest sample throughput, a computerized data handling system is highly desirable, preferably with a facility for peak comparison with a data bank of compounds, retention times and uptake rates.

5. **ACCEPTABILITY OF MONITORING DATA BASED ON DIFFUSIVE SAMPLING**

Recognition of the value of diffusive monitors will allow their proper insertion into the monitoring strategy currently under development at European Community level. The regulatory agencies agreed that subject to appropriate accuracy requirements being satisfied, monitoring data would be acceptable, irrespective of the sampling method used and that diffusive samplers can usually provide the necessary accuracy.

Among the social partners, industry has shown considerable enthusiasm for the potential use of diffusive samplers for checking compliance with exposure limits and in the context of improving hygiene conditions. Trade union representatives wished to place the question of the use of diffusive monitors in addition to compliance with exposure limits. They considered that the validity of the data was paramount. The increased availability of data through the establishment of an international data base will ge a long way towards accelerating their further wider acceptability.

6. **CONCLUSIONS AND RECOMMENDATIONS**

In recent years there has been:

- a very significant increase in the use of diffusive samplers in various circumstances (wide range of applications, including epidemiological studies)

- a continuous evolution of new designs

- an increasing interest among manufacturers for the productions of diffusive samplers.

The following conclusions have been agreed:

The theoretical basis for diffusive sampling has been confirmed by laboratory and field trials.

The follow-up analysis of diffusive samples does not appear to present specific problems.

In general there seems to be no significant difference between the accuracy and precision of diffusive sampling and those of other monitoring systems such as active pumped sampling.

Cost-benefit analysis with respect to other monitoring systems is a complex procedure due to the many parameters involved.

In routine operations when comparing with active systems diffusive sampling presents usually:
- higher costs in terms of the sampler,
- no costs in relation to pump and maintenance,
- less manpower requirements for system handling.

Active and diffusive sampling are complementary approaches each one having areas of applicability which may overlap. Each have their roles in a strategy for monitoring workers exposure to airborne pollutants.

The use of diffusive samplers may introduce flexibility in the monitoring strategy.

Validation of all sampling system is essential both in the laboratory and in the field.

For diffusive sampling, on the basis of theory and experience, the following main parameters must be included in a validation procedure:

- humidity,
- face velocity,
- temperature,
- concentration,
- exposure time,
- storage,
- shelf-life,
- interferences,
- orientations.

The protocols currently used have many features in common, which should facilitate the elaboration of a unique, simple validation protocol, adaptable to future needs. The existence of such a protocol will help users interpret performance data provided by manufacturers.

On the basis of validation protocols, acceptability criteria for the performance of diffusive samplers can be established by appropriate authorities as a function of the applications (such as compliance with limit values).

Quality assurance in the production and preparation of diffusive samplers, and in particular for the sorbent material is essential.

An increasing range of direct-reading (colour) diffusive samplers is becoming available; they are likely to play an increasing role in the future as no laboratory analysis is required.

Diffusive samplers do not interfere with worker activity and are unlikely to affect the behaviour pattern of the wearer. In this connection, at no extra inconvenience to the workers, two or more diffusive samplers with different applicabilities can be used simultaneously.

The use of these samplers, and especially direct-reading monitors, can help to lead to safer work patterns.

It was agreed that as a general principle, any method is acceptable by regulatory authorities and hygienists if used by experts within its defined limitations. This applies equally to diffusive samplers.

The following recommendations have been formulated:

A unique, simple validation protocol, adaptable to future needs, should be established with urgency.

Intercomparison programmes and the availabity of certified standards should complement the validation studies.

Quality assurance should be based on elements of the validation protocols; the quality assurance procedures and results should be available to users.

Manufacturers of diffusive samplers should provide users in addition to the validation and quality assurance data, information on applications and limitations of use.

The development of direct-reading monitors for both organic and inorganic substances should be encouraged.

The suitability of diffusive samplers for short sampling times and/or low concentrations should be further investigated.

To better identify the possible range of applications (in terms of chemicals) of the various types of diffusive samplers, a widely agreed list of suitable sorbent materials, monitor types, practical uptake rates and diffusion coefficients should be compiled. The establishment of such a list should assist with the prediction for other suitable systems and/or chemicals.

ACKNOWLEGEMENTS

The authors wish to thank all the more than 200 participants who have contributed to the success of the Symposium.

Special thanks are due to the following invited speakers, rapporteurs, chairmen and organizing committee:

J.	Auffarth (FRG),
A.	Berlin (CEC),
R.H.	Brown (UK),
J.V.	Crable (USA),
J.G.	Firth (UK),
M.	Frangiadakis (Greece),
B.	Goelzer (WHO),
J.P.	Guenier (F),
B.	Herve-Bazin (F),
H.	Ikeda (Japan),
J.	Kristensson (Sweden),
K.	Leichnitz (FRG),
R.	Lidgett (B),
P.B.	Meyer (NL),
B.	Miller (UK),
G.	Moore (UK),
L.	Pozzoli (Italy),
M.	Sarivalassis (Greece),
K.J.	Saunders (UK),
D.C.M	Squirrell (UK),
B.	Striefler (FRG),
N.	Van den Hoed (NL),
J.F.	Van der Wal (NL).

A'CAMPO C.P.
AKZO - DEPARTMENT CRL
Velperweg 76
NL-ARNHEM

ANDREJS B.
BERUFSGENOSSENSCHAFT NAHRUNGSMITTEL
UND GASTSTAETTEN
Steubenstrasse 46
D-6800 MANNHEIM 1

ANDREOLETTI D.
CLINICA DEL LAVORO LUIGI DEVOTO
UNIVERSITA DEGLI STUDI DI MILANO
Via San Barnaba 8
I-20122 MILANO

ARAVIDOU A.
DIRECTION INSPECTION DU TRAVAIL
PREFECTURE D'ATTIKI
Voulgari 2
B.P. 1108
GR-ATHENS

ARENGA J.
Rua Infanta d. Maria, Bloco 5-1
P-3000 COIMBRA

ATKINS D.H.F.
HARWELL - ENVIRONMENTAL AND MEDICAL
SCIENCES DIVISION
AERE Harwell
UK-OXFORDSHIRE OX11 ORA

AUFFARTH J.
BUNDESANSTALT FUER ARBEITSSCHUTZ
GRUPPE STOFFBESTIMMUNG
Messungen Vogelpothsweg 50-52
D-4600 DORTMUND 17

BØRRESEN E.
SINTEF-DIVISION OF APPLIED CHEMISTRY
N-7034 TRONDHEIM NTH

BARTOLUCCI G.
INSTITUTE OF OCCUPATIONAL MEDICINE
UNIVERSITY OF PADOVA
Via Facciolati 71
I-35127 PADOVA

BENVENUTI F.
ISPELS
Via Urbana 167
I-00184 ROMA

BERLIN A.
C.E.C.
HEALTH AND SAFETY DIRECTORATE, DG V/E
L-2920 LUXEMBOURG, KIRCHBERG

BERTHIER P.M.
COMMISSARIAT A L'ENERGIE ATOMIQUE
SERVICE D'HYGIENE INDUSTRIELLE
B.P. 171
F-30205 BAGNOLS CEZE/CEDEX

BERTONI G.
CONSIGLIO NAZIONALE DELLE RICERCHE
AREA DELLA RICERCA DI ROMA
ISTITUTO SULL'INQUINAMENTO
ATMOSFERICO
Via Salaria Km 29
300 CP 10
I-00016 MONTEROTONDO STAZIONE (ROMA)

BERTRAND M.
COMMISSARIAT A L'ENERGIE ATOMIQUE
SERVICE D'HYGIENE INDUSTRIELLE
B.P. 171
F-30205 BAGNOLS CEZE/CEDEX

BIENEK H.
BAYER AG. WERKSVERWALTUNG -
ARBEITSSICHERHEIT - ARBEITSSCHUTZ
Werk Dormagen
Postfach 100140
D-4047 DORMAGEN

BITTER
I.N.H.A.K. GmbH.
Im Kirchfeld 6
D-3062 BÜCKEBURG

BLOME H.
B.I.A. - BERUFSGENOSSENSCHAFTLICHES
INSTITUT FÜR ARBEITSSICHERHEIT
Lindenstrasse 80
Postfach 2043
D-5205 ST. AUGUSTIN 2

BORD B.
HEALTH AND SAFETY EXECUTIVE
MAGDALEN HOUSE
Stanley Precinct
UK-BOOTLE, LIVERPOOL

BRANDON R.W.
SIEGER LIMITED - GAS DETECTION
Nuffield Estate 31
UK-POOLE, DORSET BH17 7RZ

BREALEY G.J.
COURTAULDS FIBRES-GREAT COATES WORKS
UK-GRIMSBY, S. HUMBS DN31 2SS

BRIFFAUT G.
MINISTERE DES AFFAIRES SOCIALES ET
DE L'EMPLOI
DIRECTION DES RELATIONS DU TRAVAIL
Place de Fontenoy 1
F-75700 PARIS

BROADWAY G.M.
PERKIN - ELMER Ltd.
Post Office Lane
UK-BEACONSFIELD

BROWN R.H.
HEALTH AND SAFETY EXECUTIVE
OCCUPATIONAL MEDECINE AND HYGIENE
LABORATORY
Edgware Road 403-405
UK-LONDON NW2 6LN

BRUGNONE F.
ISTITUTO MEDICINA LAVORO
UNIVERSITA' VERONA
POLICLINICO BORGO ROMA
I-37134 VERONA

BUCHET J.P.
UNIVERSITE DE LOUVAIN
Avenue Chapelle aux Chaînes 4
B-1200 BRUXELLES

BURRI P.
FEDERAL INSTITUTE OF TECHNOLOGY
DEPARTMENT OF HYGIENE AND APPLIED
ERGONOMICS
Clausiusstrasse 21
CH-8092 ZURICH

CALVERT G.
DIGITAL EQUIPMENT Co. Limited
P.O. Box 115
UK-READING RG2 OTL

CHALVIDAN PH.
B.P. CHIMIE - USINE DE LAVERA
B.P. 6
F-13117 LAVERA

CHEVALIER B.
ROUSSEL - UCLAF
Boulevard des Invalides 35
F-75007 PARIS

CHOO YIN C.
NATIONAL SMOKELESS FUELS Ltd.
Cardiff Road
Nantgarw
UK-CARDIFF, WALES

COCHEO V.
FONDAZIONE CLINICA DEL LAVORO
Via Tassoni 8
I-35125 PADOVA

COKER D.
ESSO EUROPE INC.-ESSO RESEARCH CENTRE
UK-ABINGDON, OXON OX13 6AE

CORIGLIANO L.
MONTEPIDE PAS
Via Rosellini 15/17
I-20124 MILANO

CORNET H.
CHROMPACK INT. BV.
P.O. Box 8033
NL-4330 EA MIDDELBURG

COTTICA D.
FONDAZIONE CLINICA DEL LAVORO
Alzaia 29
I-27100 PAVIA

CRABLE J.
U.S. DEPARTMENT OF HEALTH AND HUMAN
SERVICES
DIVISION OF PHYSICAL SCIENCES AND
ENGINEERING
Columbia Parkway 4676
U.S.A.-CINCINNATI, OHIO 45226

DE BORTOLI M.
HAMLIN J.W.
B.P. CHEMICALS Ltd.
Buckingham Palace Road 76
UK-LONDON

DE ROSA E.
INSTITUTE OF OCCUPATIONAL MEDICINE
UNIVERSITY OF PADOVA
Via Facciolati 71
I-35127 PADOVA

DEHAN Y.
ETS VAN DER HEYDEN-PERKIN-ELMER
Rue du Marais 49-55
B-1000 BRUXELLES

DICESARE J.L.
PERKIN-ELMER
Main Avenue MS 20
U.S.A.-NORWACK, CT

DURKOP R.
VOLKSWAGEN AG.
SICHERHEITSTECHNIK
D-3180 WOLFSBURG

ELLWOOD P.A.
HEALTH AND SAFETY EXECUTIVE
Edgware Road 403
UK-LONDON NW2 6LN

EYRES A.R.
CONCAWE
Koninging Julianaplein 30-9
NL-2595 AA DEN HAAG

FIELDS B.
I.C.I. - PETROCHEMICALS AND PLASTICS
DIVISION
P.O. Box 90
Wilton
UK-MIDDLESBOROUGH, CLEVELAND

FIRTH J.G.
HEALTH AND SAFETY EXECUTIVE
OCCUPATIONAL HYGIENE LABORATORIES
Edgware Road 403
UK-LONDON NW2 6LN

FORBES F.
H.M. INDUSTRIAL POLLUTION
INSPECTORATE S.D.D.
Pentland House, Robb's Loan 47
UK-EDINBURGH

FRANGIADAKIS M.
MINISTRY OF LABOUR OF GREECE
Pireos Street 40
GR-ATHENS

GEUSKENS R.B.M.
M.B.L. - T.N.O.
MEDICAL BIOLOGICAL LABORATORY
Lange Kleiweg 139
P.O. Box 45
NL-2280 AA RIJKWIJK ZH

GILLHAM H.L.
THE WELLCOME FOUNDATION LIMITED
Building 140, Top fl.
Temple Hill
UK-DARTFORD, KENT DA1 5AH

GOMES ESTEVES J.F.
ADMINISTRACAO REGIONAL DE SAUDE
Praca de Londres 2-16
P-1000 LISBON

GORI G.P.
INSTITUTE OF OCCUPATIONAL MEDICINE
UNIVERSITY OF PADOVA
Via Facciolati 71
I-35127 PADOVA

GRAY W.M.
WEST OF SCOTLAND HEALTH BOARDS
DEPARTMENT OF CLINICAL PHYSICS AND
BIO-ENGINEERING
West Graham Street 11
UK-GLASGOW G4 9LF

GREEN D.
BRITISH RAILWAYS BOARDS
RAILWAY TECHNICAL CENTRE
ANALYTICAL SERVICES UNIT R.
Hartley House 110 , London Road
UK-DERBY DE2 8UP

GRIEPINK B.
C.E.C.
DIRECTORATE-GENERAL FOR SCIENCE,
RESEARCH AND DEVELOPMENT
Rue de la Loi 200
B-1049 BRUSSELS

GROSJEAN R.
MINISTERIE VAN TERWERKSTELLINGZARBEID
LAB. VOOR INDUSTRIELLE TOXICOLOGIE
Belliardstraat 51-53
B-1040 BRUXELLES

GROSSESTOVE J.
STATE UNIVERSITY,DEPARTMENT OF SAFETY
INDUSTRIAL AND ENVIRONMENTAL HYGIENE
Postbus 9500
NL-2300 RA LEIDEN

GROVES J.A.
HEALTH AND SAFETY EXECUTIVE
OCCUPATIONAL MEDICINE AND HYGIENE
LABORATORY
Edgware Road 403-405
UK-LONDON NW2 LN

GRUNENBERG D.
D.I.N. - DEUTSCHES INSTITUT FUER
NORMUNG - SECRETARIAT OF ISO ITC 146
Burggrafenstrasse 4-10
D-1000 BERLIN 30

GUARDINO X.
I.N.S.H.T. - DULCET S.N.
E-08034 BARCELONA

GUILD L.V.
S.K.C. - I.N.C.
Valley View Road 334
U.S.A.-EIGHTY FOUR, Pa

HAAN D.
DEHA INTERNATIONAL B.V.
P.O. Box 483
NL-1400 AK BUSSUM

HAEGER
BUNDESANSTALT FUER ARBEITSSCHUTZ UND
UNFALLFORSCHUNG
Vogelspothsweg 50-52
Postfach 170202
D-4600 DORTMUND 12

HAFKENSCHEID TH.L.
DIRECTORATE-GENERAL OF LABOUR
P.O. Box 69
NL-2270 MA VOORBURG

HALLBERG B.O.
NATIONAL BOARD OF OCCUPATIONAL
SAFETY AND HEALTH
RESEARCH DEPARTMENT
S-17184 SOLNA

HAMLIN J.W.
B.P. CHEMICALS Ltd.
Buckingham Palace Road 76
UK-LONDON

HAMPAKI E.
DIRECTION INSPECTION DU TRAVAIL
PREFECTURE D'ATTIKI
Voulgari 2
B.P. 1108
GR-ATHENS 10110

HARDY J.
DEPARTMENT OF CHEMISTRY
UNIVERSITY OF AKRON
U.S.A.-AKRON, OHIO OH 44325

HARLOCK L.
COURTAULDS RESEARCH
Lockhurst Lane
UK-COVENTRY

HARPER M.
DEPARTMENT OF OCCUPATIONAL HEALTH
LONDON SCHOOL OF HYGIENE AND
TROPICAL MEDICINE
Keppel Street
UK-LONDON WC1E 7HT

HEIKAMP A.
N.N.I.
NEDERLANDS NORMALISATIE INSTITUUT
P.O. Box 5059
NL-2600 GB DELFT

HENRY M.
DEPARTMENT OF LABOUR
Mespil Road
IRL-DUBLIN 4

HERVE-BAZIN B.
INSTITUT NATIONAL DE RECHERCHE ET DE
SECURITE
Avenue de Bourgogne
B.P. 27
F-54501 VANDOEUVRE CEDEX

HOUGHTON D.F.
RANK XEROX Ltd.
C.H. AND S. DEPARTMENT
ENGINEERING GROUP
Bessemer Road
UK-WELWYN GARDEN CITY, HERTS

IKEDA M.
DEPARTMENT OF ENVIRONMENTAL HEALTH
TOHOKU UNIVERSITY -SCHOOL OF MEDICINE
JAPAN-SENDAI 980

JACKSON E.S.
MOORE, BARRETT AND REDWOOD Ltd.
PETROSERVICE HOUSE
Macklin Avenue
UK-BILLINGHAM, CLEVELAND

JANSEN A.
OCCUPATIONAL HEALTH SERVICE
BGD WEST BRABANT
Lange Nieuwstraat 26
NL-4587 RJ KLOOSTERZANDE

JAROSZEWSKI M.
DIREKTORATET FOR ARBEJDSTILSYNET
Landskronagade 33-35
DK-2100 COPENHAGEN O

JONES D.C.
I.C.I. plc. - PAINTS DIVISION
Wexham Road
UK-SLOUGH

JORDAN F.H.
BERUFSGENOSSENSCHAFT DER CHEMISCHEN
INDUSTRIE BEZIRKSVERWALTUNG
HEIDELBERG
Gaisbergstrasse 11
D-6900 HEIDELBERG

JOURDAN L.
CEFIC
Avenue Louise 250
B-1040 BRUXELLES

KAALJK J.
PRINS MAURITS LABORATORY TNO
Postbus 45
NL-2280 AA RIJSWIJK

KAT W.
HOOGEVENS YMUIDEN ENVIRONMENTAL
DEPARTMENT 3D10
Postbus 1000
NL-1970 CA YMUIDEN

KELLER J.
BAYER AG, ZF-DID ing.
wiss. ABTEILUNG
D-5090 LEVERKUSEN

KELLER R.
DRÄGERWERK AG, MOISLINGER AG.
Postfach 1339
D-2400 LUEBECK 1

KENNEDY E.R.
NATIONAL INSTITUTE FOR OCCUPATIONAL
SAFETY AND HEALTH
Columbia Parkway 4676
U.S.A.-CINCINNATI, OHIO

KENNEDY M.J.
G.K.N. TECHNOLOGY
Birmingham New Road
UK-WOLVERHAMPTON

KETTRUP A.
UNIVERSITAET GESAMTHOCHSCHULE
PADERBORN - FACHBEREICH 13
CHEMIE UND CHEMIETECHNIK
Postfach 1621
D-4790 PADERBORN

KLAPMEIJER C.A.
DEHA INTERNATIONAL B.V.
P.O. Box 438
NL-1400 AK BUSSUM

KRISTENSSON J.
UNIVERSITY OF STOCKHOLM
DEPARTMENT OF ANALYTICAL CHEMISTRY
S-10691 STOCKHOLM

LABUHN P.
CIBA-GEIGY MÜNCHWILEN AG.
CH-4333 MÜNCHWILEN

LANTINGA P.H.G.
DIENST VOOR VEILIGHEID EN MILIEU
Postbus 7161
NL-1007 MC AMSTERDAM

LEANDRO D.
DIRECCAO GERAL DA HIGIENE E
SEGURANCA NO TRABALHO DO MINISTERIO
DE TRA BALHO E SEGURANGA SOCIAL
Praca de Londres 2-16
P-1000 LISBON

LEHMANN E.
BUNDESANSTALT FUER ARBEITSSCHUTZ
P.O. Box 170202
D-4600 DORTMUND

LEICHNITZ K.
IUPAC - COMMISSION ON ATMOSPHERIC
CHEMISTRY
DRÄGERWERK AG.
Moislinger AG.
Postfach 1339
D-2400 LUEBECK 1

LEINSTER P.
F.B.C. Ltd.
OCCUPATIONAL HEALTH DEPARTMENT
Hauxton
UK-CAMBRIDGE CB2 5HU

LEISSER H.
ALLGEMEINE UNFALLVERSICHERUNGSANSTALT
Abtlg. HUB
Adalbert-Stifter Strasse 65
A-1200 VIENNA

LEVIN J.O.
NATIONAL SWEDISH BOARD OF OCCUPA-
TIONAL SAFETY AND HEALTH
RESEARCH DEPARTMENT IN UMEA
P.O. Box 6104
S-90006 UMEA

LEWIS S.J.
B.P. INTERNATIONAL LIMITED
B.P. GROUP OCCUPATIONAL HEALTH CENTRE
Occam Road 10
Surrey Research Park
UK-GUILDFORD, SURREY GU2 5YQ

LIDGETT R.
FRON COTTAGE
Llandynan
UK-LLANGULLEN, CLWYD LL20 7AJ

LIGUS E.G.
NATIONAL DRAEGER INC.
Technology Drive 101
P.O. Box 120
U.S.A.-PITTSBURGH PA 15230

LINDAHL R.
NATIONAL SWEDISH BOARD OF OCCUPA-
TIONAL SAFETY AND HEALTH
RESEARCH DEPARTMENT IN UMEA
P.O. Box 6104
S-90006 UMEA

LOMENEDE B.
MINISTERE DES AFFAIRES SOCIALES ET
DE L'EMPLOI
DIRECTION DES RELATIONS DU TRAVAIL
Place de Fontenoy 1
F-75700 PARIS

LUCKAS K.H.
PERKIN-ELMER VERKAUF GmbH.
Hansa Allee 195
D-4000 DÜSSELDORF 11

LUXON S.G.
ROYAL SOCIETY OF CHEMISTRY
Burlington House
Piccadilly
UK-LONDON W1V OBN

MALVIK B.
SINTEK
DIVISION OF APPLIED CHEMISTRY
N-7034 TRONDHEIM-NTH

MARK D.
INSTITUTE OF OCCUPATIONAL MEDICINE
PHYSICS BRANCH
Roxburgh Place
UK-EDINBURGH EH8 9SU

MATTIMOE G.
DEPARTMENT OF LABOUR
Mespil Road
IRL-DUBLIN 4

MAY F.
DRÄGERWERK AG.
Moislinger AG.
Postfach 1339
D-2400 LUEBECK 1

MEIER U.
PERKIN-ELMER VERKAUF GmbH.
Hansa Allee 195
D-4000 DUESSELDORF 11

MEISSEN J.
SEREB
Rv. d. Venlaan 3A
NL-4196 PL GELDERMALSEN

MENICHINI E.
ISTITUTO SUPERIORE DI SANITA
V. le Regina Elena 299
I-00161 ROMA

MESMACQUE R.
MINISTERE DE L'EMPLOI ET DU TRAVAIL
ADMINISTRATION DE L'HYGIENE ET DE LA
MEDECINE DU TRAVAIL
Rue Belliard 53
B-1040 BRUXELLES

MEYER P.
T.N.O.
P.O. Box 214
Nl-2600 AE DELFT

MILLER B.
IMPERIAL CHEMICAL INDUSTRIES PLC
I.C.I.
MOND DIVISION, THE HEATH
P.O. Box 8
UK-RUNCORN, CHESHIRE

MOESENTHIN H.
BETRIEBSRATSMITGLIED DER BOSCH AG.
IN WAIBLINGEN
Auf der Au 41
D-7022 SCHWÄBISCH GMÜND 14

MOORE G.
S.K.C. Ltd., HAMWORTHY TRADING ESTATE
Dawkins Road
UK-POOLE, DORSET BH15 4JW

MUELLER-WILDERINK H.
BLOHM + VOSS AG.
Hermann-Blohm Strasse 3
D-2000 HAMBURG 11

MCDONALD T.
RUBENS INSTITUTE
UNIVERSITY OF SURREY
UK-GUILDFORD, SURREY

MCKEE E.
MINE SAFETY APPLIANCES CO.
P.O. Box 430
U.S.A.-PITTSBURGH, PA 15230

NEIRYNCK W.
CENTRUM VOOR TECHNOLOGISCH ONDERZOEK
Voskenslaan 270
B-9000 GENT

NEVILLE R.W.J.
THE WELLCOME FOUNDATION Ltd.,
THE WELLCOME BUILDING
Euston Road 183
P.O. Box 129
UK-LONDON NW1 2BP

NIELSEN K.E.
MILJØ-KEMI
Smedeskosvej 38
DK-8464 GALTEN

NORBÄCK D.
DEPARTMENT OF OCCUPATIONAL MEDICINE
AKADEMISKA SJUKHUSET
S-75185 UPPSALA

NORLINDER R.
DEPARTMENT OF OCCUPATIONAL MEDICINE
SAHLGREN HOSPITAL
St. Sigfridsgatan 85
S-41266 GÖTEBORG

O'SULLIVAN J.
BARNSLEY DISTRICT GENERAL HOSPITAL
QUALITY CONTROL
Gawber Road
UK-BARNSLEY S75 2EP

ONDREJKO T.
GUILD CORPORATION
Thomas 384 - Venetia Road
U.S.A.-EIGHTY FOUR, PA

OSUNA N.
MINISTERIO DE TRABAJO
Castellana N. Ministerio
E-MADRID

PANNWITZ K.H.
DRAEGERWERK AG.
Moislinger Allee 53/55
D-2400 LÜBECK

PAPENDIXK H.D.
E.C. ERDOELCHEMIE GmbH.
Postfach 752002
D-5000 KÖLN 71

PASQUIER J.L.
MINISTERE DES AFFAIRES SOCIALES ET
DE L'EMPLOI
DIRECTION DES RELATIONS DU TRAVAIL
Place de Fontenoy 1
F-75700 PARIS

PENSE R.
HOECHST AKTIENGESELLSCHAFT
REISEBÜRO C820
Postfach 800320
D-6230 FRANKFURT/MAIN 80

PERKINS B.M.
COURTAULDS RESEARCH
Lockhurst Lane
P.O. Box 111
UK-COVENTRY CV6 5RS

PFÄFFLI P.
INSTITUTE OF OCCUPATIONAL HEALTH
DEPARMENT OF INDUSTRIAL HYGIENE AND
TOXICOLOGY
Topeliuksenkatu 14a A
SF-00250 HELSINKI

PHILLIPS C.F.
SHELL INTERNATIONALE PETROLEUM
MAATSCHAPPIJ BV.
HEALTH SAFETY AND ENVIRONMENT
DIVISION - INDUSTRIAL HYGIENE
P.O. Box 162
NL-2501 DEN HAAG

POZZOLI L.
FONSAZIONE CLINICA DEL LAVORO
INSTITUTE OF OCCUPATIONAL HEALTH
Via Arzaga 19
I-27100 PAVIA

PRIHA E.
TAMPERE REGIONAL INSTITUTE OF
OCCUPATIONAL HEALTH
Ulmalankatu 1
P.O. Box 486
SF-33101 TAMPERE

RAJAN RAJADURAI R.
M.O.D. - H.M. NAVAL BASE
UK-PLYMOUTH

RAMAEKERS J.J.M.
D.S.M. RESEARCH, SECTOR FA-CO
Postbus 18
NL-GELEEN

RENTEL
HUETTEN - UND WALZWERKS-
BERUFSGENOSSENSCHAFT
TECHNISCHER AUFSICHTSDIENST
Postfach 101665
D-4300 ESSEN 1

RHEIN G.
INDUSTRIEGEWERKSCHAFT METALL
ABTEILUNG ARBEITS
UND GESUNSDHEITSSCHUTZ
Wilhelm-Leuschner Strasse 79-85
D-6000 FRANKFURT/MAIN 11

RICHARDS S.
LUCAS INDUSTRIES PLC
Great King Street
UK-BIRMINGHAM B19 2XF

RINGHAM R.
I.C.I. - PAINTS DIVISION
Wexham Road
UK-SLOUGH, BERKS

ROBERTSON A.
INSTITUTE OF OCCUPATIONAL MEDICINE
ENVIRONMENTAL BRANCH
Roxburgh Place
UK-EDINBURGH EH8 9SU

ROBERTSON S.D.
I.C.I. - PLANT PROTECTION DIVISION
Jealott's Hill Research Stat.
UK-BRACKNELL, BERKSHIRE

ROSMANITH P.
JOTUN A/S
P.O. Box 400
N-3201 SANDEFJORD

ROTHBERG M.
UUSIHAA REGIONAL INSTITUTE OF
OCCUPATIONAL HEALTH
Arinatie 3A
SF-00370 HELSINKI

ROUTLEDGE W.
MOORE, BARRETT AND REDWOOD Ltd.
PETROSERVICE HOUSE
Macklin Avenue
Billingham
UK-CLEVELAND TS23 4BY

ROWE K.L.H.
P.P.G. INDUSTRIES (UK) Ltd.
Rotton Park Street
UK-BIRMINGHAM

RUECK A.
I.I.C. - WIEDERHOLD GmbH.
Düsseldorfer Strasse 102
D-4010 HILDEN

RUME R.
DIRECTION DE LA SANTE
MINISTERE DE LA SANTE
SERVICE DE LA MEDECINE DU TRAVAIL
Rue Goethe 22
L-1637 LUXEMBOURG

SAARINEN L.
INSTITUTE OF OCCUPATIONAL HEALTH
Arinatie 3
SF-00730 HELSINKI

SALA C.
SERVIZIO DI MEDICINA DEL LAVORO DEL
L'OSPEDALE DI LECCO
Corso Promessi Sposi 1
I-22053 LECCO

SAMINI B.
GRADUATE SCHOOL OF PUBLIC HEALTH
SAN DIEGO STATE UNIVERSITY
U.S.A.-SAN DIEGO

SAPIR M.
CONFEDERATION EUROPEENNE DES
SYNDICATS
Rue Montagne aux Herbes 37-41
B-1000 BRUXELLES

SAUNDERS K.J.
THE BRITISH PETROLEUM COMPANY PLC
B.P. RESEARCH CENTRE SUNBURY
Chertsey Road
UK-SUNBURY-ON-THAMES, MIDDELSEX TW16

SCAGLIANTI G.
MONTED I PECMARG
Via Della Chimica 5
I-VENICE

SCHALLER K.H.
INSTITUT FUER ARBEITS
u. SOZIAL-MEDIZIN UND POLIKLINIK
FUER BERUFSKRANKHEITEN
DER UNIVERSITAET ERLANGEN-NUERNBERG
Schillerstrasse 25 u. 29
D-8520 ERLANGEN

SCHANDL S.
NATIONAL RESEARCH INSTITUTE OF
OCCUPATIONAL SAFETY
Ötrös Janos 1-3
H-BUDAPEST

SCHULZ H.
BAYER AG - ZENTRALBEREICH - ZENTRALE
FORSCHUNG U. ENTWICKLUNG
DID IN G. WISS. ABTEILUNG
Bayerwerk
D-5090 LEVERKUSEN

SCHUSTER A.
INSPECTION DU TRAVAIL ET DES MINES
Rue Zithe 26
L-2010 LUXEMBOURG

SCULLMAN J.
ARBETARSKYDDSSTYRELSEN
Ekelundsv 16
S-17184

SERBIN L.
LABORATORY SERVICES BR. ALBERTA
WORKERS'
HEALTH AND SAFETY
ST 10158-103
EDMONTON
CDN-ALBERTA

SHERREN A.J.
SHELL RESEARCH Ltd.
Broad Oak Road
UK-SITTINGBOURNE, KENT

SHIRTLIFFE C.J.
NATIONAL RESEARCH COUNCIL
M24/IRC
CDN-OTTAWA, ONTARIO

SLAVIN S.
PERKIN-ELMER CORPORATION
Main Avenue
Mail STN 150
U.S.A.-NORWACK, CT 06856

SOWA L.
LABORATOIRE D'ETUDE ET DE CONTROLE
DE L'ENVIRONNEMENT SIDERURGIQUE
Voie Romaine
F-57210 MAIZIERES-LES-METZ

SQUIRRELL D.C.M.
Graysfield 9, Welwyn Garden City
UK-HERTFORDSHIRE AL7 4BL

STEINHANSES J.
FRAUNHOFER-INSTITUT FUER
UMWELTCHEMIE UND OEKOTOXIKOLOGIE
Grafschaft
D-5948 SCHMALLENBERG

STENNER H.
UNIVERSITY PADERBORN
Warburgerstrasse 100
D-4790 PADERBORN

STRIEFLER B.
NIEDERSAECHSISCHE
LANDESVERWALTUNGSAMT
Bertastrasse 4-6
D-3000 HANNOVER

STUHLMANN F.
E.N.K.A.
WERK OBERNBURG ABTEILUNG PUER-CL
D-8753 OBERNBURG

SUTTER E.
S.U.V.A.
Fluhmattstrasse 1
CH-6002 LUZERN

TAYLOR T.
HEALTH AND SAFETY EXECUTIVE
MAGDALEN HOUSE
Stanley Precinct
UK-BOOTLE L20 3QZ MERSEYSIDE

TCHORZ K.
LUFTHANZA AG.
ARBEITSSICHERHEIT BODEN HAM PX2
Am Teich 16
D-2350 NEUMUENSTER

THORUD S.
INSTITUTE OF OCCUPATIONAL HEALTH
Gydas Veis 8
Box 8149 DEP
N-OSLO 1

TINDLE P.E.
SHELL U.K. LIMITED
OCCUPATIONAL HYGIENE UNIT
THORNTON RESEARCH CENTRE
P.O. Box 1
UK-CHESTER CH1 3SH

TOURRES D.
COMPAGNIE FRANCAISE DE RAFFINAGE
C.R.D. TOTAL FRANCE
B.P. 27
F-76700 HARFLEUR

TULKKI A.I.
POST AND TELEADMINISTRATION
Paasivuorenkatu 3
SF-005300 HELSINKI

TWISK J.J.
DOW CHEMICAL (NEDERLAND) BV.
P.O. Box 48
NL-4530 AA TERNEUZEN

UHLHAAS H.J.
3M DEUTSCHLAND GmbH.
HAUPTVERWALTUNG
Carl-Schurz-Strasse 1
Postfach 643
D-4040 NEUSS 1

ULFARSON U.
DEPARTMENT OF WORK SCIENCE
THE ROYAL INSTITUTE OF TECHNOLOGY
S-10044 STOCKHOLM

ULLRICH D.
INSTITUT FUER WASSER, BODEN
u. LUFTHYGIENE DES BUNDES-
GESUNDHEITSAMTES
Corrensplatz 1
D-1000 BERLIN 33

URBANUS J.
DOW CHEMICAL S.P.A.
Via Emilia N°2
I-20070 FOMBIO, MILANO

VAN DEN HOED N.
c/o KONINKLIJKE SHELL LABORATORIUM
DEPARTMENT AG/12
Postbus 3003
NL-AMSTERDAM

VAN DER WAL J.F.
M.T. - T.N.O.
DEPARTMENT OF TOXICOLOGY FOR SOCIETY
P.O. Box 21
NL-2600 AE DELFT

VAN ZAANEN A.J.
SEREB
R.V.D. VENLAAN 3A
NL-4191 PL GELDERMALSEN

VERBOEWET M.
D.S.M. RESEARCH, MVR/BO
Postbux 18
NL-GELEEN

VERGAUWE C.
DRÄGER-BELGIUM
Avenue Rodenbach 4
B-1030 BRUXELLES

VON KRAUSE J.
DEUTSCHES PRIMATENZENTRUM GmbH.
Kellnerweg 4
D-3400 GÖTTINGEN

VOSS H.
BASF AG, ABTEILUNG HRS/E - D 700
D-6700 LUDWIGSHAFEN

WALSH P.T.
HEALTH AND SAFETY EXECUTIVE
STEEL CITY HOUSE
Broad Lane
UK-SHEFFIELD S1 7HQ

WEIDHOFER J.
ALLGEMEINE UNFALLVERSICHERUNGSANSTALT
Abtlg. HUB
Adalbert-Stifter-Strasse 65
A-1200 VIENNA

WEISSKOPF V.
HESSISCHE LANDESANSTALT FUER UMWELT
Ludwig-Mond-Strasse 33b
D-3500 KASSEL

WENNIG R.
LABORATOIRE NATIONAL DE SANTE
B.P. 1102
L-1011 LUXEMBOURG

WENTRUP G.J.
TUV HANNOVER e.V.
Postfach 810740
D-3000 HANNOVER 81

WEST N.G.
HEALTH AND SAFETY EXECUTIVE
Edgware Road 403
UK-LONDON NW2 6LN

WILHARDT P.
DANISH NATIONAL INSTITUTE OF
OCCUPATIONAL
HEALTH
Baunegaardsvej 73
DK-2900 HELLERUP

WOOLFENDEN E.
PERKIN-ELMER LIMITED, CHROMATOGRAPHY
Post Office Lane
UK-BEACONSFIELD, BUCKS HP9 1QA

ZAMBELLLI S.
ZAMPELLI S.R.L.
Via Santa Rita 11-13
I-20010 BAREGGIO, MI

ZLOCZYNSTI S.
AUERGESELLSCHAFT GmbH.
Thiemannstrasse 1
D-1000 BERLIN